SOMATIC AND AUTONOMIC NERVE–MUSCLE INTERACTIONS

Research monographs in cell and tissue physiology

Volume 8

General Editors

J.T. DINGLE and J.L. GORDON

Cambridge

ELSEVIER
AMSTERDAM · NEW YORK · OXFORD

Somatic and autonomic nerve–muscle interactions

Editors

G. BURNSTOCK, R. O'BRIEN and G. VRBOVÁ

London

1983

ELSEVIER
AMSTERDAM · NEW YORK · OXFORD

© 1983 Elsevier Science Publishers, B.V.

All rights reserved. No part of this publication may be reproduced, stored in a retrieval system, or transmitted, in any form or by any means, electronic, mechanical, photocopying, recording or otherwise, without the prior permission of the copyright owner.

ISBN for the series: 0 444 80234 7
ISBN for this volume: 0 444 80458 7

Published by:

Elsevier Science Publishers, B.V.
P.O. Box 211, 1000 AE Amsterdam
The Netherlands

Sole distributors for the U.S.A. and Canada:

Elsevier Science Publishing Company Inc.
52 Vanderbilt Avenue
New York, NY 10017

Library of Congress Cataloging in Publication Data
Main entry under title:

Somatic and autonomic nerve–muscle interactions.

    (Research monographs in cell and tissue physiology; v. 8)
    Includes index.
    1. Myoneural junction – Addresses, essays, lectures.
2. Muscles – Innervation – Addresses, essays, lectures.
3. Neuromuscular diseases – Addresses, essays, lectures.
I. Burnstock, G. (Geoffrey)   II. Vrbová, Gerta.
III. O'Brien, R.   IV. Series.   [DNLM: 1. Nervous system
–Physiology. 2. Muscles–Physiology. 3. Action potentials. 4. Neuromuscular junction–Physiology.
5. Muscles–Innervation. W1 RE232GL v.8/WL 102 S6925]
QP369.5.S65   1983   599.01'88   83-5586
ISBN 0-444-80458-7

PRINTED IN THE NETHERLANDS

Preface

Interest in nerve–muscle interactions has grown during recent years, particularly in terms of the mechanisms associated with the formation of neuromuscular junctions during development, such as the neurochemical differentiation of transmitters, the specificity of neuromuscular connections, synapse elimination, and the influence of the target organ on neuronal death. In the mature animal too, questions have been raised about long-term plasticity of neuromuscular connections and of the effects of such factors as impulse activity, hormones, disease and ageing. The mechanisms involved in regeneration of neuromuscular junctions in adults have also been investigated. For historical reasons and the influence of certain dominant minds, these studies have followed two quite separate lines, dictated largely by the neuromuscular model employed, namely the skeletal neuromuscular junction and the autonomic nerve-smooth or cardiac muscle junction. Each system has particular features and advantages for examining particular questions, but there has been little overlap or even cross reference between work in these two models even when the starting questions were identical.

The emphasis in this book is to combine under one cover for the first time studies of nerve–muscle interactions in both models, so that there is an opportunity to see where there are common answers and where there are different answers to the same questions in the two systems. Perhaps in this way, new questions will arise and new directions of research will be initiated. The first part of the book deals with autonomic and somatic nerve–muscle interactions during embryonic and perinatal development and makes comparisons with interactions in both systems that can be so conveniently observed in tissue culture. The role of nerve growth factors in neurotrophic functions is discussed. A series of chapters follow that are concerned with nerve–muscle interactions in the adult, including the influence of nerves on both smooth and skeletal muscle function and the influence of target organs on the development of nerves. Finally the significance of disturbances of nerve–muscle interactions in disease are briefly explored.

Contents

Preface v

Chapter 1.   Embryonic development of the autonomic system, by Nicole M. Le Douarin and Philippe Cochard   1

1. Introduction   1
2. Migration pathways followed by neural crest cells during the ontogeny of peripheral ganglia   2
   2.1. Truncal region   2
      2.1.1. Migration of neural crest cells in the cervico-dorsal region from somites 7 to 28   2
      2.1.2. Migration of neural crest cells in the lumbosacral region of the embryo   8
   2.2. Vagal region   9
3. Fate map of the autonomic structures in the neural crest   10
4. Developmental capabilities of the crest cells at various levels of the neural axis   10
5. Expression of the adrenergic phenotype in vivo   12
6. Differentiating potentialities of peripheral ganglionic cells during development   18
   6.1. Graft of autonomic and dorsal root ganglia   18
   6.2. Back-transplantation of the nodose ganglion   22
7. In vitro studies of neural crest cell differentiation   23
   7.1. Culture of crest cells in serum-containing medium   23
   7.2. Cholinergic traits in the migrating neural crest   25
   7.3. Neuronal differentiation in cultured neural crest cells in serum-deprived medium   28
8. Concluding remarks   30

Chapter 2.   Development of smooth muscle and autonomic nerves in culture, by Julie H. Chamley-Campbell and Gordon R. Campbell   35

1. Influence of sympathetic nerves on the development of smooth musle in culture   35
   1.1. Development of smooth muscle   35
   1.2. Phenotypic modulation of smooth muscle   36
   1.3. Influence of sympathetic nerves on development and phenotypic modulation of smooth muscle   39
2. Influence of effector organs on the development of autonomic nerves in culture   40
   2.1. Expression of differentiated functions of autonomic nerves in culture   40
      2.1.1. Sympathetic neurons   40

	2.1.2. Parasympathetic neurons	43
	2.1.3. Enteric neurons	43
2.2.	Influence of autonomic effectors on neuronal transmitter	44

3. Tropic influence of autonomic effector organs on nerve fibre growth — 45

4. Formation of functional junctions in culture — 48
    4.1. Interactions between sympathetic nerves and isolated smooth and cardiac muscle cells — 48
    4.2. Ultrastructure of nerve–muscle associations — 51
    4.3. Functional junctions — 52

5. Concluding remarks — 53

Chapter 3.   *Aspects of naturally-occurring motoneuron death in the chick spinal cord during embryonic development, by Ronald W. Oppenheim and I.-Wu Chu-Wang*   57

1. Introduction — 57

2. Evidence that significant cell death occurs in spinal motoneurons — 59

3. Are spinal motoneurons preprogrammed to die? — 60
    3.1. Induced cell death — 61
    3.2. Peripheral or target innervation prior to the onset of cell death — 66

4. Normal development of anterior (ALD) and posterior (PLD) latissimus dorsi muscles and their innervation — 68
    4.1. Motoneurons — 68
    4.2. Peripheral nerves — 69
    4.3. Neuromuscular junction — 72
    4.4. Muscle development — 75

5. Prevention of naturally-occurring loss in spinal motoneurons — 79
    5.1. Increased target size — 79
    5.2. Modification of motoneuron death by chronic treatment with pharmacological agents that affect neuromuscular activity — 81
    5.3. Development of ALD and PLD muscles and their innervation following neuromuscular blockade
        5.3.1. Motoneurons — 88
        5.3.2. Peripheral nerve — 88
        5.3.3. Muscle development — 89
        5.3.4. Innervation and the neuromuscular junction — 89

6. How does chronic neuromuscular blockade prevent the death of spinal motoneurons? A Model — 92

7. Cell death in a population of spinal visceral efferent neurons: the sympathetic preganglionic column of Terni — 96

8. Function of naturally-occurring neuronal death — 102

Chapter 4. *Neuromuscular development in tissue culture, by L.L. Rubin and K.F. Barald*   109

1. Introduction   109
2. Myogenesis in vitro   109
   2.1. Overview   109
   2.2. Primary cultures and cell lines   110
   2.3. Cell cycle and fusion   111
   2.4. Advantages of culture systems for studies of myogenesis   112
   2.5. Cell–cell interactions among prefusion myoblasts   113
   2.6. Early cell–cell interactions in the fusion process   113
   2.7. Gap junctions and cell–cell communication   113
   2.8. Involvement of membrane lipids in the fusion event   114
   2.9. Lipid interactions in fusion   115
   2.10. Effects of other membrane alterations or perturbations on the fusion process   116
      2.10.1. Lections as mediators of muscle cell fusion   116
      2.10.2. Other effects on cell surface-mediated events   117
      2.10.3. Effects of viral transformation on muscle cell fusion   117
   2.11. Potential triggers of myogenesis   118
      2.11.1. The involvement of the substratum   118
      2.11.2. Growth factors   119
   2.12. Cell fusion and production of muscle-specific components   120
3. Synapse formation   121
   3.1. Initial nerve–muscle contacts   121
      3.1.1. Experimental systems   121
      3.1.2. Growth of spinal cord neuronal processes   122
      3.1.3. Specificity of synapse formation   123
      3.1.4. Onset of synaptic transmission   125
      3.1.5. Ultrastructure of early synapses   126
   3.2. AChR Clustering at newly-formed synapses   127
      3.2.1. Nerve–muscle synapse formation causes the appearance of new AChR clusters   127
      3.2.2. Is AChR clustering a invariable consequence of nerve–muscle synapse formation?   129
      3.2.3. Redistribution versus local insertion of synaptic AChRs   130
      3.2.4. Further modification of synaptic AChRs   130
   3.3. AChE appearance at newly-formed nerve–muscle synapses   132
      3.3.1. Chick nerve–muscle cultures   132
      3.3.2. Other types of cultures   133
   3.4. Regulatory events in nerve–muscle synapse formation: dependence on synaptic activity   135
      3.4.1. Onset of synaptic transmission   135
      3.4.2. AChR clustering at synapses   135
      3.4.3. AChE appearance at newly formed synapses   136
   3.5. Molecular mechanisms underlying accumulation of AChE and AChRs at synapses   138
      3.5.1. Activity-independent events   138
      3.5.2. Activity-dependent events   143
4. Conclusions   145

Chapter 5. Postnatal development of the innervation of mammalian skeletal muscle, by R.A.D. O'Brien   153

1. Introduction   153

2. Development of muscle fibres and their innervation   155
   2.1. Muscle fibre development   155
   2.2. The first nerve–muscle contact   156
   2.3. Development of polyneuronal innervation   158

3. The elimination of polyneuronal innervation   160
   3.1. Time course of elimination   163
   3.2. Effect of activity on elimination   165
   3.3. Nerve regeneration in neonatal muscles   167
   3.4. Differences between surviving and eliminated nerve terminals   168

4. The mechanism of elimination   171
   4.1. A hypothetical model   174

5. Summary and conclusions   176

6. Appendix: Techniques for examining polyneuronal innervation   176
   6.1. Contraction measurements   177
   6.2. Electrophysiology   178
   6.3. Microscopy   180
      6.3.1. Light histology   180
      6.3.2. Electron microscopy   181

Chapter 6. Influence of preganglionic fibres and peripheral target organs on autonomic neuronal development, by Caryl E. Hill and Ian H. Hendry   185

1. General introduction   185

2. Development of sympathetic neurones   186
   2.1. Normal biochemical development of sympathetic ganglia   186
   2.2. Influence of preganglionic fibres   187
      2.2.1. Development in the absence of the preganglionic fibres   187
      2.2.2. Agents responsible for preganglionic influences   187
      2.2.3. Trophic effects of sympathetic neurones on preganglionic fibres   188
      2.2.4. Preganglionic fibres and sympathetic cholinergic neurones   189
   2.3. Influence of tissues on sympathetic neuronal development   189
      2.3.1. Neurite outgrowth   189
      2.3.2. Factors other than NGF influencing neurite outgrowth   190
      2.3.3. Development of the innervation pattern in target organs   191
      2.3.4. Factors influencing the innervation pattern in target organs   191
      2.3.5. Influence of target organs on neuronal survival. A role for NGF   192
      2.3.6. Transport of NGF to the nerve cell soma   194
      2.3.7. Is NGF in target organs?   195
      2.3.8. Effects of innervation on NGF release from target organs   196

2.3.9. Factors other than NGF influencing neuronal survival	197
2.3.10. Developmental changes in requirements for survival factors	198
2.4. Regulation of synaptic connections	199
2.4.1. Effect of the loss of synapses with target organs on preganglionic innervation	199
2.4.2. Effect of the loss of preganglionic synapses on peripheral innervation	201
2.4.3. Effect of spinal transection on the development of sympathetic neurones	201
3. Development of parasympathetic neurones	202
3.1. The avian ciliary ganglion	202
3.2. Influence of preganglionic fibres	202
3.2.1. Trophic effects of parasympathetic neurones on preganglionic neurones	203
3.3. Influence of target organs	203
3.3.1. Neurite outgrowth	203
3.3.2. Target tissue derived factors influencing neurite outgrowth	203
3.3.3. Development of the innervation pattern in target organs	204
3.3.4. Influence of target organs on neuronal survival	204
3.3.5. Factors influencing neuronal survival	205
3.3.6. Effects of synapse formation on nerve and muscle maturation	206
4. General discussion	206

Chapter 7. Nerve Growth Factor and the neuronotrophic functions, by Silvio Varon and Stephan D. Skaper   213

1. Introduction	213
1.1. Sources of neuronotrophic factors	214
1.1.1. Target-derived factors	214
1.1.2. Other peripheral sources and glial sources	215
1.2. Nerve Growth Factor as a model factor	216
1.2.1. NGF sources	216
1.2.2. NGF-responsive cells	218
2. NGF effects and roles	220
2.1. Neuronal survival and general growth	220
2.1.1. NGF effects	220
2.1.2. NGF supplements and NGF surrogates	221
2.2. Neurite growth	222
2.2.1. NGF effects on PC12 and primary neurons	222
2.2.2. Substratum-bound neurite promoting factors	223
2.2.3. Neurite directional guidance	224
3. Modes of action of NGF	224
3.1. NGF receptors	225
3.2. NGF internalization	226
3.3. RNA and protein synthesis	227
3.3.1. NGF effects on general synthesis of RNA and/or protein	227
3.4. NGF intermediate effects	228
3.4.1. Selected proteins	228
3.4.2. Membrane properties	229
3.4.3. Cytoskeletal elements	229
3.4.4. Cyclic AMP involvements	230
3.4.5. Phosphorylation effects	230

4. The ionic hypothesis for NGF 231
  4.1. Concepts and strategies 232
  4.2. The ionic responses to NGF 233
    4.2.1. Features of the ionic responses 233
    4.2.2. Correlations between ionic and traditional effects of NGF 234
  4.3. Molecular aspects of the ionic effects of NGF 236
    4.3.1. From receptor binding to ionic pump control 236
    4.3.2. From ionic pump control to neuronal survival 238
  4.4. General aspects of the ionic hypothesis 240
    4.4.1. Multiple consequences of NGF action 241
    4.4.2. Ionic involvements in cell regulations 242

Chapter 8. *The effects of autonomic nerves on smooth muscle properties, by Mark D. Dibner* 253

1. Introduction 253

2. Chemical interactions between nerve and muscle 254
  2.1. Cholinergic interactions 254
  2.2. Adrenergic interactions 256
  2.3. Cross interactions 258
  2.4. Non-adrenergic, non-cholinergic autonomic effects 259
  2.5. Histamine effects 259
  2.6. Peptides and hormone effects 259

3. Changes in smooth muscle under autonomic influence 261
  3.1. Ion changes 261
  3.2. Cyclic nucleotides 263
  3.3. Phenotypic properties 264

4. Regulation of smooth muscle responsiveness 265
  4.1. Supersensitivity 267
  4.2. Desensitization 269

5. Conclusions 271

Chapter 9. *The influence of the motor nerve on the contractile properties of mammalian skeletal muscle, by A.J. Buller* 277

1. Introduction 277

2. Contractile responses 279
  2.1. General 279
  2.2. Isometric contractions 280
  2.3. Isotonic studies 282

3. Biochemical correlates 284

4. Patterns of nerve activity 284
  4.1. Introduction 284
  4.2. Normal activity patterns 285
  4.3. Stimulation experiments 286

5. Conclusion 286

Chapter 10. Dependence of peripheral nerves on their target organs, by Tessa Gordon — 289

1. Introduction — 289
2. Effects of axotomy — 290
   2.1. Wallerian degeneration in the distal nerve stump — 290
   2.2. Metabolic response of axotomized neurones — 292
      2.2.1. Chromatolysis and the metabolic response — 292
      2.2.2. Two state model of the motoneurone: growing or transmitting — 294
      2.2.3. Fast axonal transport in axotomized neurones — 295
      2.2.4. Cytoskeleton: slow axoplasmic flow in axotomized neurones — 298
   2.3. Physiological parameters of axotomy — 301
      2.3.1. Synaptic transmission — 301
      2.3.2. Conduction velocity and nerve fibre diameter — 303
      2.3.3. Electrophysiological changes in axotomized motoneurons — 306
3. Regeneration and reinnervation of target organs — 307
   3.1. Two stage regeneration: elongation and increase in diameter — 307
   3.2. Outgrowth and mechanical barriers for regenerating fibres — 310
      3.2.1. Suture line — 311
      3.2.2. Conditions in peripheral stumps — 311
      3.2.3. Condition of the denervated end-organs — 312
   3.3. Functional connections with end-organs and maturation — 312
      3.3.1. Rematching of nerve and muscle properties after reinnervation — 315
      3.3.2. Influence of muscle on nerve — 316

Chapter 11. Pathology of autonomic nerve–smooth muscle mechanisms in the gut of man, by E.R. Howard and J.R. Garrett — 327

1. Anatomy of autonomic innervation — 328
   1.1. Intrinsic nerves — 328
   1.2. Extrinsic nerves — 329
      1.2.1. Sympethatic — 329
      1.2.2. Parasympathetic — 330
   1.3 Other types of innervation — 330
2. Oesophagus — 331
   2.1. Anatomy — 331
   2.2. Normal function — 332
   2.3. Oesophageal pathology — 333
      2.3.1. Lower sphincter incompetence (reflux) — 333
      2.3.2. Achalasiaa of the cardia (cardiospasm) — 333
      2.3.3. Tumour induced achalasia — 335
      2.3.4. American trypanosomiasis (Chagas' disease) — 335
      2.3.5. Diffuse oesophageal spasm — 336
      2.3.6. Systemic sclerosis (scleroderma) — 336
3. Stomach and pylorus — 337
   3.1. Stomach — 337
      3.1.1. Peptic ulceration — 337
      3.1.2. American trypanosomiasis (Chagas' disease) — 338

Contents

3.2.	Pyloric innervation	338
3.3.	Pyloric pathology	338
	3.3.1. Hypertrophic pyloric stenosis of infancy	338
	3.3.2. Hypertrophic pyloric stenosis of adults	340

4. The small intestine — 340
   4.1. Duodenum — 341
      4.1.1. Megaduodenum — 341
      4.1.2. Congenital aganglionosis (Hirschsprung's disease) — 341
      4.1.3. Systemic sclerosis (scleroderma) — 341
      4.1.4. American trypanosomiasis (Chagas' disease) — 341
      4.1.5. Autonomic neuropathy of unknown origin — 341
   4.2. Jejuno-ileum — 342
      4.2.1. Chronic idiopathic intestinal obstruction — 342
      4.2.2. Segmental dilatation — 343
      4.2.3. Familial neuronal disease — 343
      4.2.4. Crohn's disease — 343

5. Large bowel and anal sphincter — 344
   5.1. Developmental anomalies (neuronal dysplasias) — 345
      5.1.1. Congenital aganglionosis (Hirschsprung's disease) — 345
      5.1.2. Hypoganglionosis — 347
      5.1.3. Hyperganglionosis — 348
      5.1.4. Neurofibromatosis of the colon — 348
   5.2. Acquired disorders — 348
      5.2.1. American trypanosomiasis (Chagas' disease) — 348
      5.2.2. Ulcerative colitis — 349
      5.2.3. Irritable colon — 349
      5.2.4. Pathology of central nervous system — 349
      5.2.5. Autonomic neuropathy in diabetes mellitus — 350

6. Drugs and the myenteric plexus — 350

7. Conclusion — 351

Chapter 12. Neuromuscular diseases viewed as a disturbance of nerve–muscle interactions, by Gerta Vrbová — 359

1. Introduction — 359

2. Spinal muscular atrophy of the Werdnig–Hoffman type — 360
   2.1. Clinical features — 360
   2.2. Histological changes in muscles and CNS — 361
   2.3. Abnormalities of EMG — 363
   2.4. Motoneurone disease viewed as a disturbance of nerve–muscle interaction — 366

3. Duchenne dystrophy — 370
   3.1. Brief description of clinical features — 370
   3.2. Histological and physiological changes — 371
   3.3. Changes of EMG — 374
   3.4. Current ideas on the pathogenesis of Duchenne dystrophy — 375
   3.5. Duchenne dystrophy viewed as a disturbance of nerve–muscle interaction — 376

4. Conclusions — 380

Subject index — *385*

CHAPTER 1

Embryonic development of the autonomic system

NICOLE M. LE DOUARIN and PHILIPPE COCHARD

Institut d'Embryologie du CNRS et du Collège de France, 49bis, Avenue de la Belle-Gabrielle, 94130 Nogent-sur-Marne, France

1. Introduction

Considerable effort has been expended in recent years in endeavors to elucidate the mechanisms underlying the development of the peripheral nervous system (PNS). Many aspects of this problem made it particularly attractive since, in the vertebrate embryo, the PNS is derived from transient structures, the neural crest and the ectodermal placodes, the cells of which have to migrate throughout the embryo before reaching their destination in ganglia and plexuses where they differentiate into a variety of cell types. The precise migration pattern of neural crest cells in the developing body has for long intrigued investigators who have tried to understand how crest cells find their way to their destination and when the segregation of the different cell lines arising from this structure occurs.

Because they are available for experimentation during the whole period of embryogenesis, amphibians and birds have been often used in these studies, whereas in mammals, for a long time, only descriptive studies were performed. Research on neural crest cell ontogeny in the avian embryo has recently become particularly active owing to the availability of a stable cell marking technique which allows crest cells to be followed during the course of their migration and until they have reached a completely differentiated state.

This technique based on the structural differences that exist between the nucleus of two species of birds, the quail and the chick, has been previously described by one of us (Le Douarin, 1969, 1973) and exploited in a number of studies devoted to the ontogeny of neural crest derivatives (cf., Le Douarin, 1980, 1981 for reviews). In this chapter, we will describe the more recent advances that our own group and others have achieved in the study of the early development of the autonomic nervous system (ANS). Since differentiation of the sensory ganglia of the rachidian and cranial nerves is closely related to that of the autonomic ganglia and plexuses, references to the peripheral nervous system (PNS) as a whole will be made in many instances. We will first consider the migratory behavior of the precursor cells of the peripheral ganglia and the way in which the entire ganglionic complex is built in embryogenesis.

Finally by describing in vivo and in vitro experiments we will consider the problem of the differentiation of various cellular components that go to make up the ganglia and plexuses of the PNS.

2. Migration pathways followed by neural crest cells during the ontogeny of peripheral ganglia

The migration pathways of ganglion precursor cells have been identified through the use of three different methods: (i) the quail-chick marker system (Le Douarin, 1969, 1973); (ii) the histochemical revelation of acetylcholinesterase (AChE), which is present in neural crest cells through most of the migration process (Drews, 1975; Cochard and Coltey, 1983); and (iii) the immunocytochemical staining of one of the major components of the crest cell migration pathway, fibronectin (Duband and Thiéry, 1982; Thiéry et al., 1982).

2.1. Truncal region

2.1.1. Migration of neural crest cells in the cervico-dorsal region from somites 7 to 28

At the level of the trunk, early neural crest cell migration proceeds in three streams. One, dorsolateral between ectoderm and somites, concerns the melanoblasts which eventually home to the skin and differentiate into pigment cells. The two others are dorsoventral and give rise to the dorsal root ganglia (DRG) and the sympathetic chains and plexuses. They pass both between the neural tube and the inner face of the somites (Fig. 1) and in the space located between two consecutive somites (Figs. 2 and 3).

Observations of Tosney (1978) using scanning electron microscopy, as well as ours, have shown that during the first phase of the migration, the crest cell

---

Fig. 1. Dorso-ventral migration of neural crest cells in the cervico-dorsal region (somites 7 to 28). (a) Histochemical demonstration of AChE on a transverse section at the level of somite 9 of a 21-somite chick embryo. Neural crest cells, characterized by a strong AChE activity leave the top of the neural tube (arrow heads) and migrate bilaterally (arrows) between the wall of the neural tube (NT) and the inner face of the somites (S). (b) Chimaeric quail-chick embryo at the 18-somite stage. The neural primordium of a 13-somite chick embryo has been removed between somites 5 and 10 and replaced by a similar fragment from a quail embryo at the same stage. This transverse section at the level of somite 7 shows neural crest cells (arrows) identified by their large mass of ▶

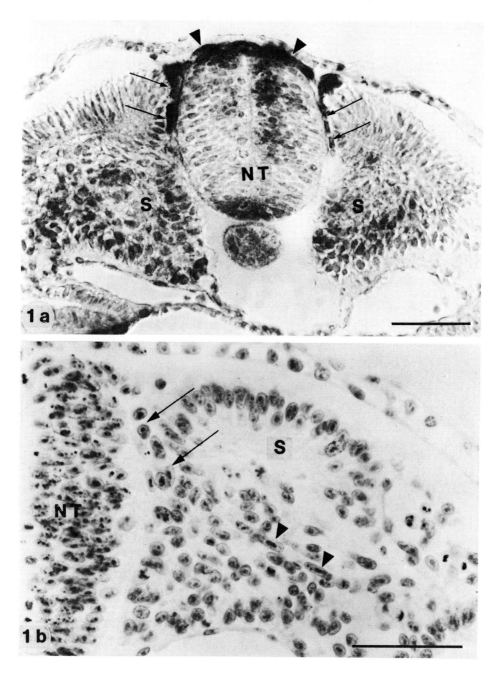

heterochromatin, migrating between neural tube (NT) and somite (S) while others (arrow heads) progress within the somite itself in the space located between the dermomyotome and the sclerotome. Bars represent 50 μm.

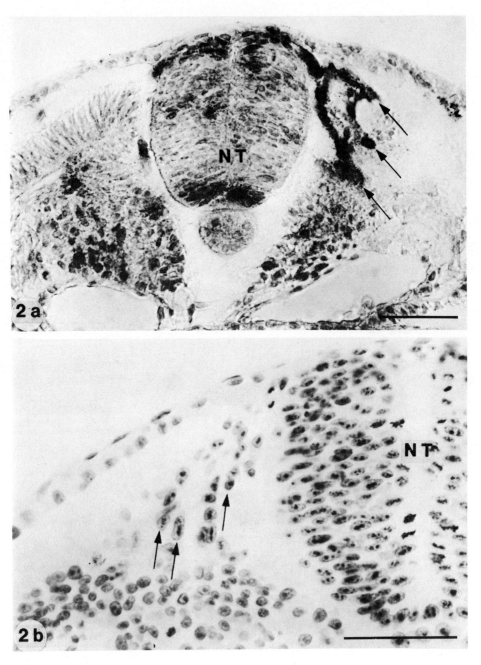

Fig. 2. Intersomitic dorso-ventral migration of neural crest cells in the cervico-dorsal region (somites 7 to 28). (*a*) Histochemical demonstration of AChE on a transverse section between the 8th and the 9th somite (right side) of a 21-somite chick embryo, slightly rostral to that shown in Fig. 1a. AChE-positive crest cells (arrows) migrating in the extracellular space between 2 consecutive ▶

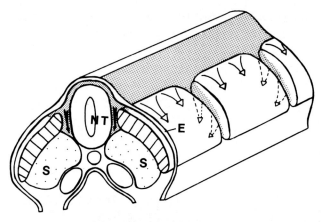

Fig. 3. Schematic representation of the dorsal region of a 2-day-old chick embryo illustrating the early stages of crest cell migration at trunk levels. Crest cell migration proceeds along three different pathways: (1) dorsoventrally in the space located between 2 consecutive somites (dashed arrows); (2) dorsoventrally between the neural tube (NT) and the inner faces of the somites (black arrows); and (3) dorsolaterally (white arrows) between ectoderm (E) and somites (S).

population expands laterally to cover approximately the dorsal half of the neural tube. Then they are stopped for a while, probably because the migration pathways are too narrow as this stage for them to proceed further (Figs. 4b and 5).

The first pathway to appear is that located between two consecutive somites (Fig. 2). The crest cells migrate rapidly through it and become distributed along the dorsal aorta roots where they aggregate to form the primary sympathetic chains.

About 10 h after the initiation of the migration, the dorsoventral route between somites and neural tube opens up, probably by hydration of the hyaluronic acid and glycosaminoglycans present at this level. It is only somewhat later that the dorsolateral pathway taken by the presumptive skin melanocytes materializes (Thiéry et al., 1982).

It is interesting to follow the fate of the cells which have taken the dorsoventral route at the level of a given somite. Between the 2nd and the 3rd days of development, and according to a cranio-caudal gradient, the somites become separated into two regions, the dermomyotome and the sclerotome. At that time, sclerotomal cells start to dissociate and to surround the axial structures, notochord and neural tube, eventually forming the primordium of the vertebra. The

---

somites divide into three streams. Note that the section is slightly oblique and that the somite itself is cut on the left side. (b) Chimaeric quail-chick embryo at the 36-somite stage. Orthotopic and isochronic grafting of a fragment of a quail neural primordium at the posterior level of a 22-somite chick host. As in Fig. 2a, this transverse section from the 31st to 32nd somite level shows quail neural crest cells (arrows) migrating in streams from the top of the neural tube (NT) in the intersomitic space. Bars represent 50 μm.

dorsoventral pathway seems then to become obstructed, resulting in the arrest and accumulation of the neural crest cells, which thereafter give rise to the primordium of the sensory ganglion (Figs. 4b and 5).

In the quail-chick chimaeras, some cells are found within the somites, first between the dermomyotome and the sclerotome, and later on intermingled with the sclerotomal cells (Fig. 1b). Whether they can reach the primary sympathetic chains by this way is not clear, but cannot be ruled out.

Fibronectin which previously surrounded the migrating crest cells, disappears completely as the hyaluronic acid level drops (Toole, 1972; Weston et al., 1978) at the site where the ganglia are forming (Fig. 5a). In contrast the neural cell

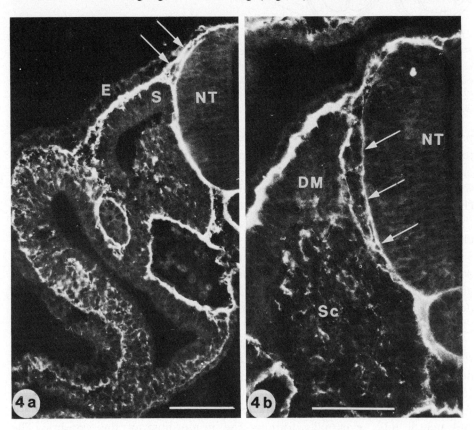

Fig. 4. Dorsoventral migration of neural crest cells in the cervicodorsal region (15th somite level) of the chick embryo. Demonstration of crest cell migration pathways by the immunohistochemical detection of fibronectin (FN). (a) Transverse section at the 25-somite stage. Crest cells (arrows) accumulate for a while at the apex of the somite (S). They are surrounded by the FN-rich basement membranes of ectoderm (E), neural tube (NT) and somite. (b) Transverse section at the 32-somite stage. Crest cells (arrows) accumulate in a FN-rich area between dermomyotome (DM), sclerotome (Sc) and neural tube (NT). They do not progress further ventrally since the ventral pathway seems to become obstructed by sclerotomal cells. Bars represent 50 μm.

adhesion molecule (N-CAM) (Thiéry et al., 1977) is abundantly produced by the developing ganglionic cells (Fig. 5b).

One ends up with the migration pattern described in Fig. 6 showing that the cells migrating between two somites form the sympathetic structures whilst those which migrate between the somites and the neural tube form the DRG. Whether some dispersed cells might reach the ventral sites where the primary sympathetic chain is formed by following a pathway located between the sclerotome and the dermomyotome is a possibility which requires further investigation.

At the cervico-dorsal level, no crest cells, except those lining the nerves as Schwann cells and the melanoblasts which seed the peritoneal epithelium, penetrate the dorsal mesentery and migrate in the gut or in the somatopleure.

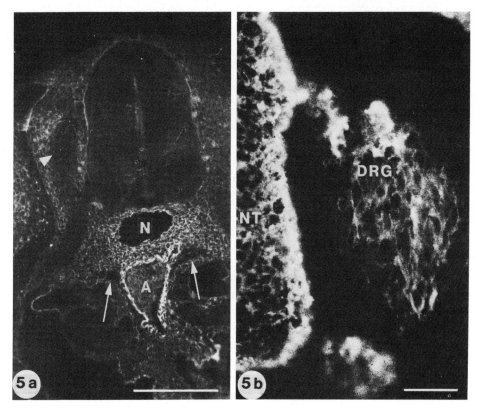

Fig. 5. Immunohistochemical demonstration of fibronectin (FN) (*a*) and neural cell adhesion molecule (N-CAM) (*b*) after the migration of neural crest cells in the cervico-dorsal region of the chick embryo (45-somite stage, 15-somite level). (*a*) The sclerotome has expanded ventrally around the notochord (N) and part of the aorta (A), and dorsally around the dorsal root ganglion (arrow head). The aortic region contains the highest concentration of FN. Note the absence of FN within the sympathetic (arrows) and dorsal root ganglia. Bar represents 200 μm. (*b*) In contrast at the same stage and same level the N-CAM is abundantly produced by the neural tube (NT) and dorsal root ganglion (DRG) cells. Bar represents 30 μm.

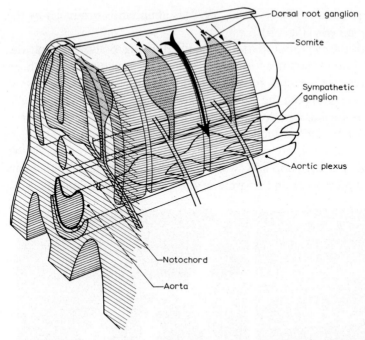

Fig. 6. Schematic representation of the dorsal trunk region of a 5-day-old chick embryo illustrating the migration pathways of neural crest cells that give rise to sympathetic and dorsal root ganglia. Cells migrating between two somites (large arrow) form the sympathetic structures (ganglia and plexuses) whereas those which migrate between the somites and the neural tube (small arrows) form the dorsal root ganglia.

2.1.2. Migration of neural crest cells in the lumbosacral region of the embryo

In this region, the neural crest cells follow the early pathways described above. However, their migration at this level is not strictly restricted to the dorsal area. In birds, some crest cells penetrate the mesentery where they form the ganglion of Remak. Cells of the primordium of Remak ganglion migrate caudocranially within the mesentery and colonize the mesentery up to the level of the duodenojejunal junction.

Through the use of the quail-chick marker system, the lumbosacral neural crest was also shown to participate in the formation of the enteric ganglia of the post-umbilical gut. This participation is not massive but involves only a small number of neuronal and satellite cells of the two plexuses whose precursors stay for a while in the anlage of the Remak ganglion before pursuing their dorsoventral migration from the neural crest to the gut (Le Douarin and Teillet, 1973; Teillet, 1978).

In the caudal region, which is devoid of coelomic cavity, the adrenergic sympathetic ganglia of the pelvic plexus are closely associated with the parasympathetic structures of the ganglion of Remak. However, the use of the formol-induced fluorescence method of Falck (1962) which allows catecholamine (CA)-containing cells to be identified, shows that the ganglion of Remak and the adrenergic ganglia remain well distinct entities except in the very posterior cloacal region, where groups of cells with and without characteristic CA fluorescence are very closely associated (see Teillet, 1978). More cranially, the region of the Remak ganglion located along the hind-gut does not show CA-containing cells, but adrenergic fibres appear in this region from day 6 in the chick and day 5.5 in the quail and remain a constant component of the ganglion.

2.2. Vagal region

Isotopic and isochronic grafts of the neural primordium between quail and chick embryos (Le Douarin and Teillet, 1973) have shown that most of the contribution

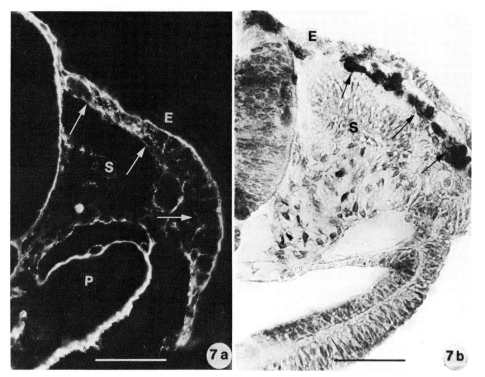

Fig. 7. Dorsolateral migration of neural crest cells at the vagal level (somites 1 to 5), illustrated by the immunohistochemical demonstration of fibronectin (FN) and the histochemical reaction for AChE. (a) Transverse section at the level of somite 2 of an 18-somite chick embryo. Neural crest cells (arrows) follow the FN-rich migration pathway located between the superficial ectoderm (E) and the dorsal aspect of the somite (S) and reach the pharyngeal region (P). (b) Transverse section at the level of somite 3 of a 13-somite chick embryo. Neural crest cells (arrows) migrating dorsolaterally are characterized by an intense AChE reaction, while the neighboring ectoderm (E) and somite (S) are almost completely devoid of enzyme activity. Bars represent 50 μm.

of the neural crest to the enteric ganglia arises from the level of the neural axis corresponding to somites 1 to 7.

At the level of somites 1 to 5, the dorsoventral pathways do not seem to play a major role in crest cell distribution as they do in more caudal regions of the body. In fact, in this area the dispersion of the somitic cells is concomitant with the onset of crest cell migration and prevents the crest cells from infiltrating significantly the spaces between two consecutive somites and between the somites and the neural tube. Most of the migration flow takes the pathway located underneath the superficial ectoderm and thereafter reaches the pharyngeal region (Fig. 7). The precursors of the enteric ganglia then become incorporated in the mesodermal wall of the gut within which they will undergo a cranio-caudal migration.

At the level of somites 6 and 7, the dorsolateral pathway is not the only one taken by neural crest cells. The latter also penetrate between the somites and the neural tube as already described for more caudal regions. It is interesting to note that the first DRG is formed as the level of somite 6, as though DRG formation were strictly related to the availability of the dorsoventral pathway between somite and neural tube.

3. Fate map of the autonomic structures in the neural crest

Isotopic and isochronic grafts of the neural primordium between quail and chick embryos, allowed recognition of the levels of the neural axis from which the various autonomic ganglia and paraganglia arise in normal development. Figure 8 shows that the neural crest is precisely regionalized in areas whose cells have a different migratory behavior and a different fate in terms of transmitter functions. Some migrate into the splanchnopleural mesoderm while others have a migration pattern restricted to the dorsal trunk structures (from somites 8 to 28). It is also clear that the cells which seed the gut differentiate into cholinergic, peptidergic and serotoninergic neurons (see Le Douarin, 1982, for a review) to the exclusion of adrenergic cells. The latter, in contrast, develop in the autonomic structures which are formed in the dorsal-truncal mesoderm derived from the somites.

4. Developmental capabilities of the crest cells at various levels of the neural axis

In order to see whether the migration pattern and the embryonic site where the crest cells become localized influence their differentiation, we modified their initial position along the neural axis before they started migrating. For example, we transplanted the cephalic or vagal neural crest at the level of somites 18 to 24 (*adrenomedullary* level of the crest); conversely, we transferred the adreno-

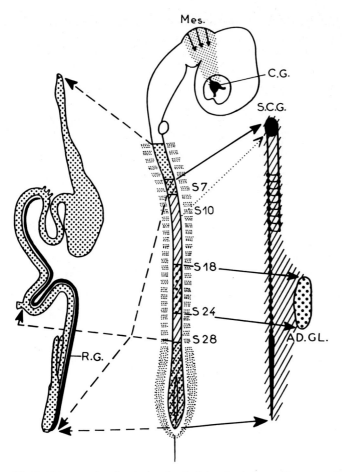

Fig. 8. Diagram showing the levels of origin of enteric ganglia, sympathetic chains and plexuses and ciliary ganglion on the neural crest. The vagal level of the neural crest (from somites 1 to 7) provides all the enteric ganglia of the pre-umbilical gut and contributes to the innervation of the post-umbilical gut. The lumbosacral level of the neural crest, posterior to the level of the 28th somite, gives rise to the ganglion of Remak and some ganglion cells of the post-umbilical gut. The ciliary ganglion arises from the mesencephalic crest. The sympathetic chain and plexuses are derived from the entire length of the neural crest posterior to the 5th somite, the superior cervical ganglion arising more precisely from the level of somites 5 to 10, and the adrenomedullary cells originate from the level of somites 18 to 24. ADGL, adrenal gland; SCG, superior cervical ganglion; S, somite; RG, ganglion of Remak; CG, ciliary ganglion; Mes, mesencephalic crest.

medullary level of the crest to the vagal level. Since these heterotopic grafting experiments were done between quail and chick embryos with quail as donor and chick as host, we could recognize the grafted crest cells and identify their differentiated phenotype. Both the cephalic and the vagal neural crest, transplanted at the level of somites 18 to 24, provided adrenomedullary-like cells for the suprarenal gland. Conversely, the cervico-truncal neural crest, grafted into the vagal region, colonized the gut and gave rise to cholinergic enteric ganglia (Le Douarin and Teillet, 1974; Le Douarin et al., 1975) and normally distributed peptidergic neurons (Fontaine-Pérus et al., 1981, 1982).

These results indicate that the migratory behavior of crest cells depends on the pathways available when they leave the neural primordium rather than on some specificity related to their origin in the neural axis. The vagal and the adrenomedullary regions of the embryo provide preferential routes leading crest cells to the gut and to the suprarenal gland, respectively. There, the phenotypic expression of the crest cell population is regulated by environmental factors, which elicit either cholinergic or adrenergic cell differentiation, irrespective of their fate in normal development.

It can be concluded that the potentialities for giving rise to cholinergic and peptidergic enteric neurons, adrenergic sympathetic ganglia and adrenomedullary paraganglia are not locally restricted but are present practically in the entire crest.

The pluripotentiality of neural crest was also shown by Noden (1978), through heterotopic transplantations of crest from the cranial level. Forebrain crest normally does not give rise to neurons, but when grafted at the midbrain region, crest cells emigrated from their new position and formed normal ciliary and trigeminal ganglia.

Similarly, the capacity for prosencephalic crest cells to differentiate into neurons was further demonstrated by the graft of quail prosencephalic crest cells in the migration pathway of the trunk neural crest (cf., Fig. 11). Such an experiment resulted in the differentiation of quail neurons and glia both in the sensory and in the sympathetic host ganglia (Le Douarin et al., 1979; Le Lièvre et al., 1980 and unpublished results).

5. Expression of the adrenergic phenotype in vivo

The experiments reported above clearly indicate the critical importance of the cellular environment in regulating the choice of neurotransmitter metabolism, adrenergic or cholinergic, in neural crest cells irrespective of their normal fate during development. However, the nature of the cellular interactions that govern neurotransmitter phenotype expression is still unknown. In recent years a growing number of studies have focused on adrenergic cell differentiation, thanks to the availability of very specific and sensitive histofluorescence and immunocytochemical techniques. Using these methods, the normal ontogenetic appearance and the development of adrenergic traits in the autonomic nervous system have been described. Researches carried out on avian as well as mammalian embryos have shown that neither CA (Allan and Newgreen, 1977; Cochard et al., 1978, 1979) nor the enzymes responsible for their biosynthesis (Cochard et al., 1978,

1979; Teitelman et al., 1979) are detectable in neural crest cells during their initial dorsoventral migration. They are found first, soon after crest cells have coalesced to form the primordia of paravertebral ganglia, at 3.5 days of incubation in chick embryos (Enemar et al., 1965; Allan and Newgreen, 1977) and 11 to 11.5 days of gestation in rat embryos (Cochard et al., 1978, 1979; Teitelman et al., 1979). In the rat embryo, there is a remarkable degree of synchrony in the appearance of tyrosine hydroxylase (TOH) and dopamine-β-hydroxylase (DBH), two enzymes of CA biosynthesis, showing that as soon as the enzymes are present in detectable quantities they are metabolically active. This also indicates that the CA detected are the product of an actual synthesis rather than an accumulation, via an uptake system, of molecules that would come from other sources (such as maternal in mammals, vitelline in birds).

From the above, it appears that adrenergic differentiation occurs only when presumptive sympathetic neurons have reached their destination at the dorsolateral aspect of the aorta, within the sclerotomal part of the somite. Then the question arises as to whether an inductive process, responsible for the expression of the adrenergic phenotype, acts while crest cells are migrating dorsoventrally or when they are in their final location. By associating the neural crest with trunk structures, Cohen (1972) reached the conclusion that crest cells receive differentiating signals on their route, the ventral neural tube, notochord and somitic mesenchyme being of decisive importance in promoting the expression of the adrenergic phenotype. In addition, he also concluded that the specificity of the somitic mesenchyme in this process, since cardiac or limb-bud mesenchymes were unable to promote the appearance of CA-containing cells.

This question was further examined by Norr (1973) in a series of experiments involving in vitro organotypic cultures of dorsal trunk structures. The development of the catecholaminergic phenotype was found to depend upon cellular contacts between neural crest cells and the somitic mesenchyme. However, this last structure acquired its inductive capacity only after being previously conditioned by the neural tube and the notochord.

These results, and the fact that sympathoblasts first develop in the vicinity of the notochord, prompted us to investigate whether this structure could, by itself, promote adrenergic cell differentiation. We used, as substratum for crest cell development, the aneural gut mesenchyme. As previously described, in the chick embryo, enteric neuron precursor cells from the vagal and lumbo-sacral levels of the neural crest reach the hind-gut around the 7th day of incubation only (Le Douarin and Teillet, 1973). Thus it is possible to grow the colorectum in a totally aneural state when it is removed from the embryo before this stage. In various series of experiments, the colorectum was explanted at 5 days of incubation and associated with the quail neural primordium from either the vagal or the trunk level, in the presence or absence of the notochord and cultured on the chick chorio-allantoic membrane for 2 to 10 days.

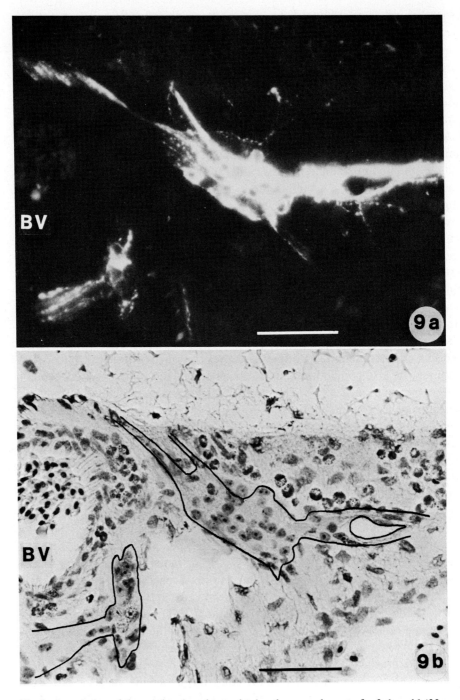

Fig. 9. Association of the cervico-dorsal neural tube plus neural crest of a 2-day-old (25 somites) quail embryo with the entire notochord of the same embryo and the hind gut rudiment of a ▶

In all cases, crest cells migrated into the gut and differentiated into neurons that were shown to express cholinergic and peptidergic traits (Smith et al., 1977; Fontaine-Pérus et al., 1982). In the absence of the notochord the adrenergic phenotype was never expressed in the cells which had seeded the gut. In some cases, however, CA-containing cells developed at some distance of the gut wall, along blood vessels of the chorio-allantoic membrane from some erratic neural crest cells which migrated out of the explant (Fig. 9) (Teillet et al., 1978).

In contrast, when the notochord was included in the explant groups of catecholaminergic cells developed in the site of the myenteric plexus and occasionally also in the submucous plexus (Teillet et al., 1978). These cells were already present after 2 to 3 days of grafting and could still be evidenced 10 days after the beginning of the graft (Fig. 10). We observed that there was a direct relationship between the occurrence of CA-containing cells and the amount of notochordal material included in the graft. In initial experiments, a single fragment of notochord had been included in the explant. In such cases adrenergic cells developed in the gut wall only in 25% of the grafts (Teillet et al., 1978). Additional experiments in which 2 or 3 fragments of notochord were associated with the gut resulted in the development of adrenergic cells in 80% of the explants observed (Cochard et al., unpublished results). These findings do not confirm the exclusive ability of somitic derived mesenchymes to induce adrenergic cell differentiation, as previously claimed (Cohen, 1972; Norr, 1973) since this process can occur in the splanchnopleural mesenchyme of the gut and in the chorio-allantoic membrane. The fundamental role of the notochord is demonstrated by its ability to promote the appearance of the adrenergic phenotype in an ectopic environment such as the gut mesenchyme, where it is not normally expressed.

In addition, the differentiation of adrenergic cells along blood vessels of the chorio-allantoic membrane, even in the absence of the notochord, is an interesting observation. It suggests that a factor similar to that produced by the notochord might also be present in the environment of the chorio-allantoic membrane blood vessels. In fact, the chorio-allantoic membrane is not the only vascular environment which is able to trigger CA synthesis in neural crest cells. Recent experiments in our laboratory have shown that catecholaminergic cells can also develop ectopically along arterial walls in the umbilical cord, thus suggesting that the wall of the arteries per se might be responsible for this phenomenon.

Additional support for the decisive role played by environmental factors in the regulation of adrenergic phenotype expression is also provided by certain observations of the normal development in the mammalian embryo.

4–5-day-old chick embryo. Culture for 8 days on the chorio-allantoic membrane. (a) Two groups of CA-containing cells, evidenced by the formaldehyde-induced fluorescence technique, are found outside the explant in the vicinity of blood vessels (BV). Note that these brightly fluorescent cells extend long CA-containing processes. (b) Feulgen-Rossenbeck staining of the same section showing that both fluorescent cell clusters are made up of quail cells. Bars represent 50 μm.

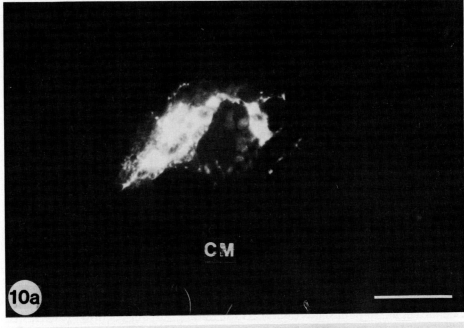
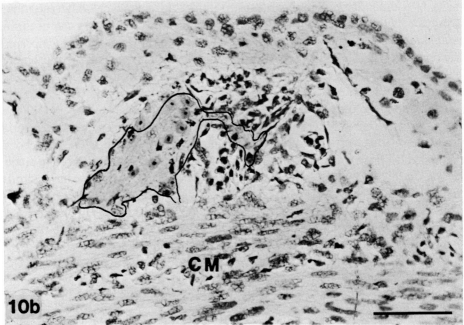

Fig. 10. Same experiment as in Fig. 9. (*a*) A group of CA-containing, brightly fluorescent cells is found near the developing circular muscle layer of the gut (CM), in the myenteric plexus. (*b*) Feulgen-Rossenbeck staining of the same section. CA-containing cells of the ganglion are of quail origin. Bars represent 50 μm.

As described above, in the rat embryo, CA and their synthesizing enzymes first appear in cells of the developing sympathetic ganglia at 11.5 days of gestation. In addition, cells immunoreactive for TOH and DBH, that also contain CA, are present at the same stage in ectopic position within the gut mesenchyme. Their morphological and biochemical features are identical to those of sympathoblasts developing at the same stage along the aorta. Moreover, their time of appearance, location and the recent evidence that at least some of them elaborate neurofilament proteins (Cochard, Paulin and Le Douarin, unpublished results) suggest that they are of neural crest origin, and possibly represent precursor cells of enteric neurons, although this remains to be demonstrated.

In any case, the adrenergic phenotype is only transiently expressed in the gut by these cells, since they are not detected after the 14th gestational day (Cochard et al., 1978, 1979; Teitelman et al., 1979). In addition to CA synthetic abilities, these transient catecholaminergic gut cells also exhibit a high affinity uptake process specific for noradrenaline (Jonakait et al., 1979). However, this last noradrenergic character does not disappear together with transmitter synthesizing ability and can still be demonstrated at later stages, up to 17th day of gestation. This fact strongly suggests that the disappearance of catecholaminergic cells at 14 days of gestation is not due to cell death but rather to the loss of some adrenergic traits, namely the capacity to synthesize CA, while the ability to take up noradrenaline is retained for several additional days.

Transient adrenergic cells with a similar developmental time course have also been found in the gut of the mouse embryo (Teitelman et al., 1981) but so far we have been unable to detect such cells in the avian gut (Cochard and Le Douarin, unpublished observations).

The resemblance to sympathoblasts of the transient catecholaminergic cells in the gut is further accentuated by the fact that they are able to proliferate (Teitelman et al., 1981) and that they respond to nerve growth factor (Kessler et al., 1979) and glucocorticoids (Jonakait et al., 1980) by expressing their CA synthetic activity at a higher level and for longer periods of time.

If these cells are actually derived from the neural crest, it is thus conceivable that they express adrenergic traits similar to those displayed by sympathoblasts as a result of interactions with the somite-notochord–neural tube complex that they meet during their dorsoventral migration. However, the expression of adrenergic phenotype could not be maintained in cells which home to the gut, possibly because of the lack of appropriate stimulations.

Another example of plasticity in transmitter choice during normal development is provided by the differentiation of the sympathetic *cholinergic* neurons, which in cats and rats, innervate the eccrine sweat glands. This cholinergic innervation is provided by neurons located in the coeliac ganglion. Recent evidence, based on electron microscope observations, indicates that during the course of development the coeliac sympathetic fibers that reach the eccrine sweat glands convert from the adrenergic to the cholinergic-peptidergic phenotype (Landis and Keefe, 1980).

Although the mechanisms responsible for this apparent shift in neurotransmitter expression remain unclear, these observations substantiate the experimental demonstrations, both in vivo and in vitro, of mutability in transmitter phenotypic expression (see Patterson, 1978; Le Douarin, 1981) and show that in normal development as well the expression of the adrenergic phenotype is not always a stable and irreversible process. This in turn suggests that initial expression and further maintenance of adrenergic differentiation are two relatively independent events that might be regulated by different factors.

6. Differentiating potentialities of peripheral ganglionic cells during development

From the experiments described above two notions arise. One is that as far as the ontogeny of the peripheral nervous system is concerned, the neural crest population is multipotential, at practically all levels of the neural axis. The other is that the expression of a given phenotype in the crest-derived structures is selected by the environment to which the crest cells are subjected in the tissues where they differentiate.

The problem of the mechanisms by which the selection of only some of the developing potentialities of neural crest cells occur in the ganglia was therefore raised.

Experiments involving the back-transplantation of developing quail ganglia into the neural crest migration pathway of a 2-day-old chick host were carried out. They showed that potentialities that are not expressed by a given ganglion can be elicited when cells of such ganglia are led to develop in different environments (Le Douarin et al., 1978, 1979; Le Lièvre et al., 1980).

6.1. Graft of autonomic and dorsal root ganglia

Fragments of developing quail autonomic and sensory ganglia of about 2000 cells were implanted into 2-day-old chick embryos at the adrenomedullary level of the neuraxis, between the somites and the neural primordium (Fig. 11).

As a control experiment, pieces of neural crest taken from the truncal level of quail embryos at the 15- to 25-somite stage were inserted as supernumerary grafts into chick hosts by the same procedure as for the ganglia. This was done to see whether the crest cells became localized in their normal sites of arrest in these types of grafts. The distribution of crest cells in the host was similar to that observed in isotopic-isochronic grafting experiments in which the host neural primordium was replaced by the equivalent tissue from the donor (Le Douarin, 1976): quail cells participated in the formation of the host DRG, the sympathetic

Embryonic development of the autonomic system | 19

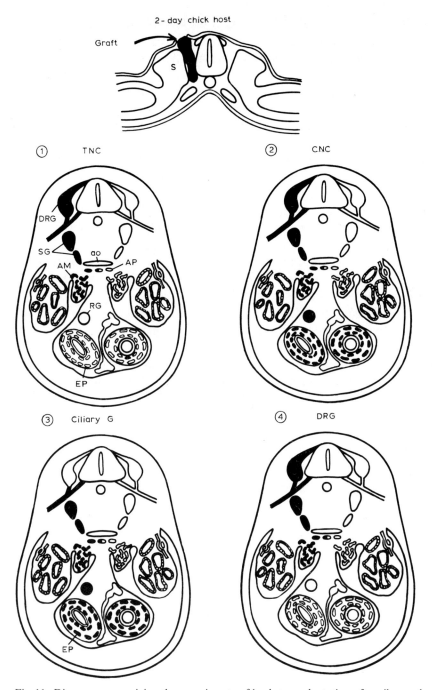

Fig. 11. Diagram summarizing the experiments of back-transplantation of quail neural crest cells or crest derivates at the adrenomedullary level of a 2-day-old chick host. The top schematic drawing illustrates the experimental procedure: the grafted tissue is inserted between the somite (S) and the ▶

(Legend continues on page 20)

chain and plexuses and the adrenal medulla at the level of the operation but they did not penetrate the dorsal mesentery. This indicates that grafting of neural crest or neural crest derivatives in addition to the normal host structures does not prevent the grafted cells from finding the sites of arrest that they would have reached under normal conditions. Interestingly, when cephalic or vagal neural crest cells were grafted as supernumerary tissues, their migration extended to the enteric plexuses and to the ganglion of Remak (Fig. 11). The ganglia chosen for the experiments were autonomic (ciliary and sympathetic) and sensory (DRG), taken from quail donors at 4.5 to 6 days of incubation. The fate of the grafted cells was followed at various times after the operation. Irrespective of the nature of the grafted ganglion, its peripheral cells detached from the bulk of the explant, mainly on the side facing the neural tube and the ventral root of the aorta. However, the pattern of disaggregation of the implanted tissue was not identical for the different types of ganglia grafted. Namely, cell death of grafted cells was much more noticeable following DRG implantation than it was after grafting of autonomic ganglion cells. From 6 days onwards the definitive localization of implanted cells was established for all types of grafts.

For autonomic ganglia, quail cells were found: (i) lining the fibres of the rachidian (spinal) and autonomic nerves; (ii) in the sympathetic ganglia; (iii) in the aortic and adrenal plexuses (as neurons and glia); and (iv) in the adrenomedullary cords of the suprarenal glands. Grafted sympathetic ganglia did not contribute to the enteric plexuses, whereas ciliary ganglion cells migrated into the ganglion of Remak and into Auerbach's and Meissner's plexuses of the mid and hind gut (Fig. 11). It is important to emphasize that during the first 2 postoperative days, quail cells were often seen in the dorsal somitic region at the sites of formation of the DRG but, subsequently, were absent from these structures except in a few instances where they were found as scattered satellite cells among chick neurons. In any case, they never showed the neuronal phenotype characterized by large nuclei and neurofibrils.

---

neural tube of the host, at the level of somites 18 to 24 and at a stage preceding the migration of host crest cells. (1 and 2) Supernumerary crest cells from the truncal level (TNC) (1) or the cephalic level (CNC) (2) become localized in the normal sites of arrest of the host trunk crest cells: quail cells participate in the formation of the dorsal root ganglia (DRG), the sympathetic ganglia (SG), the aortic plexuses (AP) and the adrenal medulla (AM). In addition, cephalic crest cells (2), penetrate the dorsal mesentery and populate the ganglion of Remak (RG) and the enteric plexuses (EP). (3 and 4) Results from the graft of a fragment of 5- to 7-day ciliary ganglion (3) or DRG (4). In both types of grafts, quail cells are found as Schwann cells along the rachidian nerves, as supportive cells of the sympathetic structures and as CA-containing cells in the sympathetic ganglia, aortic plexuses and adrenal medulla. However, two major differences are observed between the two kinds of grafts: Ciliary ganglion cells (3) penetrate the dorsal mesentery and colonize the cholinergic Remark and enteric ganglia, as already seen for the graft of supernumerary cephalic crest cells (2), but they never give rise to DRG cells. In marked contrast, grafted DRG cells (4) participate in the formation of host sensory ganglia but do not penetrate the dorsal mesentery.

When the grafted DRG were taken from the quail embryo at 5 to 7 days of incubation, cells from the graft were found in the host DRG, sympathetic ganglia and adrenal medulla. In the DRG, the majority of the quail cells aggregated at the ventrolateral aspect of the host ganglion, where they appeared as large neurons and supporting cells (Fig. 11). In none of the cases observed was the host DRG predominantly made up of quail cells, as it was when a piece of trunk crest was implanted. When the sensory ganglion originated from a host older than 7 days (from 7 to 10 days), no sensory neurons were ever found in the host DRG (unpublished results).

Associated formol-induced fluorescence and Feulgen-Rossenbeck techniques applied to the host embryos showed fluorescent quail cells in the sympathetic ganglia and plexuses, and in the adrenal medulla, whether the graft was of sensory or autonomic origin. Under the electron microscope the chimeric suprarenal adrenal gland exhibited cells with both the typical DNA-rich quail nucleolus and characteristic CA secretory granules.

These results led to the following conclusions. Firstly, the observed distribution of grafted ganglion cells in the host is the result of a cell–cell recognition process between host and grafted cells. Autonomic ganglion neurons and satellite cells become localized in the autonomic structures of the host (i.e., the ciliary ganglion cells become localized in the sympathetic ganglia and plexuses, adrenal medulla and enteric ganglia). In contrast, the grafted sensory neurons (of the DRG) colonize the host DRG in addition to the sympathetic and adrenomedullary structures in the host.

Therefore, at 5- to 7-days of incubation, the developmental capabilities of DRG and autonomic ganglia are different: those of the DRG cells are broader since, in the back-transplantation system, they differentiate along both 'sensory' and 'autonomic' cell lines. In contrast, autonomic ganglia of any kind, when back-transplanted, give rise only to autonomic ganglion cells. From the time all neurons of the DRG become post-mitotic (Carr and Simpso, 1978) onward, the developing capabilities of the ganglion cells are more restricted. After 7 days of incubation, quail DRG transplants give rise only to autonomic derivatives and sensory neurons no longer develop from the graft.

Since our experiments involve transplantation of a heterogeneous population of cells, the problem was to identify in the graft the particular cell types able to respond to new developmental cues by expressing phenotypes that would have been repressed in normal conditions.

At the time of transplantation, peripheral ganglia still contain undifferentiated cells that are possible candidates for phenotypic reorientation. Such undifferentiated cells belong to the small-sized cell population of the ganglia. The problem of whether neurons can develop from this small-sized cell population, which normally gives rise only to satellite cells, was approached through the use of the nodose ganglion.

6.2. Back-transplantation of the nodose ganglion

Because of its dual origin both from the neural crest and the placodal ectoderm (Narayanan and Narayanan, 1980), the nodose ganglion of the vagus nerve appeared to be an interesting model to further analyze the respective behavior of the developing neurons and of the non-neuronal cell population of a given ganglion in the situation provided by back-transplantation into a younger host.

By grafting a quail rhombencephalic primordium isotopically into a chick host, before crest cell migration had begun at this level, it was possible to construct a chimeric nodose ganglion in which the satellite cells were of the quail type and the neurons of the chick type. Pieces of such a chimeric nodose ganglion taken between 5.5 and 9 days were subsequently implanted at the adrenomedullary level of 2-day-old chick embryos as indicated in Fig. 11.

In these conditions, the fate of the presumptive satellite cells could be selectively followed in the host because they were the sole elements to be labelled by the quail nucleolus. Interestingly, when the host was observed at 6 to 10 days of development, certain quail cells were found in the sympathetic ganglia and in the adrenal medulla and plexuses. In some of them, CA content could be revealed with the formol-induced fluorescence and Feulgen-Rossenbeck associated techniques. In addition, a massive colonization of the gut took place, namely in the jejunum and enteric neurons with the quail marker developed in the host myenteric and submucosal plexuses (Fig. 12).

The reverse type of chimeric nodose ganglion could also be constructed by taking a quail embryo as the host and implanting isotopically a chick rhombencephalon at the appropriate stage of development. In this instance, the neurons were of the quail type while the presumptive satellite population belonged to the chick donor. Grafting of pieces of this chimeric tissue into a 2-day-old chick in the neural crest cell migration pathway allowed the fate of the placode derived neuronal cells to be studied.

It appeared that only a few (3 to 4) days after the graft, quail neurons could no longer be detected in the chick host tissue. In fact, the grafted post-mitotic neurons were unable to adjust to the new environment with which they were confronted and died soon after the implantation (Ayer-Le Lièvre and Le Douarin, 1982).

These experiments certainly demonstrate that neuronal potentialities exist in the small-sized population of the developing nodose ganglion although normally they do not seem to be expressed at all. One can speculate that such potentialities are repressed under the influence of the placodal neurons. In the conditions of back-transplantation into a 2-day-old embryo, dispersion of the grafted cells takes place and the placodal elements cannot exert their inhibitory effect on the neural crest-derived cells which are subjected in the host to new developmental cues.

Although the model provided by the nodose is somewhat different from that

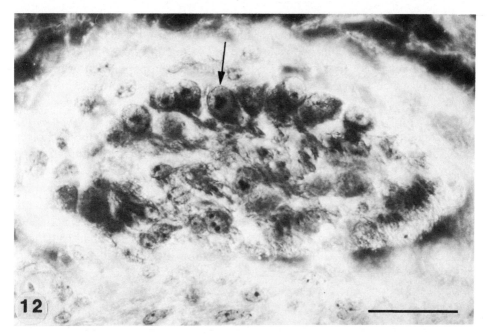

Fig. 12. Silver stain of a chimaeric enteric ganglion resulting from the back-transplantation of a 5-day chimaeric nodose ganglion into the trunk region of a 2-day-old chick host as explained in Fig. 11. Cells of the small-sized population of the nodose ganglion, selectively labelled by the quail cell marker (arrow) are found in the enteric plexuses and differentiate into neurons as evidenced by their affinity for silver salts. Bar represents 20 μm.

of the other peripheral ganglia because of its mixed neural crest–placodal origin, it suggests that in all types of experiments the ganglion cells which respond to the differentiation cues of the young host belong to the non-neuronal population. The latter probably contains a reserve of undifferentiated (or reversibly committed) cells which can be programmed toward differentiation pathways which are not normally expressed in the donor ganglion. In contrast, the postmitotic neurons of the grafted ganglia very likely do not survive in graft.

7. In vitro studies of neural crest cell differentiation

7.1. Culture of crest cells in serum-containing medium

Although in vivo studies permit the identification of the tissues that influence autonomic neuronal development, only an in vitro approach can lead to the isolation and characterization of the factor(s) involved. That is why a number of workers, including ourselves, have turned their attention to the culture of neural crest. The migration of crest cells away from a neural primordium explanted in vitro, already described by Dorris (1936), has been exploited by Cohen and Konigsberg (1975) to obtain viable primary cultures of quail neural crest. Crest cell migration in vitro begins almost at once and 48 h after explantation, the neural tube can be removed, thus leaving on the culture dish a population of cells, the majority of which can be considered as being of neural crest origin. Cultures of trunk crest prepared in this way were shown by Cohen (1977) to develop adrenergic properties spontaneously in medium supplemented with horse serum and with embryo extract. Similarly, Greenberg and Schrier (1977) have reported the differentiation of cholinergic properties in mesencephalic crest cultures derived in vitro from whole mesencephalic primordium.

This method has also been successfully exploited in studies of the behavior and of the role of the substratum in the migration process of crest cells in vitro (Newgreen and Thiéry, 1980; Rovasio et al., 1983). However, in experiments dealing with the differentiation of neuronal cells, we considered that excision of crest cells as free as possible from contamination by other cell types prior to cultivation would provide better conditions for analyzing their potentialities of autodifferentiation. The even momentary presence in culture of the neural tube may have an influence on crest cell differentiation by producing extracellular material. Furthermore, some contamination of the crest cell culture by neuroblasts of the spinal cord cannot be completely ruled out when the whole neural primordium is explanted. By microdissection, we could remove the crest both from the trunk and the mesencephalic levels at the neural fold stage. In addition, the mesencephalic crest is a particularly convenient subject for such experiments, since at the 8- to 12-somite stage, crest cells migrate as a multilayered sheet of cells underneath the ectoderm and can be easily removed in a pure state. About 800 and 2000 cells are obtained on each side of the mesencephalon at the neural fold and at the migrating stages respectively.

Biochemical determination of neurotransmitter synthesis as well as cellular morphology were used as indexes of neuroblast differentiation (Ziller et al., 1979; Fauquet et al., 1981). We were able to show that mesencephalic crest (a region from which the ciliary ganglion and also part of the trigeminal ganglion arise and which does not normally participate in adrenergic ganglion formation), cultured for seven days in Dulbecco's Modified Eagle Medium (DMEM) supplemented with 15% fetal calf serum, developed both cholinergic and adrenergic traits (Table 1). Explants of trunk neural fold always differentiated along the cholinergic pathway, but only rarely (one case out of nine cultures) were they found with CA-synthesizing activity (Table 1). In contrast, when cultures were prepared from the total trunk-level primordium (i.e., by explantation of neural tube plus neural fold – cf., Cohen and Konigsberg, 1975), adrenergic differentiation was observed in just over half the cases (Table 1). A possible explanation for these results is that an early, albeit brief, contact between neural crest cells and the neural tube, or extracellular material produced by the neural tube, is sufficient to induce catecholaminergic differentiation in vitro.

TABLE 1
Synthesis of neurotransmitters by neural crest before and after culture

Origin of neural crest	Culture conditions	Number of samples of cultures synthesizing		Quantity of neurotransmitter synthesized (fmol)	
		CA	ACh	CA	ACh
Mesencephalic	Non-cultured	0/6	8/8	0	48 ± 14
Mesencephalic	Cultured 7 d + HS	4/11	11/11	9 ± 5	553 ± 83
Mesencephalic	Cultured 7 d + FCS	9/10	10/10	20 ± 4	165 ± 37
Trunk	Cultured 7 d + HS	0/10	10/10	0	30 ± 7
Trunk	Cultured 7 d + FCS	1/9	9/9	2 ± 2	43 ± 6
Trunk (from total neural primordium)	Cultured 7 d + FCS	5/9	9/9	12 ± 5	74 ± 18

Cultures of trunk and mesencephalic crest, and freshly removed fragments of the latter, were incubated with [^{3}H]tyrosine and [^{3}H]choline for 4 h. The synthesis of catecholamine (CA) and acetylcholine (ACh), determined essentially as described by Mains and Patterson (1973), is expressed as fmoles (mean ± SEM) formed per dish during the incubation period. Each culture of excised crest contained 14 crest explants; trunk crest obtained from total neural primordium was derived from 7 neural tubes. In experiments with non-cultured crest, 14 fragments (about 2×10^4 cells) were used for each incorporation. Note that horse serum (HS) is more favourable than foetal calf serum (FCS) to ACh synthesis in cultured neural crest cells. In contrast with FCS both the number of cultures synthesizing CA as well as the level of CA-synthesis are enhanced with respect to the cultures in HS.

The above described results led us to orientate our investigations in two directions:

(i) The occurrence of acetylcholine(ACh)-synthesizing activity in cell cultures of neural crest arising from any level of the neuraxis prompted us to examine its possible onset earlier in development, i.e., before the precursors of the peripheral neurons have aggregated into ganglia and started to differentiate into neurons.

(ii) In addition, although neural crest cultures in DMEM + 15% serum synthesize neurotransmitters, it is striking to see that no morphological differentiation of neuronal cells is apparent. This is why we undertook a series of studies in which we manipulated the culture medium in order to induce neurite outgrowth by the crest cells in culture.

7.2. Cholinergic traits in the migrating neural crest

The particular features of the neural crest at the level of the head (the abundance of cells and the migration as a sheet underneath the ectoderm), which allow a pure population of cells in the process of migration to be obtained, prompted us to see whether ACh-synthesizing ability already exists in the crest before explantation.

We were able to show that isolated 'migrating' mesencephalic crest contained choline acetyltransferase (CAT) and could synthesize ACh (Smith et al., 1979). The quantities measured, although small, were significant; what is more, crest removed from mesencephalic neural folds (when migration is only just beginning) could also make ACh. Neither type of crest could convert [^{3}H]tyrosine to CA (Table 2). Thus, even before the mesencephalic crest population is subjected to

TABLE 2
Neurotransmitter synthesis by trunk neural crest cells included in the sclerotomal moiety of the somite, before and after culture

Sclerotome	Quantity of neurotransmitter synthesized		
	Catecholamines (total, fmol)	Acetylcholine (total, fmol)	Molar ratio CA/ACh
Non-cultured	0 (6)	103 ± 35 (5)	0.000 (5)
Cultured 24 h	82 ± 25 (8)	47 ± 9 (8)	2.142 ± 0.757 (8)
Cultured 7 d	949 ± 113 (16)	604 ± 130 (16)	2.473 ± 0.372 (16)

The conversion of [^{3}H]tyrosine and [^{3}H]choline to catecholamine (CA) and acetylcholine (ACh) respectively, by preparations of sclerotomes before and after culture in medium supplemented with fetal calf serum. Synthetic activity is expressed as fmoles (mean ± S.E.M.) transmitter produced in 4 h per dish (each containing 20 sclerotome explants or non-cultured rudiments). Number of determinations in parentheses. (From Fauquet et al., 1981.)

any influences from tissues encountered on its route to the target site, it already contained cholinergic properties. It is obviously difficult to determine whether all or only a fraction of the crest possesses cholinergic activity, but it is likely that the cells responsible include the presumptive neuroblasts of the ciliary ganglion.

At the trunk level, neural crest cells are not abundant enough to permit the biochemical assay for ACh or CA synthesis to be carried out. However, if the trunk crest cells are taken from the embryo during the migration process when they are included in the sclerotomal moiety of the somites (at 3 days of incubation in either chick or quail embryos) it is possible to demonstrate that they can synthesize ACh (Table 1).

CA synthesis, in contrast, could be evidenced only after the cells had been cultured for a certain period of time (Table 2). It is interesting to notice that when removed from the embryo during the 3rd day of incubation together with the sclerotomal mesenchyme, the trunk neural crest cells differentiate into neurons in which CA storage can be easily evidenced by cytochemistry as well as at the electron microscope level (Fauquet et al., 1981).

As mentioned earlier in this article, the ACh degradating enzyme AChE has been demonstrated in crest cells from all axial levels, before and during their

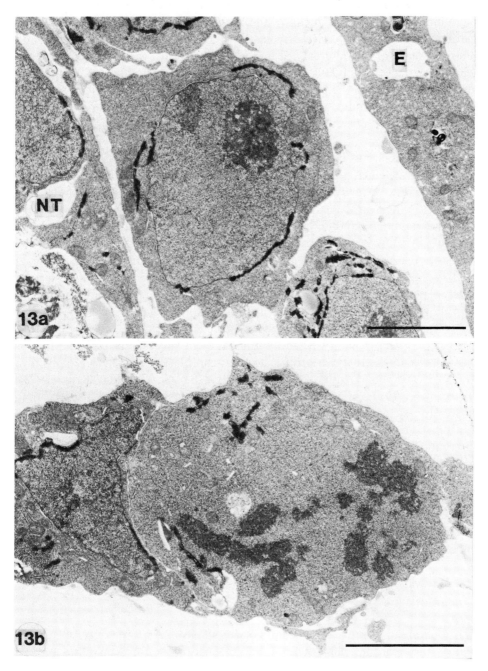

Fig. 13. Electron micrographs showing the intracellular distribution of the AChE-cytochemical reaction product in migrating neural crest cells of the chick embryo. (a) 9-Somite embryo, middle-rhombencephalic level. AChE reaction product is restricted to perinuclear cisternae and endoplasmic reticulum and is never found on the plasma membrane. E: ectoderm; NT: neural tube. (b) 24-Somite embryo, cervical level. AChE reaction product is also found in a migrating crest cell undergoing mitosis. Bars represent 3 μm.

migration (Drews, 1975; Cochard and Coltey, 1983). A careful study of the distribution of AChE-positive cells in the chick embryo at different levels and different stages of neural crest cell migration has shown that about 90% of the neural crest cells exhibit this enzymatic activity. Only small groups of negative cells were observed in the trunk and the cephalic crest. It is interesting to notice that the reaction product, as detected at the EM level, was always present inside the ergastoplasmic reticulum and not on the plasma membrane (Fig. 13) (Cochard and Coltey, 1983).

The impossibility to evidence CAT activity cytochemically allows a doubt to subsist as to whether AChE activity (although located intracellularly) is always associated with ACh synthesis in all neural crest cells. Whether cholinergic characters in the neural crest cells are the sign of their early determination toward the peripheral neuronal differentiation pathway is also a question mark. One can speculate that low levels of cholinergic metabolism could be a trait linked to the ability of crest cells to migrate actively. Furthermore a number of observations have shown that cholinergic characters are present during development in a variety of peripheral ganglia which do not necessarily become functional cholinergic neurons when they have reached a fully differentiated state.

For example CAT activity has been detected in early mouse sympathetic ganglia (Coughlin et al., 1978) and a number of in vitro studies indicate that at least part of this activity is intrinsic to the neuroblasts themselves (Johnson et al., 1976; Hill and Hendry, 1977; Ross et al., 1977; Coughlin et al., 1978). Furthermore, CAT, ACh and AChE have been transiently found in developing sensory (dorsal root) ganglia (Strumia and Baima-Bollone, 1964; Marchisio and Consolo, 1968; Pannese et al., 1971; Karczmar et al., 1980; and our own observations). Consequently, a cholinergic metabolism might well be an early feature of the presumptive neuroblasts contained in the neural crest all along the neural axis. This metabolism would then be retained during the early stages of neuronal differentiation, irrespective of the ultimate neurotransmitter synthesized by the various peripheral ganglion neurons. Its subsequent stabilization or disappearance in cholinergic and non-cholinergic neurons respectively, would then depend on external cues.

7.3. Neuronal differentiation in cultured neural crest cells in serum-deprived medium

Our attempts to improve the culture conditions so that neural crest cells could express the neuronal phenotype in the absence of non-neuronal tissue were successful in various respects. First they led to the demonstration that determined neuronal precursors are present in the neural crest as early as the neural fold stage (at the mesencephalic level) (Ziller et al., 1981, 1983). If the mesencephalic neural crest is cultured in a medium totally devoid of serum but containing a mixture of hormones and growth factors (Basic Brazeau's Medium – as described in Brazeau et al., 1981) a number of cells with neurite outgrowths differentiate within the first 24 h and can survive several weeks (Ziller et al., 1981) (Fig. 14).

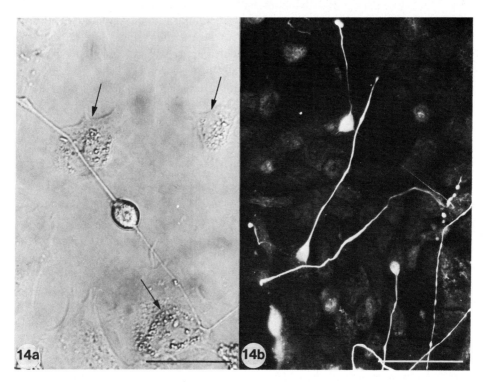

Fig. 14. Mesencephalic neural crest cells cultivated for 7 days in serum-deprived medium. (a) Within 24 h a number of neural crest cells differentiate into neuron-like cells and extend long fine processes. Other crest cells affect a flat fibroblastic-like morphology (arrows) (Nomarski optics). (b) The neuronal nature of the process-bearing cells is controlled here by the presence of the 200 kilodalton subunit of the neurofilament protein detected by a specific antibody in indirect immunofluorescence. Bars represent 30 μm.

Various cytochemical markers were used to control that the process-bearing cells were really of the neuronal type. Such were for example tetanus toxin binding, neurofilament protein synthesis (Fig. 14b) and silver staining (Ziller et al., 1983). In addition, [^{3}H]thymidine incorporation by the explanted crest cells at various times during the culture period showed that no cell division occurred in the neuronal precursors prior to their differentiation. In contrast, the non-neuronal cells of the neural crest were found to incorporate [^{3}H]thymidine at practically all stages of the culture period.

The presence of some inducible neuronal precursors among the neural crest cells demonstrates the heterogeneity of the crest population in which only certain cells are determined toward the neuronal differentiation pathway and are able to respond to serum deprivation by readily extending neurites. A study of the dynamics of cell division in the neural crest cells of the mesencephalon from the time they start migrating to the stage when the ciliary ganglion neurons become post-mitotic is presently in progress. Our preliminary results indicate that most crest cells undergo several mitotic cycles during their migration and in the developing ciliary ganglion. Whether the neuronal precursors, whose presence we demonstrate in the crest, normally generate only neurons or, alternatively, give rise to neurons plus satellite cells is an interesting question to which our experiments do not bring an answer at the present time.

8. Concluding remarks

The experiments reported in this article were aimed at elucidating the early ontogenetic events which lead from the neural crest to the ganglia of the peripheral nervous system. The migratory behaviour of the crest cells was shown to be directed by the general morphogenesis of the non-neuronal structures of the embryo. The development of the somites in particular appeared to play a very decisive role in the differentiating events leading to the formation of both the DRG and the sympathetic ganglia, plexuses and paraganglia. The development of the enteric innervation in contrast is apparently relatively independent of that of the axial mesodermal structures and also of the establishment of early connections with the central nervous system. This was attested by the differentiation of apparently normal cholinergic and peptidergic enteric plexuses in isolated gut fragments associated with neural crest cells from either the vagal or the truncal levels of the neural axis.

Although the normal fate of the neural crest-derived cells varies according to their initial position on the neural primordium, their developmental capabilities are fairly equally distributed at all levels of the crest as far as the cell types encountered in the PNS are concerned. This could be concluded from the fact that enteric ganglia, DRG and sympathetic structures as well, could be obtained from any level of the crest provided that appropriate environmental conditions were encountered by the cells for the expression of each type of differentiation.

Back-transplantation experiments of developing peripheral ganglia into the migration pathway of younger host embryos have been informative in many respects. In particular, they showed that each type of ganglion contains a larger variety of differentiating potencies than those which are actually expressed during the normal course of ontogeny. This can be interpreted as meaning that those extra-potentialities are either repressed in the developing ganglia or that the proper differentiation cues are lacking.

The nodose ganglion transplantation experiment teaches us that although only glial cell will normally arise from its constitutive neural crest cell population, the latter contains a large range of capacities for neuronal phenotypic expression going from sympathetic neurones and adrenomedullary-like cells to enteric ganglia.

So far, the factors originating from the non-neuronal cells which influence their fate in terms of expression of either the neuronal phenotypic morphology or of neurotransmitter synthesis have not been isolated and therefore their nature has not been biochemically defined. Attempts made to investigate such problems in in vitro systems permit to hope that significant advances will emerge soon. Progress is being made in elucidating the nature of the factor produced by glial cells and various other cell types which elicit the switch from adrenergic to cholinergic biosynthetic activities in cultured neurones of the superior cervical ganglion of the newborn rat (Hawrot, 1980; Fukada, 1980; Weber, 1981). In addition, our recent observation that serum-deprivation of the culture medium induces neurite outgrowth also opens up new avenues of research to study the extrinsic influences triggering neuronal differentiation within the developing nervous system.

Acknowledgements

This work was supported by the Centre National de la Recherche Scientifique and by NIH research grant R01 DEO 4257 03 CBY.

References

Allan, I.J. and Newgreen, D.F. (1977) Am. J. Anat. *149*, 413–421.
Ayer-Le Lièvre, C.S. and Le Douarin, N.M. (1982) Dev. Biol., *94*, 291–310.
Brazeau, P., Ling, N., Esch, F., Bohlen, P., Benoit, R. and Guillemin, R. (1981) Regulatory Peptides *1*, 255–264.
Carr, V.McM. and Simpson, S.B. (1978) J. Comp. Neurol. *182*, 727–740.
Cochard, P. and Coltey, P. (1983) Dev. Biol., in press.
Cochard, P., Goldstein, M. and Black, I.B. (1978) Proc. Natl. Acad. Sci. USA, *75*, 2986–2990.
Cochard, P., Goldstein, M. and Black, I.B. (1979) Dev. Biol. *71*, 100–114.
Cohen, A.M. (1972) J. Exp. Zool. *179*, 167–182.
Cohen, A.M. (1977) Proc. Natl. Acad. Sci. USA *74*, 2899–2903.
Cohen, A.M. and Konigsberg, I.R. (1975) Dev. Biol. *46*, 262–280.
Coughlin, M.D., Dibner, M.D., Boyer, D.M. and Black, I.B. (1978) Dev. Biol. *66*, 513–528.
Dorris, F. (1936) Proc. Soc. Exp. Biol. Med. *34*, 448–449.
Drews, U. (1975) Prog. Histochem. Cytochem. *7*, 1–52.
Duband, J.-L. and Thiéry, J.P. (1982) Dev. Biol. *93*, 308–323.
Enemar, A., Falck, B. and Håkanson, R. (1965) Dev. Biol. *11*, 268–283.
Falck, B. (1962) Acta Physiol. Scand. 56, Suppl. *197*, 1–25.
Fauquet, M., Smith, J., Ziller, C. and Le Douarin, N.M. (1981) J. Neurosci. *1*, 478–492.
Fontaine-Pérus, J., Chanconie, M., Polak, J.M. and Le Douarin, N.M. (1981) Histochemistry *71*, 313–323.

Fontaine-Pérus, J., Chanconie, M. and Le Douarin, N.M. (1982) Cell Differ. *11*, 183–193.
Fukada, K. (1980) Nature (London) *287*, 553–555.
Greenberg, J.H. and Schrier, B.K. (1977) Dev. Biol. *61*, 86–93.
Hawrot, E. (1980) Dev. Biol. *74*, 136–151.
Hill, C.E. and Hendry, I.A. (1977) Neuroscience *2*, 741–750.
Johnson, M., Ross, D., Meyers, M., Rees, R., Bunge, R., Wakshull, E. and Burton, H. (1976) Nature (London) *262*, 308–310.
Jonakait, G.M., Wolf, J., Cochard, P., Goldstein, M. and Black, I.B. (1979) Proc. Natl. Acad. Sci. USA *76*, 4683–4686.
Jonakait, G.M., Bohn, M.C. and Black, I.B. (1980) Science *210*, 551–553.
Karczmar, A.G., Nishi, S., Minota, S. and Kindel, G. (1980) Gen. Pharmacol. *11*, 127–134.
Kessler, J.A., Cochard, P. and Black, I.B. (1979) Nature (London) *280*, 141–142.
Landis, S.C. and Keefe, D. (1980) Soc. Neurosci. Abstr. *6*, 379.
Le Douarin, N.M. (1969) Bull. Biol. Fr. Belg. *103*, 435–452.
Le Douarin, N.M. (1973) Dev. Biol. *30*, 217–222.
Le Douarin, N.M. (1976) in Embryogenesis in Mammals, Ciba Foundation Symposium, pp. 71–101, Elsevier, Amsterdam.
Le Douarin, N.M. (1980) Curr. Top. Dev. Biol. *16*, 31–85.
Le Douarin, N.M. (1981) in Development of the Autonomic Nervous System, Ciba Foundation Symposium 83, pp. 19–50, Pitman Medical, London.
Le Douarin, N.M. (1982) The Neural Crest, Cambridge University Press, Cambridge.
Le Douarin, N.M. and Teillet, M.-A. (1973) J. Embryol. Exp. Morphol. *30*, 31–48.
Le Douarin, N.M. and Teillet, M.-A. (1974) Dev. Biol. *41*, 162–184.
Le Douarin, N.M., Renaud, D., Teillet, M.-A. and Le Douarin, G.H. (1975) Proc. Natl. Acad. Sci. USA *72*, 728–732.
Le Douarin, N.M., Teillet, M.-A., Ziller, C. and Smith, J. (1978) Proc. Natl. Acad. Sci. USA, *75*, 2030–2034.
Le Douarin, N.M., Le Lièvre, C.S., Schweizer, G. and Ziller, C.M. (1979) in Cell Lineage, Stem Cells and Cell Determination (Le Douarin, N., ed.), pp. 353–365, Elsevier, Amsterdam.
Le Lièvre, C.S., Schweizer, G.G., Ziller, C.M. and Le Douarin, N.M. (1980) Dev. Biol. *77*, 362–378.
Mains, R.E. and Patterson, P.H. (1973) J. Cell Biol. *59*, 329–345.
Marchisio, P.C. and Consolo, S. (1968) J. Neurochem. *15*, 759–764.
Narayanan, C.H. and Narayanan, Y. (1980) Anat. Rec. *196*, 71–82.
Newgreen, D. and Thiéry, J.P. (1980) Cell Tiss. Res. *211*, 269–291.
Noden, D.M. (1978) Dev. Biol. *67*, 313–329.
Norr, S.C. (1973) Dev. Biol. *34*, 16–38.
Pannese, E., Luciano, L., Iurato, S. and Reale, E. (1971) J. Ultrastruct. Res. *36*, 46–67.
Patterson, P.H. (1978) Ann. Rev. Neurosci. *1*, 1–17.
Ross, D., Johnson, M. and Bunge, R. (1977) Nature (London) *267*, 536–539.
Rovasio, R.A., Delouvée, A., Yamada, K., Timpl, R. and Thiéry, J.P. (1983) J. Cell Biol., *96*, in press.
Smith, J., Cochard, P. and Le Douarin, N.M. (1977) Cell Differ. *6*, 199–216.
Smith, J., Fauquet, M., Ziller, C. and Le Douarin, N.M. (1979) Nature (London) *282*, 853–855.
Strumia, E. and Baima-Bollone, P.L. (1964) Acta Anat. *57*, 281–293.
Teillet, M.-A. (1978) Roux's Arch. Dev. Biol. *184*, 251–268.
Teillet, M.-A., Cochard, P. and Le Douarin, N.M. (1978) Zoon *6*, 115–122.
Teitelman, G., Baker, H., Joh, T.H. and Reis, D.J. (1979) Proc. Natl. Acad. Sci. USA *76*, 509–513.
Teitelman, G., Gershon, M.D., Rothman, T.P., Joh, T.H. and Reis, D.J. (1981) Dev. Biol. *86*, 348–355.
Thiéry, J.P., Brackenbury, R., Rutishauser, U. and Edelman, G.M. (1977) J. Biol. Chem. *252*, 6841–6845.

Thiéry, J.P., Duband, J.-L. and Delouvée, A. (1982) Dev. Biol. *93*, 324–343.
Toole, B.P. (1972) Dev. Biol. *29*, 321–329.
Tosney, K.W. (1978) Dev. Biol. *62*, 317–333.
Weber, M.J. (1981) J. Biol. Chem. *256*, 3447–3453.
Weston, J.A., Derby, M.A. and Pintar, J.E. (1978) Zoon *6*, 103–113.
Ziller, C., Smith, J., Fauquet, M. and Le Douarin, N.M. (1979) Prog. Brain Res. *51*, 59–74.
Ziller, C., Le Douarin, N.M. and Brazeau, P. (1981) C.R. Acad. Sci. *292*, 1215–1219.
Ziller, C., Dupin, E., Brazeau, P., Paulin, D. and Le Douarin, N.M. (1983) Cell, in press.

CHAPTER 2

Development of smooth muscle and autonomic nerves in culture

JULIE H. CHAMLEY-CAMPBELL[1] and GORDON R. CAMPBELL[2]

[1]Cell Biology Laboratory, Baker Medical Research Institute, Commercial Rd, Prahran, Victoria 3185, Australia, and [2]Department of Anatomy, University of Melbourne, Parkville, Victoria 3052, Australia

1. Influence of sympathetic nerves on the development of smooth muscle in culture

1.1. Development of smooth muscle

The ultrastructural changes occurring during the development of smooth muscle in vivo are well documented (Bennett and Cobb, 1969; Gonzalez-Crussi, 1971; Leeson and Leeson, 1965; Yamamoto, 1961). In the initial stages of development the cells are morphologically very similar to fibroblasts with their cytoplasm containing a large number of organelles, particularly free ribosomes, rough endoplasmic reticulum, mitochondria and Golgi complexes; few filaments are present at this stage. With increasing age there is a progressive reduction in the nucleocytoplasmic ratio and an increase in the number of bundles of thin (50–80 Å) actin-containing filaments with associated dark bodies (Yamauchi and Burnstock, 1969). Thick (120–180 Å) myosin-containing filaments are apparent in these bundles of filaments at a later stage and there is a concomitant decrease in organelles as the size of the filament bundles increases. Other characteristic features of smooth muscle such as dense areas along the cell membrane, plasmalemmal vesicles and basal lamina appear at about the same time or a little later than the thick myofilaments (Imaizumi and Kuwabura, 1971).

In cell culture, isolated smooth muscle cells from the 10-day-old chicken embryo gizzard undergo intense proliferation until a confluent monolayer forms. Differentiation of the cells then proceeds along the same lines, although somewhat slower, as in vivo (Campbell et al., 1974). Thin myofilaments and dark bodies are first noted in the cultured cells at 3 days, a few plasmalemmal vesicles and dense areas at 6 days, and thick myofilaments at 11 days. At 11 days there is also an increase in the amount of basal lamina and plasmalemmal vesicles. By 14 days in culture the cells appear of equivalent morphological development as those from the 18-day-old embryo with most of the features of the adult gizzard. With phase contrast microscopy the same cells are observed to change shape with time in culture from polyhedral to spindle, ribbon-spindle, ribbon and finally to ribbon with tapered ends (fusiform). These changes in shape appear to parallel those of the cells from the developing gizzard in vivo, as cells from the 8, 12, 17 and 19-day-old embryos in the first 24 h of culture show a similar gradation in shape as the age of the donor embryo is increased.

1.2. Phenotypic modulation of smooth muscle

Cell differentiation describes the characteristics of cells in terms of their morphologic structure, the nature of the extracellular matrix they synthesize, and their ability to retain normal function, e.g., in the case of muscle, contract. When cells lose these characteristics they are often called 'dedifferentiated'. However, the ability of a cell to retain its differentiated phenotype can be reversibly dependent on the environmental conditions. We therefore refer to the loss of phenotypic markers of differentiation as 'phenotypic modulation', as distinct from the reversion of cells to a less differentiated state from which they could potentially redifferentiate in a different direction.

This concept is particularly important in the case of smooth muscle of the mammalian artery wall since it is the only cell type present within the media (Pease and Paule, 1960), and must therefore be responsible for maintaining tension via contraction–relaxation and arterial integrity by profileration and synthesis of connective tissue elements (Wissler, 1968). To accomplish this multiplicity of functions the smooth muscle cell is capable of expressing a range of phenotypes (Chamley-Campbell et al., 1979; Chamley-Campbell et al., 1982c; Chamley-Campbell and Campbell, 1981). At one end of the spectrum is the smooth muscle cell whose major function is contraction (contractile phenotype) with its cytoplasm filled with thick and thin myofilaments (Chamley et al. 1974; Chamley et al., 1977; Campbell et al., 1981). The vast majority of smooth muscle cells in the media of an adult aorta are in this phenotype. At the other end of the spectrum is the smooth muscle cell whose cytoplasm lacks thick filaments but contains scattered bundles of thin filaments and large amounts of rough endoplasmic reticulum, free ribosomes and Golgi elements, which are organelles associated with synthesis of both extracellular and intracellular material (Chamley-Campbell et al., 1979). These cells are in a synthetic state (hence the term synthetic phenotype) and predominate in developing and regenerating smooth muscle tissues (Poole et al., 1971; Imai et al., 1970; Campbell et al., 1971).

When mature, contractile smooth muscle is enzyme-dispersed into single cells and placed in primary culture, the cells phenotypically modulate to the synthetic state if a confluent monolayer of background cells is not present during the first few days (Chamley and Campbell, 1975a,b; Chamley et al., 1974; Chamley et al., 1977; Chamley-Campbell et al., 1979, 1981b; Chamley-Campbell and Campbell, 1981). Thus in cultures of newborn guinea pig vas deferens and ureter, adult monkey, pig and rabbit aorta, and human saphenous vein, isolated smooth muscle cells modulate morphologically and functionally after 5–8 days (depending on the source) losing their capacity for spontaneous or induced contraction and becoming capable of migration and intense proliferation (Fig. 1). Synthetic state smooth muscle cells are broader and less phase dense than contractile state cells and have a larger nucleus and often more nucleoli. They show an increase in the number of organelles, particularly rough endoplasmic reticulum and free ribosomes. Bundles of thin filaments tend to be located toward the periphery of the cell and there are no thick filaments. During the first 5–7 days in culture when the cells are in the contractile functional state, only 1–3% incorporate [^{3}H]thymidine into DNA in the presence of 5% whole blood serum and there are few cell divisions (Chamley-Campbell et al., 1981b). However, on the same day that modulation to the synthetic state has occurred, the percentage of cells incorporating [^{3}H]thymidine increases to 10–20%. This percentage increases linearly with time in culture as progressively more cells are recruited into the cell cycle such that 3–5 days after modulation up to 60% of the cell population is actively synthesizing DNA during the 4 h [^{3}H]thymidine pulse period. When the cells of sister cultures are counted in a haemocytometer each day, the numbers remain constant until 1–2 days after modulation of phenotype and increased incorporation of [^{3}H]thymidine. The number of cells per dish then increases logarithmically until confluence is achieved at which time cell proliferation ceases.

If the contractile state cells are seeded sufficiently densely (5×10^4–1×10^5/ml) that a confluent monolayer results after less than 5 or 6 days of intense proliferation following modulation on days 5–8, then the cells rapidly regain the ultrastructural appearance of contractile state smooth muscle. With the guinea pig vas deferens, monkey, pig and rabbit aorta, and human saphenous vein, the cells in the confluent monolayer spontaneously aggregate into clumps over the entire culture area. The clumps are up to 3 mm × 0.5 mm and often consist of two layers of cells at right angles to each other. Shortly after the clumps form, the constituent cells resume a ribbon or fusiform shape and sometimes recommence spontaneous contractions. Foci of synchronous contraction also sometimes appear in groups of 10–100 cells (Chamley et al., 1974; Chamley et al., 1977). With the guinea pig ureter, smooth muscle cells regain the contractile phenotype and form muscle chains after the culture becomes confluent. They gradually join with other chains, producing meshworks under a fibroblast and epithelial cell layer. Spontaneous asynchronous contractions (2–4/min) develop in these cells 14–16 days after chain formation, increasing in rate and strength with time in culture until after 6 weeks the contraction rate becomes steady at about 10/min (Chamley and Campbell, 1975b).

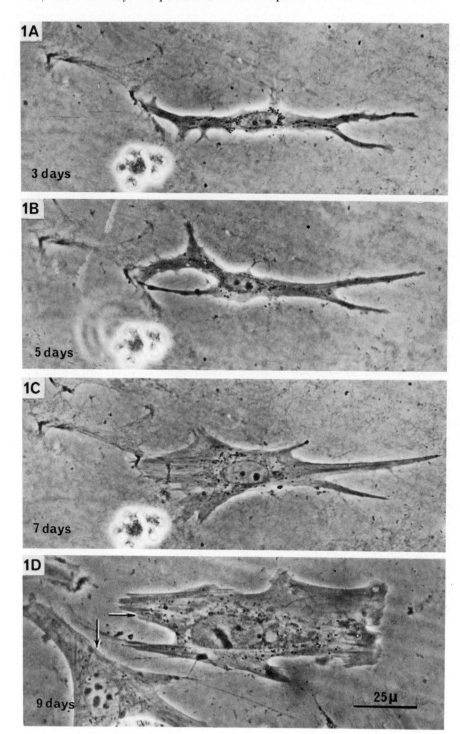

If the contractile state cells are seeded so densely (1×10^6/ml) that a confluent monolayer is present from day 1 in culture, they do not undergo a change in phenotype to the synthetic state on days 5–8, but remain in the contractile state (Chamley-Campbell et al., 1982; Chamley-Campbell and Campbell, 1981). However, if the contractile state cells are seeded so sparsely (1×10^3–5×10^3/ml) that following modulation on days 5–8 a confluent monolayer does not result until after 21–24 days of proliferation then the cells do not return to the contractile state but appear permanently in the synthetic state.

1.3. Influence of sympathetic nerves on development and phenotypic modulation of smooth muscle

There are several reports suggesting a trophic influence of sympathetic nerves on smooth muscle in vitro. Thus, in primary cultures of dissociated newborn guinea pig ureter, long processes grow from intrinsic non-adrenergic neurons to form close and long-lasting associations with re-formed muscle chains (Chamley and Campbell, 1975b). Where this has occurred the muscle chains commence spontaneous contraction about 13 days in advance of muscle chains without nerve fibres, and their contractions are considerably stronger and more rapid. Nerve fibre growth from extrinsic sympathetic ganglia into clumps of partially dispersed guinea pig vas deferens stimulates the formation of gap junctions between muscle cells and thus the formation of foci of synchronous contractions (Chamley et al., 1974); the rate and strength of contraction in explants of mouse intestine in culture are enhanced by the presence of extrinsic sympathetic ganglion neurons (Cook and Peterson, 1974); and denervation atrophy of skeletal muscle is prevented by the presence, but not necessarily innervation from, explants of sympathetic ganglia (Crain and Peterson, 1974).

A trophic influence of autonomic nerves on smooth muscle in culture is further demonstrated by the delay of 3–7 days in spontaneous phenotypic modulation to the synthetic state and subsequent proliferation of isolated contractile smooth muscle cells from the guinea pig vas deferens and rabbit aorta and ear artery (Chamley and Campbell, 1976; Chamley et al., 1974). A similar delay in modulation and proliferation occurs when the cells are grown in the presence of a crude extract of homogenized sympathetic ganglia, indicating that the trophic influence of sympathetic neurons on smooth muscle cells is chemically mediated (Chamley and Campbell, 1975a, 1976). There is little or no effect with

◀ Fig. 1. The same smooth muscle cell from the newborn guinea pig vas deferens in culture. (*A*) After 3 days. (*B*) After 5 days. (*C*) After 7 days. (*D*) After 9 days. Note that the smooth muscle cell has undergone a change in phenotype from the contractile to synthetic state and has undergone division (arrows). From Chamley et al. (1974).

approximately equivalent amounts of homogenized spinal cord and liver or noradrenaline or acetylcholine at $10^{-7} \to 10^{-5}$ g/ml.

In vivo it has long been recognized that motor nerves exert a trophic influence on skeletal muscle (see Gutmann, 1976), but there are few reports that sympathetic nerves can exert a trophic influence on the structure and function of smooth muscle. Essex and co-workers (1943) have reported medial hypertrophy in arterioles of the hind limb of a dog 10 years after left lumbar sympathetic ganglionectomy, and Campbell et al. (1977) found that denervation of the expansor secundariorum muscle (a totally adrenergically innervated piloerector smooth muscle in the wing) of 2-week-old and adult chickens, results in an approximate 2-fold increase in dry weight over 8 weeks due to hyperplasia of the smooth muscle cells. However, denervation of the rabbit ear artery appears to have the opposite effect. Six weeks after unilateral postganglionic sympathetic denervation the ear artery is lighter, thinner and stiffer than its control (Bevan and Tsuru, 1979). Although supersensitive to noradrenaline the denervated vessel develops a lower maximum contractile force which is reduced proportionately more than tissue weight. Hart et al. (1980) also found attenuation of cerebral vessels of rats after superior cervical ganglionectomy.

2. Influence of effector organs on the development of autonomic nerves in culture

2.1. Expression of differentiated functions of autonomic nerves in culture

2.1.1. Sympathetic neurons

Two classes of neurons have been distinguished in explant cultures of rat and guinea pig sympathetic ganglia (Fig. 2) (Chamley et al., 1972a; Perry et al., 1975). Type II neurons, which are in the clear majority, correspond to mature (principal)

Fig. 2. (A) Type I neuron with high nucleo-cytoplasmic ratio, clumped granular material in the cytoplasm and oval nucleus containing pale nucleoli and chromatin clumps. Note absence of satellite cells around the cell body and absence of Schwann cells on processes. From 18-day-old rat superior cervical ganglion, 9 days in vitro. A inset. Same cell as 1A taken a few minutes later showing small, pale nucleoli and clumped chromatin. (B) Type II neuron with low nucleocytoplasmic ratio, evenly distributed granular material in the cytoplasm and round nucleus containing a single large, dark nucleolus. Note satellite cells (arrows) closely applied to the cell body. From 7-day-old rat sympathetic chain, 21 days in vitro. (C) Type II neuron of normal appearance with no closely applied satellite cells. From 5-day-old rat sympathetic chain, enzyme-dispersed, 21 days in vitro. (D) Type II neurons in explant. No type I neurons can be distinguished. From newborn guinea pig sympathetic chain, 7 days in vitro. (E) Neuron with characteristics of both types I and II. Note oval nucleus with chromatin clumps, absence of satelite cells and presence of neuropil around the cell body-type I; low nucleo-cytoplasmic ratio, evenly distributed granular material in the cytoplasm, single large, dark nucleolus-type II. From 5-day-old rat sympathetic chain, 7 days in vitro. From Chamley et al. (1972a).

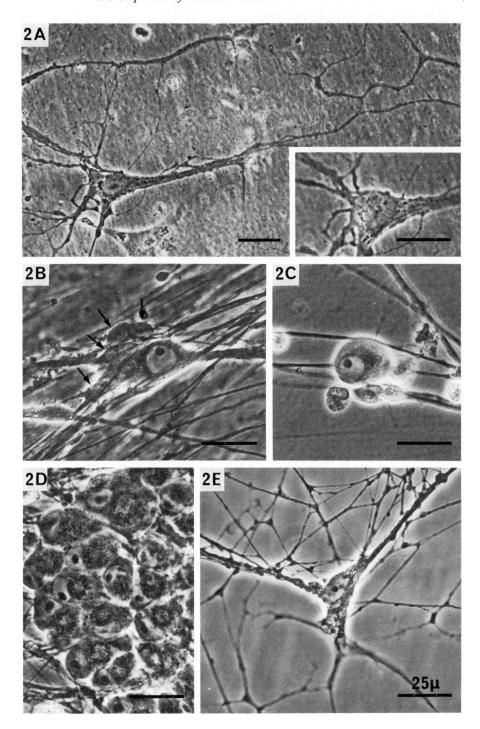

sympathetic neurons described in situ by most histologists. Their processes are varicose and have histochemical and ultrastructural features of normal regenerating sympathetic nerves (Burnstock and Bell, 1974). Type I neurons are small and migratory and their status is not so clear. In situ small neurons with a high nucleo–cytoplasmic ratio and chromatin clumps in the nucleus have been considered as immature neurons capable of proliferation and differentiation or retarded elements which never reach the degree of differentiation attained by other neurons (Ping, 1921; de Castro, 1932; Botár, 1966; Zaimis, 1971). Whether some or all of these early reports are describing SIF (small intensely fluorescent) cells is unknown, however, in culture the two are distinct, as SIF cells are uni- or bipolar, weakly migratory, and usually smaller and more intensely fluorescent than the multipolar and strongly migratory type I neurons (Chamley et al., 1972b).

In recent years, much interest has been shown in the culture of dissociated sympathetic neurons. Thus, isolated neurons from adult human trigeminal and superior cervical ganglia have been shown to be normal ultrastructurally and electrophysiologically (Kim et al., 1979), and those from 11-day-old chick sympathetic ganglia to possess functional muscarinic cholinergic receptors as well as the ability to release noradrenaline via a stimulus-secretion coupling mechanism (Green and Rein, 1978). Dissociated neurons from the newborn rat superior cervical ganglion, grown for 3 weeks or longer in the presence of non-neuronal cells, have resting potentials, passive electrical properties and action potentials generally reported for principal neurons of the superior cervical ganglion of adult rats (O'Lague et al., 1978). Frequently the action potential is followed by a prolonged after-hyperpolarization whose properties suggest the presence of potassium channels controlled by calcium ions. When the neurons are grown in the absence of non-neuronal cells the action potentials are similar but the prolonged after-hyperpolarization is rarely seen and the neurons usually discharge repetitively in response to a steady depolarization. Obata (1974) has shown that iontophoretically applied acetylcholine on dissociated newborn rat sympathetic neurons produces the classical depolarizing effects which are blocked by α-tubocurarine, and that there is no effect with noradrenaline. Alpha-bungarotoxin binding has also been used to demonstrate acetylcholine receptors on dissociated sympathetic neurons (Greene et al., 1973).

Much research has been devoted to the role of Nerve Growth Factor (NGF) on the survival, growth and differentiation of sympathetic neurons in culture (see Levi-Montalcini and Angeletti, 1968a,b; Varon and Adler, 1980). Like other growth factors and hormones, NGF induces rapid and delayed responses. The rapid responses include the uptake of metabolites such as uridine and glucose (Horii and Varon, 1977; Skaper et al., 1979; Thoenen et al., 1971), while the delayed responses include polymerization of microtubules, stimulation of two enzymes involved in catecholamine biosynthesis (Thoenen et al., 1971), production of neurites, and long-term maintenance of cell viability (Levi-Montalcini, 1976).

It is currently believed that NGF initially interacts with plasma membrane receptors, becomes internalized and attaches to a second type of receptor located in the nucleus (Andres et al., 1977; Yanker and Shooter, 1979), with the rapid responses of NGF a consequence of the binding to the plasma membrane, and the internalized growth factor and nuclear receptors playing a role in the activation of the delayed processes (see Levi et al., 1980).

Dissociated newborn rat sympathetic neurons in the absence of NGF and other cell types, do not survive for more than 1 day in culture, however in the presence of increasing levels of NGF there is increased survival with saturation at 0.5 μg/ml 7 S NGF (Chun and Patterson, 1977). Low concentrations of NGF (such as occur in plasma) elicit maximum axon formation from these cells, but it is the high concentrations (as occur in effector organs) that induce maximal cell survival and tyrosine hydroxylase activity (Hill and Hendry, 1976). It has therefore been suggested that in vivo NGF from the two sources may subserve different regulatory functions in the developing neuron. That is, circulating NGF may stimulate neurite elaboration, while target organ NGF may cause increased survival.

A variety of artificial and cell-derived substrata have also been examined for their effects on the survival and neurite extension of dissociated rat sympathetic neurons (Hawrot, 1980). Compared to dried collagen films both three-dimensional hydrated collagen gels and surfaces coated with basic polymers (polylysine, polyornithine) provide a highly effective substratum, as do killed non-neuronal cells or an extracellular, substrate-associated material produced by non-neuronal cells. Adult rat brain astrocytes promote the in vitro survival not only of NGF-sensitive (superior cervical ganglion) but also NGF-insensitive (nodose ganglion) autonomic neurons (Lindsay, 1979).

2.1.2. Parasympathetic neurons
The survival and development of parasympathetic nerves in culture require co-culture with target tissues (Coughlin, 1975a,b; Hooisma et al., 1975; Nishi and Berg, 1977; White and Bennett, 1979) or growth in target-organ conditioned medium (Bennett and Nurcombe, 1979; Helfand et al., 1976; Varon et al., 1979) or with tissue homogenates (Ebendal et al., 1979; McLennan and Hendry, 1978; Tuttle et al., 1980). A macromolecule which supports the survival of dissociated embryonic chick ciliary ganglion neurons in culture has recently been isolated from ox cardiac muscle (Bonyhady et al., 1980).

2.1.3. Enteric neurons
Auerbach's plexus from the taenia libera or taenia mesocolica of the caecum of newborn guinea pigs has been isolated completely free of smooth muscle by the action of collagenase and careful dissection (Jessen et al., 1978). Intracellular recordings following impalement of the neurons with microelectrodes reveal the

presence of two types of neurons: one showing action potentials which resemble those of neurons from other autonomic ganglia, and another whose action potentials are followed by a long after-hyperpolarization similar to that found in Auerbach's plexus in situ (Hirst et al., 1974).

2.2. Influence of autonomic effectors on neuronal transmitter

Each neuron possesses the potential ability to synthesize the complete enzymatic machinery for all transmitter substances. The particular transmitter that appears at the end of differentiation seems to be determined by external trophic factors released from certain cells which trigger the expression of the appropriate genetic programme (see Burnstock, 1976). Thus it is possible to influence the transmitter produced in populations of developing autonomic neurons in vivo by transplantation of neural crest cells to foreign sites (Le Douarin and Teillet, 1974; Le Douarin et al., 1975). Presumptive adrenergic neuroblasts from the adrenomedullary neural crest area transplanted into the vagal neural crest area are induced by splanchnic mesoderm to become fully differentiated cholinergic neurons when they reach the gut. Conversely when presumptive cholinergic neuroblasts from the cephalic neural crest are transplanted into the adrenomedullary area the cells migrate into the suprarenal glands and differentiate into catecholamine-containing cells.

In cell culture, sympathetic neurons dissociated from the superior cervical ganglia of newborn rats and grown in the absence of non-neuronal cells are predominantly adrenergic, and synthesize, store, release and take up noradrenaline while acetylcholine synthesis is negligible (Mains and Patterson, 1973). However, if the same neurons are grown in the presence of non-neuronal cells from ganglia or the heart or in medium conditioned by them, the cultures synthesize significant amounts of acetylcholine as well as catecholamines. If sufficiently high concentrations of conditioned medium are used then the cultures become predominantly cholinergic (Patterson and Chun, 1974, 1977; Johnson et al., 1976; Landis, 1980). Differential survival and selection of subclasses of the neuronal population have been ruled out indicating that the conditioned medium acts by altering the differentiated fate of individual neurons and that many of the neurons isolated from ganglia of newborn rats are plastic with respect to transmitter functions. However, the development of cholinergic characteristics in adrenergic neurons is age dependent, with a decrease in the ability of sympathetic ganglion explants taken from rats older than 2–5 days to develop choline acetyltransferase activity in culture (Hill and Hendry, 1977; Ross et al., 1977; Johnson et al., 1980b). This suggests that most sympathetic neurons may become committed to a certain neurotransmitter at a critical stage and lose their ability to respond to alternative instructions after that time.

A major question is why in culture is it possible through conditioning with non-neuronal cells from ganglia or from target tissues to induce virtually all of the neurons to become cholinergic, whereas in vivo, despite the potential presence of a cholinergic-inducing factor in non-neuronal cells in the ganglia and in tissues to which nerve processes grow, only about 5% are cholinergic? Cholinergic-inducing signals selectively localized in a target which receives sympathetic cholinergic innervation would explain why only some neurons become cholinergic, and indeed it has been found that targets normally receiving cholinergic innervation (skeletal and heart muscle) are far better at 'cholinergic-conditioning' the culture medium than a target receiving only adrenergic inervation (liver) (see Patterson, 1978). However, cell types not known to receive synapses at all (fibroblasts, glia) can also cholinergically condition medium. NGF from either the blood supply to the ganglia or transported retrogradely from the target tissue cannot be involved as NGF stimulates both adrenergic and cholinergic differentiation (Coughlin et al., 1978).

One possibility is that the influence of the preganglionic synapses is to prevent the neurons of the sympathetic ganglia from becoming cholinergic. Cholinergic synapses form between more than 25% of sympathetic neurons cultured in the presence of spinal cord explants, which suggests that preganglionic nerves have no effect on the induction of adrenergic properties (Ko et al., 1976b). However, only 20% of the sympathetic neurons in these cultures were synaptically coupled with the spinal nerves (Ko et al., 1976a). Prolonged depolarization of sympathetic neurons in culture by elevated $[K^+]$, veratridine or direct electrical stimulation prevents the large increase in acetylcholine/catecholamine ratio caused by conditioned medium from non-neuronal cells (Walicke et al., 1977). However, transection of the preganglionic nerve trunk of the young rat in vivo does not result in an increase in the intrinsic choline acetyltransferase activity of the ganglion (Hill and Hendry, 1979). Therefore the role of preganglionic input remains equivocal.

Recently it has been shown that the synthetic glucocorticoid dexamethasone, and the naturally occurring glucocorticoid in rats, corticosterone, prevent the development of cholinergic properties in sympathetic neurons in culture (McLennan et al., 1980). The influence of the glucocorticoids is apparently on the non-neuronal cells to alter selectively the production or liberation of a factor that induces neuronal cholinergic development, rather than a direct effect on the nerves (Fukada, 1980). Reduction of circulating glucocorticoids in vivo by adrenalectomy does not lead to an increase in cholinergic properties in the superior cervical ganglion (Hill et al., 1980).

3. Tropic influence of autonomic effector organs on nerve fibre growth

The term 'trophic' as applied to the autonomic system denotes an influence on

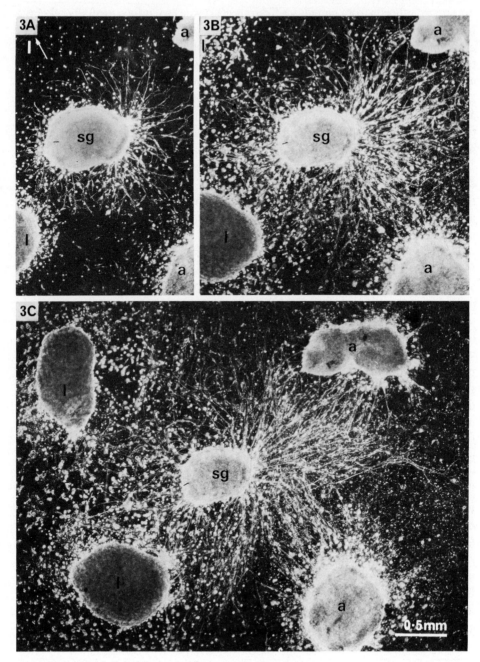

Fig. 3. Nerve fibre growth from a sympathetic ganglion explant (sg) to explants of lung (l) and atrium (a). 5-day-old rat. (A) After 2 days. (B) After 3 days. (C) After 5 days. At all times a greater number of nerve fibres grew on the side of the sympathetic ganglion explant nearer the atrium than ▶

maintenance or survival, whereas 'tropic' implies an influence on directional growth. While its contribution in vivo is not understood, it is clear that at least in culture the growth and direction of both sympathetic and parasympathetic nerves can be influenced by gradients of diffusible agents. The best studied case is the influence of NGF on sympathetic and sensory nerve fibre growth. Thus regenerating neurites in culture grow preferentially towards a capillary tube filled with NGF (Charlwood et al., 1972; Ebendal and Jacobson, 1977a) and processes from dissociated neurons grow preferentially into adjacent chamber compartments containing NGF (Campenot, 1977). With semisolid matrices of agar and culture medium, orientation of nerve tips from dissociated neurons occurs up a defined gradient of NGF as far as 23 mm from the NGF source (Letourneau, 1978). Using a tissue source of NGF, Levi-Montalcini et al. (1954) have found growth stimulation of sympathetic and sensory nerves on the side of the ganglion explant opposite an explant of NGF-producing mouse sarcoma 1–2 mm distant. A similar, although not as potent, growth stimulation occurs towards explants of mouse heart.

As an extension of these observations, it has been shown that sympathetic nerve fibre growth is stimulated towards explants of tissue normally densely innervated by sympathetic nerves in vivo (atrium, vas deferens) but not towards explants of tissues normally sparsely innervated (lung, kidney medulla, uterus, distal ureter) (Chamley et al., 1973a; Chamley and Dowel, 1975; see also Ebendal and Jacobson, 1977a; Coughlin et al., 1978). The 'attraction' effect is evident as soon as the nerve fibres emerge from the sympathetic ganglion explant and occurs over a distance of 2 mm (Fig. 3). With the chick, the maximum neurite-stimulating activity occurs with explants from 16–18-day-old embryo heart, reflecting the phase of sympathetic innervation in ovo (Ebendal, 1979). Whether a NGF gradient is responsible for this directed outgrowth is uncertain, since the amount of NGF is higher in tissues which become densely innervated by sympathetic nerves (Johnson et al., 1971), but the 'attraction' and growth stimulatory effects in culture are not diminished by the presence of antibodies to NGF (Ebendal and Jacobson, 1977a). Also, there is a growth stimulatory effect of heart on ciliary ganglia neurons which are not responsive to NGF (Ebendal and Jacobson, 1977a,b; Ebendal, 1979), but heart has no effect on spinal cord neurons which are not responsive to NGF (Chamley and Dowel, 1975).

A tropic effect of NGF in vivo was suggested by the original experiments of Bueker (1948) and of Levi-Montalcini (1952) and her colleagues (Levi-Montalcini and Hamburger, 1953). More recently, it has been shown that intracerebral injection of NGF causes a highly abnormal growth of sympathetic axons into the

---

lung explants. The nerve fibre growth to the atrium explants appeared to be directed, many fibres changing course to penetrate atrium explants. Nerve fibre growth to lung explants always appeared random. From Chamley et al. (1973b).

spinal cord of developing mice and rats, with postganglionic axons from thoracic and cervical sympathetic ganglia running cephalad in the spinal cord and brain stem toward the region of NGF injection (Menesini Chen et al., 1978). This leaves no doubt that NGF can exert a tropic influence on sympathetic nerves.

While most attention has been paid to a tropic influence on sympathetic nerve fibres, there is some evidence to suggest that parasympathetic nerves are 'attracted' by their normal targets in vivo and in vitro (Coughlin 1975a,b; Coughlin and Rathbone, 1977; Helfand et al., 1976; Ebendal, 1979). The agent of this influence is not known, but it is possible that it is the same substance responsible for the survival and development of parasympathetic neurons in culture (see above; Bonyhady et al., 1980).

It remains to be shown that NGF or other diffusible chemical agents contribute a tropic stimulus in normal development. Other possible mechanisms for eliciting directed growth in vivo can be inferred from the fact that growing axons in vitro are very sensitive to mechanical factors (Weiss, 1934) and may be affected by differential adhesiveness of the surfaces over which they grow (Letourneau, 1975a,b; Sidman and Wessells, 1975). Finally, there is some evidence for recognition of specific surface molecules by developing neurons (Moscona, 1976).

4. Formation of functional junctions in culture

4.1. Interactions between sympathetic nerves and isolated smooth and cardiac muscle cells

The initial interactions between sympathetic nerve fibres and isolated smooth and cardiac muscle cells from rats, guinea pigs and rabbits have been observed with time-lapse microcinematography (Mark et al., 1973; Chamley et al., 1973a; Chamley and Campbell, 1975b, 1976). There is no obvious 'attraction' from a distance of the nerve fibres to the single cells, with chance contacts occurring through random growth. Once a nerve growth cone encounters a cell, it halts in its growth and palpates the cell for about 50 min. The nerve fibre then appears to be able to distinguish whether or not the cell is a suitable target to innervate (Fig. 4). That is, contact with a fibroblast, immature cardiac muscle cell, or

---

Fig. 4. Abstract from a time-lapse microcinematography film. (*A*) Nerve fibre bundle (n) from a sympathetic ganglion explant grows towards a contracting smooth muscle cell (m) and a fibroblast (f) from the newborn guinea pig vas deferens. (*B,C,D*) The nerve fibre bundle pauses in its growth equidistant from the muscle cell and fibroblast, repeatedly palpating both cells (arrow). (*E,F,G*) After a pause of 2 h 47 min some of the nerve fibres (arrow) begin to grow again, with a definite change in direction towards the muscle cell. The remaining fibres in the nerve bundle continue to palpate the fibroblast. (*H*) After 23 h the nerve fibres in contact with the muscle cell have increased in number and have stopped growing. From Chamley-Campbell et al. (1981a).

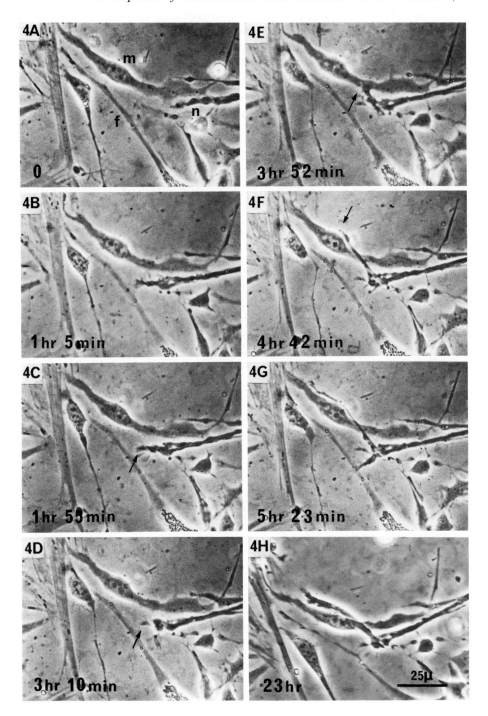

smooth muscle cell from a normally sparsely-innervated organ (ureter, aorta) leads to a transitory relationship lasting no longer than a few hours, whereas contact with a mature cardiac muscle cell or smooth muscle cell from a normally densely-innervated organ (vas deferens, ear artery) leads to an association that lasts for days or weeks. The association may be complex, with many fine fibres plexusing over the cell (Fig. 5.), or may consist of a single varicose fibre running along the length of the cell. Receptor sites for noradrenaline or acetylcholine on

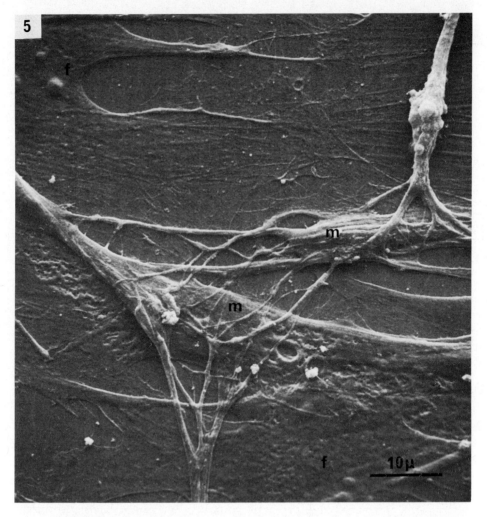

Fig. 5. Close and extensive associations between sympathetic nerve fibres and two smooth muscle cells (m) from the newborn guinea pig vas deferens in culture as observed with scanning electron-microscopy. Note that the nerve fibres do not form associations with the background fibroblasts (f). From Chamley, J.H. et al. (1973a).

Development of smooth muscle and autonomic nerves in culture | 51

the muscle cells do not appear to be involved in the process of nerve–muscle recognition since the presence of propranolol or hyoscine does not inhibit the formation of nerve–cardiac muscle associations in culture (Campbell et al., 1978).

4.2. Ultrastructure of nerve–muscle associations

The associations that form between sympathetic nerve fibres and smooth muscle cells from the guinea pig vas deferens closely resemble the in vivo situation with a separation of 100 Å and no specialization of either nerve or muscle (Fig. 6.) (Chamley et al., 1973a). The separation between sympathetic nerves and isolated smooth muscle cells from the rabbit ear artery is 100–500 Å (Chamley and Campbell, 1976). This is considerably less than the minimum 5000 Å separation that occurs in vivo with this tissue (Bevan et al., 1972) and may be due to the

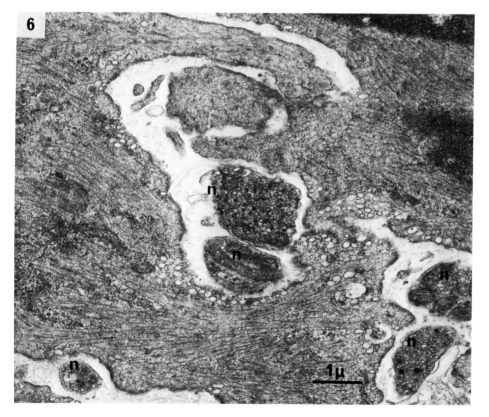

Fig. 6. Five nerve fibre varicosities (n) in association with a single smooth muscle cell from the newborn guinea pig vas deferens in culture. Note the presence of both granular and agranular vesicles. From Chamley et al. (1973a).

absence of connective tissue elements in the culture situation. The associations that form between sympathetic nerve fibres and smooth muscle cells in mouse intestinal explants have a separation of 200–300 Å (Cook and Peterson, 1974), whereas those between nerves and muscle cells of rat iris explants are generally no closer than 3000 Å (Hill et al., 1976).

4.3. Functional junctions

Nerve fibres from explants of sympathetic and ciliary ganglia form functional junctions in culture with explants and single cells of rat cardiac muscle and smooth muscle from the rat iris sphincter pupillae and guinea pig vas deferens and taenia coli (Purves et al., 1974; Hill et al., 1976). Stimulation of the ganglia evoke contractions of the smooth muscle and either excitatory or inhibitory responses in the cardiac muscle. Hyoscine abolishes contractile responses of the smooth muscle cells and blocks the inhibitory response of the cardiac muscle, indicating that in these cases the innervation is cholinergic.

Electrophysiological studies performed on microcultures (300–500 μm in diameter) of solitary sympathetic neurons plated on cardiac muscle from the newborn rat have shown that some neurons inhibit, some excite, and others first inhibit then excite the cardiac muscle (Furshpan et al., 1976). Application of drugs has provided evidence for secretion of acetylcholine by the first group, catecholamines by the second, and both acetylcholine and catecholamines by the third. Ultrastructurally, the cholinergic endings contain small agranular vesicles, the adrenergic endings small granular vesicles and the dual function endings mainly agranular vesicles with an occasional granular vesicle (Landis, 1976; see also Johnson et al., 1976; Johnson et al., 1980a). Excitatory cholinergic synapses form between dissociated sympathetic neurons in culture in a medium permissive for proliferation of non-neuronal cells (O'Lague et al., 1974), and between dissociated sympathetic neurons and rat skeletal myotubes (Nurse and O'Lague, 1975).

Ciliary ganglia from newborn rabbits (which normally make exclusively cholinergic muscarinic connections with smooth muscle cells) form cholinergic nicotinic junctions with chick skeletal muscle in culture, indicating that the character of the muscle and not that of the neuron determines the pharmacological nature of the neuromuscular junctions (Stevens et al., 1978). However, unlike spinal cord–skeletal muscle junctions in culture, innervation of skeletal muscle by the ciliary ganglia does not provide the support to maintain cross-striations, suggesting that recognition of a cell as an appropriate target and its functional innervation does not automatically lead to adequate neurotrophic support of the cell.

5. Concluding remarks

The numerous and diverse studies on the autonomic system in culture have raised or emphasized more questions than they have answered. For example, what is the identity or nature of the tropic agent supplied by sympathetic neurons to control the development and maintain the contractile phenotype of smooth muscle? What are the substances released by target organs that control the directional growth of sympathetic and parasympathetic nerves and aid in their maintenance? What controls which neurotransmitter will be synthesized? What controls whether a cell will accept innervation and from which nerves? The answers to these basic and important questions can be best found using the culture system and this ensures a vigorous interest in the culture of the autonomic system for many years to come.

Acknowledgments

The authors are supported by the National Health and Medical Research Council of Australia, the Life Insurance Medical Research Fund of Australia and New Zealand and the National Heart Foundation of Australia.

References

Andres, R., Jeng, I. and Bradshaw, R. (1977) Proc. Natl. Acad. Sci. USA 74, 2785–2789.
Bennett, M.R. and Nurcombe, V. (1979) Brain Res. 173, 543–548.
Bennett, T. and Cobb, J.L.S. (1969) Z. Zellforsch. Mikrosk. Anat. 98, 599–621.
Bevan, J.A., Bevan, R.D., Purdy, R.E., Robinson, C.P., Su, C. and Waterson, J.G. (1972) Circ. Res. 30, 541–548.
Bevan, R.D. and Tsuru, H. Blood Vessels 16, 109–112.
Bonyhady, R.E., Hendry, I.A., Hill, C.E. and McLennan, I.S. (1980) Neurosci. Lett. 18, 197–201.
Botár, J. (1966) The Autonomic Nervous System. pp. 183–238, Akadémiai Kiadó, Budapest.
Bueker, E.D. (1948) Anat. Rec. 102, 369–370.
Burnstock, G. (1976) Neuroscience 1, 239–248.
Burnstock G. and Bell, C. (1974) in The Peripheral Nervous System (Hubbard, J.I. ed.) pp. 277–327, Plenum Press, New York.
Campbell, G.R., Uehara, Y., Malmfors, T. and Burnstock, G. (1971) Z. Zellforsch. Mikrosk. Anat. 117, 155–175.
Campbell, G.R., Chamley, J.H. and Burnstock, G. (1974) J. Anat. 117, 295–312.
Campbell, G.R., Gibbins, I., Allan, I. and Gannon, B. (1977) Cell Tiss. Res. 176, 143–156.
Campbell, G.R., Chamley, J.H. and Burnstock, G. (1978) Cell Tiss. Res. 187, 551–553.
Campbell, G.R., Chamley-Campbell, J.H. and Burnstock, G. (1981) in Structure and Function of the Circulation (Schwartz, C.J., Werthessen, N.T. and Wolf, S., eds.), pp. 357–399, Vol. III, Plenum Press, New York.
Campenot, R.B. (1977) Proc. Natl. Acad. Sci. USA 74, 4516–4519.
Chamley-Campbell, J.H. and Campbell, G.R. (1981) Atherosclerosis 40, 347–357.

Chamley-Campbell, J.H., Campbell, G.R. and Ross, R. (1979) Physiol. Rev. 59, 1–61.
Chamley-Campbell, J.H., Campbell, G.R. and Burnstock, G. (1981a) in Structure and Function of the Circulation (Schwartz, C.J., Werthessen, N.T. and Wolf, S., eds.) Vol. III, pp. 401–425, Plenum Press, New York.
Chamley-Campbell, J.H., Campbell, G.R. and Ross, R. (1981b) J. Cell Biol. 89, 379–383.
Chamley-Campbell, J.H., Nestel, P. and Campbell, G.R. (1982) in Nato Advanced Study Institute on Formation and Regression of the Atherosclerotic Plaque. Belgirate, (Born, G.R.V., Catapano, R.L. and Paoletti, R., eds.), Series A, Vol. 51, pp. 115–124, Plenum Press, New York.
Chamley, J.H. and Campbell, G.R. (1975a) Cell Tiss. Res. 161, 497–510.
Chamley, J.H. and Campbell, G.R. (1975b) Cytobiology 11, 358–365.
Chamley, J.H. and Campbell, G.R. (1976) in Vascular Neuroeffector Mechanisms. Proc. 2nd Intern. Symp., Odense (Bevan, J.A., Burnstock, G., Johansson, B., Maxwell, R.A. and Nedergaard, O.A. eds.) pp. 10–18, Karger, Basel.
Chamley, J.H. and Dowel, J.J. (1975) Exp. Cell Res. 90, 1–7.
Chamley, J.H., Mark, G.E., Campbell, G.R. and Burnstock, G. (1972a) Z. Zellforsch. Mikrosk. Anat. 135, 287–314.
Chamley, J.H., Mark, G.E. and Burnstock, G. (1972b) Z. Zellforsch. Mikrosk. Anat. 135, 315–327.
Chamley, J.H., Campbell, G.R. and Burnstock, G. (1973a) Devel. Biol. 33, 344–361.
Chamley, J.H., Goller, I. and Burnstock, G. (1973a) Devel. Biol. 31, 362–379.
Chamley, J.H., Campbell, G.R. and Burnstock, G. (1974) J. Embryol. Exp. Morph. 32, 297–323.
Chamley, J.H., Campbell, G.R., McConnell, J.D. and Gröschel-Stewart, U. (1977) Cell Tiss. Res. 177, 503–522.
Charlwood, K.A., Lamont, D.M. and Banks, B.E.C. (1972) in Nerve Growth Factor and its Antiserum (Zaimis, E., ed.) pp. 102–107, Athlone Press, London.
Chun, L.L.Y. and Patterson, P.H. (1977) J. Cell Biol. 75, 694–704.
Cook, R.D. and Peterson, E.R. (1974) J. Neurol. Sci. 22, 25–38.
Coughlin, M.D. (1975a) Devel. Biol. 43, 123–139.
Coughlin, M.D. (1975b) Devel. Biol. 43, 140–158.
Coughlin, M.D. and Rathbone, M.P. (1977) Devel. Biol. 61, 131–139.
Coughlin, M.D., Dibner, M.D., Boyer, D.M. and Black, I.B. (1978) Devel. Biol. 66, 513–528.
Crain, S.M. and Peterson, E.R. (1974) Ann. N.Y. Acad. Sci. 228, 6–35.
De Castro, F. (1932) in Cytology and Cellular Pathology of the Nervous System (Penfield, W., ed.) pp. 317–379, P.B. Hoeber, New York.
Ebendal, T. (1979) Devel. Biol. 72, 276–290.
Ebendal, T. and Jacobson, C.-O. (1977a) Brain Res. 131, 373–378.
Ebendal, T. and Jacobson, C.-O. (1977b) Exp. Cell. Res. 105, 379–387.
Ebendal, T., Belew, M., Jacobson, C.-O. and Porath, J. (1979) Neurosci. Lett. 14, 91–95.
Essex, H.E., Herrick, J.F., Baldes, E.J. and Mann, F.C. (1943) Am. J. Physiol. 139, 351–370.
Fukada, K. (1980) Nature 287, 553–555.
Furshpan, E.J., McLeish, P.R., O'Lague, P.H. and Potter, D.D. (1976) Proc. Natl. Acad. Sci. USA 73, 4225–4229.
Gonzalez-Crussi, F. (1971) Am. J. Anat. 130, 441–460.
Green, L.A. and Rein, G. (1978) J. Neurochem. 30, 579–586.
Greene, L., Sytkowski, A., Vogel, Z. and Nierenberg, N. (1973) Nature (London) 243, 163–164.
Gutmann, E. (1976) Ann. Rev. Physiol. 38, 177–217.
Hart, M.N., Heistad, D.D. and Brody, M.J. (1980) Hypertension 2, 419–423.
Hawrot, E. (1980) Devel. Biol. 74, 136–151.
Helfand, S.L., Smith, G.A. and Wessells, N.K. (1976) Devel. Biol. 50 541–547.
Hill, C.E. and Hendry, I.A. (1976) Neuroscience 1, 489–496.
Hill, C.E. and Hendry, I.A. (1977) Neuroscience 2, 741–749.

Hill, C.E. and Hendry, I.A. (1979) Neurosci. Lett. *13*, 133–139.
Hill, C.E., Purves, R.D., Watanabe, H. and Burnstock, G. (1976) Pflügers Archiv. *361*, 127–134.
Hill, C.E., Hendry, I.A. and McLennan, I.S. (1980) in Histochemistry and Cell Biology of Autonomic Neurons, SIF cells and paraneurons (Eränkö, O., Soinila, S. and Päivärinta, H., eds.) pp. 69–74, Raven Press, New York.
Hirst, G.D.S., Holman, M.E. and Spence, I. (1974) J. Physiol. (London) *236*, 303–326.
Hooisma, J., Slaaf, D.W., Meeter, E. and Stevens, W.F. (1975) Brain Res. *85*, 79–85.
Horii, Z.-I. and Varon, S. (1977) Brain Res. *124*, 121–133.
Imai, H., Lee, K.J., Lee, S.K., Lee, K.T., O'Neal, R.M. and Thomas, W.A. (1970) Lab. Invest. *23*, 401–415.
Imaizumi, M. and Kuwabura, T. (1971) Invest. Ophthalmol. *10*, 733–744.
Jessen, K.R., McConnell, J.D., Purves, R.D., Burnstock, G. and Chamley-Campbell, J. (1978) Brain Res. *152*, 573–579.
Johnson, D.G., Gorden, P. and Kopin, I.J. (1971) J. Neurochem. *18*, 2355–2362.
Johnson, M., Ross, C.D., Meyers, M., Rees, R. Bunge, R., Wakshull, E. and Burton, H. (1976) Nature *262*, 308–310.
Johnson, M.I., Ross, C.D., Meyers, M., Spitznagel, E.L. and Bunge, R.P. (1980) J. Cell Biol. *84*, 680–691.
Johnson, M.I., Ross, C.D. and Bunge, R.P. (1980) J. Cell Biol. *84*, 692–704.
Kim, S.U., Warren, K.G. and Kalia, M. (1979) Neurosci. Lett. *11*, 137–141.
Ko, C.-P., Burton, H. and Bunge, R.P. (1976) Brain Res. *117*, 437–460.
Ko, C.-P., Burton, H., Johnson, M.I., and Bunge, R.P. (1976) Brain Res. *117*, 461–485.
Landis, S.C. (1976) Proc. Natl. Acad. Sci. USA *73*, 4220–4224.
Landis, S.C. (1980) Devel. Biol. *77*, 349–361.
Le Douarin, N.M. and Teillet, M.A. (1974) Devel. Biol. *41*, 162–184.
Le Douarin, N.M., Renaud, D., Teillet, M.A. and Le Douarin, G.H. (1975) Proc. Natl. Acad. Sci. USA *72*, 728–732.
Leeson, T.S. and Leeson, C.R. (1965) Acta Anat. *62*, 60–79.
Letourneau, P.C. (1975a) Devel. Biol. *44*, 77–91.
Letourneau, P.C. (1975b) Devel. Biol. *44*, 92–101.
Letourneau, P.C. (1978) Devel. Biol. *66*, 183–196.
Levi, A., Shechter, Y., Neufield, E.J. and Schlessinger, J. (1980) Proc. Natl. Acad. Sci. USA *77*, 3469–3473.
Levi-Montalcini, R. (1952) Ann. N.Y. Acad. Sci. *55*, 330–343.
Levi-Montalcini, R. (1976) Prog. Brain Res. *45*, 235–258.
Levi-Montalcini, R. and Angeletti, P.U. (1968a) Physiol. Rev. *48*, 534–569.
Levi-Montalcini, R. and Angeletti, P.U. (1968b) in Growth of the Nervous System. A Ciba Foundation Symposium (Wolstenholme, G.E.W. and O'Connor, M., eds.) pp. 126–147, Churchill, London.
Levi-Montalcini, R. and Hamburger, V. (1953) J. Exp. Zool. *123*, 223–278.
Levi-Montalcini, R., Meyer, H. and Hamburger, V. (1954) Cancer Res. *14*, 49–57.
Lindsay, R.M. (1979) Nature (London) *282*, 80–82.
Mains, R.E. and Patterson, P.H. (1973) J. Cell Biol. *59*, 329–366.
Mark, G.E., Chamley, J.H. and Burnstock, G. (1973) Devel. Biol. *32*, 194–200.
McLennan, I.S. and Hendry, I.A. (1978) Neurosci. Lett. *10*, 269–273.
McLennan, I.S., Hill, C.E. and Hendry, I.A. (1980) Nature (London) *283*, 206–207.
Menesini Chen, M.G., Chen, J.S. and Levi-Montalcini, R. (1978) Archo. Ital. Biol. *116*, 53–84.
Moscona, A.A. (1976) in Neuronal Recognition (Barondes, S.H., ed.) pp. 205–226, Plenum Press, New York.
Nishi, R. and Berg, D.K. (1977) Proc. Natl. Acad. Sci. USA *74*, 5171–5175.

Nurse, C.A. and O'Lague, P.H. (1975) Proc. Natl. Acad. Sci. USA 72, 1955–1959.
Obata, K. (1974) Brain Res. 73, 71–88.
O'Lague, P.H., Obata, K., Claude, P., Furshpan, E.J. and Potter, D.D. (1974) Proc. Natl. Acad. Sci. USA 71, 3602–3606.
O'Lague, P.H., Potter, D.D. and Furspan, E.J. (1978) Devel. Biol. 67, 384–403.
Patterson, P.H. (1978) Trends Neurosci. 11, 126–128.
Patterson, P.H. and Chun, L.L.Y. (1974) Proc. Natl. Acad. Sci. USA 71, 3607–3610.
Patterson, P.H. and Chun, L.L.Y. (1977) Devel. Biol. 56, 263–280.
Pease, D.C. and Paule, W.J. (1960) J. Ultrastruct. Res. 3, 469–483.
Perry, R., Chamley, J.H. and Robinson, P.M. (1975) J. Anat. 119, 505–516.
Ping, C. (1921) J. Comp. Neurol. 33, 281–312.
Poole, J.C.F., Cromwell, S.B. and Benditt, E.P. (1971) Am. J. Pathol. 62, 391–414.
Purves, R.D., Hill, C.E., Chamley, J.H., Mark, G.E., Fry, D.M. and Burnstock, G. (1974) Pflügers Arch. 350, 1–7.
Ross, D., Johnson, M. and Bunge, R. (1977) Nature (London) 267, 536–539.
Sidman, R.L. and Wessells, N.K. (1975) Exp. Neurol. 48, 237–251.
Skaper, S.D., Bottenstein, J.E. and Varon, S. (1979) J. Neurochem. 32, 1845–1851.
Stevens, W.F., Slaaf, D.W., Hooisma, J., Magchielse, T. and Meeter, E. (1978) Biochem. Soc. Trans. 6, 487–490.
Thoenen, H., Angeletti, P.U., Levi-Montalcini, R. and Kettler, R. (1971) Proc. Natl. Acad. Sci. USA 68, 1598–1602.
Tuttle, J.B., Suszkiw, J.W. and Ard, M. (1980) Brain Res. 183, 161–180.
Varon, S. and Adler, R. (1980) Curr. Top. Devel. Biol. 16, 207–252.
Varon, S., Manthorpe, M. and Adler, R. (1979) Brain Res. 173, 29–45.
Walicke, P.A., Campenot, R.B. and Patterson, P.H. (1977) Proc. Natl. Acad. Sci. USA 74, 5767–5771.
Weiss, P. (1934) J. Exp. Zool. 68, 393–448.
White, W.B. and Bennett, M.R. (1979) Brain Res. 159, 379–384.
Wissler, R.W. (1968) J. Atheroscler. Res. 8, 201–213.
Yamamato, I. (1961) J. Electron Microsc. 10, 145–160.
Yamauchi, A. and Burnstock, G. (1969) J. Anat. 104, 1–16.
Yankner, B.A. and Shooter, E.M. (1979) Proc. Natl. Acad. Sci. USA 76, 1269–1273.
Zaimis, E. (1971) Proc. Physiol. Soc. C15, 16–17.

CHAPTER 3

Aspects of naturally-occurring motoneuron death in the chick spinal cord during embryonic development

RONALD W. OPPENHEIM and I-WU CHU-WANG

Neuroembryology Laboratory, Research Section, North Carolina Division of Mental Health, Anderson Hall, Dorothea Dix Hospital, Raleigh, NC 27611, USA

1. Introduction

A careful study of the intellectual history of science reveals that for most disciplines a remarkably large number of the fundamental questions, issues and phenomena in a field were already recognized – often to a very great extent – by the early pioneers and that later efforts frequently represent elaboration and modifications on a set of original themes. By-and-large this appears to be the case with neuroembryology. Not only were most of the major events in the ontogenetic organization of the nervous system – such as the proliferation, determination, migration, growth and differentiation of neurons – recognized by embryologists such as W. His, W. Roux, H. Driesch, H. Spemann. R. Harrison and S. Ramon y Cajal, but a number of the important questions concerning the specific mechanisms involved in neurogenesis were also rather clearly understood by these and other of our intellectual predecessors of the late 19th and early 20th century (Jacobson, 1978).

In neuroembryology, however, there is a striking exception to this general historical observation; namely, the phenomenon of massive naturally-occurring neuronal death in the central and peripheral nervous system. Although it is true that the devoted antiquarian can point to a few isolated instances early in this century in which degenerating neurons were observed in apparently normal embryos (Glücksmann, 1951), these reports were either ignored or considered exceptional and thus failed to have any lasting influence on conceptions of neurogenesis (Oppenheim, 1981a). Consequently, in all textbooks that were written prior to 1950 and which deal with development of the nervous system, no mention is made of the naturally-occurring loss of neurons during embryogenesis. The year 1950 is significant as a watershed as it was in 1949 and 1950 that Viktor

Hamburger and Rita Levi-Montalcini reported two cases of a normal massive neuronal degeneration, first in the dorsal root ganglia (DRG) and then in the cervical motoneurons of the chick embryo (Hamburger and Levi-Montalcini, 1949; Levi-Montalcini, 1950).

Despite the compelling nature of these pioneer observations, it took more than another decade before naturally-occurring neuronal death began to be gradually recognized as a potentially widespread and fundamental feature of normal vertebrate neurogenesis (Cowan, 1973; Oppenheim, 1981a). Since it is now known that massive neuronal death, often involving 40 to 60% of a given population of cells, does, in fact, occur in a wide variety of regions and cell types in the central and peripheral nervous system it seems curious, in retrospect, that such a striking phenomenon was overlooked for so long. Part of the reason for this probably lies in the mundane, but important observation, that embryonic neurons degenerate and are phagocytized exceedingly rapidly (see below). Consequently, unless one is deliberately searching for dying cells they can be easily missed or their magnitude greatly underestimated. Until investigators began to utilize quantitative methods, in which all healthy and/or degenerating cells in a population were counted at closely spaced intervals during embryogenesis, the significance of massive neuronal degeneration was overlooked. Another apparent reason for this oversight, and one which has previously been discussed in more detail (Oppenheim, 1981a), is that the occurrence of a large scale cellular degeneration or regression, during what is generally considered an exemplar of a constructive or progressive period in the life-history of an organism (i.e., embryogenesis), is conceptually counter-intuitive. As Saunders so aptly put it in his important review on non-neural embryonic cell death, 'One confronts less than comfortably the notion that cellular death has a place in embryonic development; for why should the embryo, progressing towards an ever more improbable state, squander in death those resources of energy and information which it has won from a less ordered environment?' (Saunders, 1966, p. 604).

Since one of us has recently written a relatively comprehensive review of the history and current status of neuronal cell death (Oppenheim, 1981a), our aim in the present account is to focus more narrowly on recent progress in our understanding of the mechanisms regulating cell death, and on the putative adaptive significance of this phenomenon, in spinal motoneurons of the chick. Largely because of the sheer amount of information available on somatic motoneurons, the major emphasis will be on this population of neurons; however, cell death in a specific population of *visceral* efferent (visceromotor) preganglionic neurons will also be briefly discussed.

2. Evidence that significant cell death occurs in spinal motoneurons

In light of the relatively long period during which naturally-occurring neuronal cell death was overlooked, it seems appropriate to begin this review with a brief discussion of the evidence now available supporting the existence of this phenomenon for spinal motoneurons. Theoretically, a number of different lines of evidence could be viewed as consistent with the proposition that a temporally-restricted cell loss occurs in a defined population of neurons: (a) following the cessation of proliferation and migration, counts of healthy cells should decline over time; (b) in the same situation, counts of pyknotic or dying neurons should first increase and then decrease over time; (c) total DNA content in a given region as measured biochemically could decrease over time; and finally, (d) shortly after the completion of migration, the number of cells previously labelled with a [^{3}H]thymidine marker might be higher than at later, more mature stages. In practice, the first two procedures have been the most widely used and are generally considered the most acceptable criteria for demonstrating the occurrence of cell death. Although a decrease in the number of healthy cells is generally considered to be a valid index of naturally-occurring cell death, it must be remembered that in some populations, differentiating neurons undergo a secondary migration, in which some proportion of an apparently homogeneous neuronal population move to a new location (e.g., Levi-Montalcini, 1950). Although this is probably a relatively uncommon phenomemon, one needs to be aware of such a possibility when studying neuronal death in the embryo.

In a brief footnote to a report on the induced cell death of limb-innervating motoneurons following early limb-bud removal in the chick, Hamburger (1958) suggested, based on counts of degenerating neurons, that there may also be a naturally-occurring cell death in this population. However, it wasn't until almost 20 years later that rigorous, quantitative data were reported by Hamburger (1975) confirming this suggestion and showing that at least 40% of the motoneurons in the lumbar lateral motor column (LMC) of the chick are lost between day 5 and day 10 of incubation. Perhaps the strongest support for a substantial cell death in this population is the combination of quantitative data on both healthy *and* degenerating cells in which the two sets of data are inversely related over time (Fig. 1). In other words, in this case, the loss of healthy cells can probably be accounted for by actual cell death as against a secondary migration or by later, in situ, phenotypic changes (e.g., by apparent early motoneurons becoming transformed into interneurons or glia). Moreover, recent ultrastructural studies have shown conclusively that the pyknotic nuclei seen in the light microscope reflect young neurons undergoing a process of degeneration and subsequent phagocytosis (Chu-Wang and Oppenheim, 1978a). Axon counts in the ventral roots of limb-innervating segments of the chick spinal cord also show a decline during the period in which motoneuron numbers are decreasing (Chu-Wang and

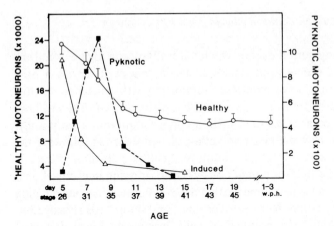

Fig. 1. The number of healthy and pyknotic LMC motoneurons in lumbar segments 23 to 30. Counts were made from every 10th section. *Induced* represents the number of surviving motoneurons in the deprived LMC following limb-bud removal on day 2. Sample size = 8 to 12 animals per age for healthy and pyknotic, and 3 to 5 for induced.

Oppenheim, 1978b). Since similar kinds of evidence are now available for preganglionic visceral motoneurons in the chick spinal autonomic system (see Section 7), there can be no question that naturally-occurring cell death is a real phenomenon in spinal motor as well as in many other neuronal systems.

3. Are spinal motoneurons preprogrammed to die?

There is considerable evidence (reviewed by Saunders, 1966) showing that genetic factors may determine the occurrence and temporal schedule of cell degeneration among non-neuronal cells during embryogenesis. In such cases actualization of the normal phenotype apparently depends upon the genetic and epigenetic control of patterns of degeneration, indicating that the death of cells is programmed as a normal morphogenetic event. A classic example of this phenomenon is the loss of mesenchyme cells in the posterior necrotic zone (PNZ) of the chick wing bud, in which case the loss of cells is an integral part of the sculpturing of normal wing morphology. Isolation of prospective stage 17 PNZ tissue in organ culture or by transplantation to heterotopic sites in vivo results in the death of the PNZ cells at the time at which they would normally die in situ (i.e., stage 24). However, prior to stage 22 such cells can be rescued by the appropriate environmental manipulation. Thus, as has been demonstrated over and over again for constructive events in embryogenesis, prior to their final determination, regressive processes such as cell death can also be reversed by modifying cellular interactions; by contrast, after the time of final determination their fate is virtually sealed, hence the designation of *programmed cell death*. These observations raise the obvious question of whether the death of neurons is also regulated by similar mechanisms. Before embarking on a discussion of this issue, however, we wish to point out that we will be dealing here exclusively with the death of differentiating neurons or what has been termed *histogenetic* cell death. Massive, as well as a more sporadic death of exceedingly immature postmitotic neurons, has also been described (Carr and Simpson, 1978; Hamburger et al., 1981), but it seems possible that cell death during this early phase of neurogenesis is regulated by different mechanisms and serves a different function than the later histogenetic neuronal death.

As far as can be determined the death of spinal motoneurons is not restricted to a specific, spatially restricted group of cells, as is the case with the cells of the PNZ. Rather the cells which die are apparently randomly intermingled with those that survive. Moreover, quantitative data on cell or nuclear size in the chick lumbar LMC prior to the onset of cell death indicate that initially there is only a single, relatively uniform population of immature motoneurons (Fig. 2). Similarly, ultrastructural studies of the LMC prior to cell death have failed to reveal a sub-population of neurons that might be considered to reflect signs of an impending degeneration (Oppenheim et al., 1978). Despite these negative findings, however, it is still conceivable that there is a sub-population of motoneurons that are genetically programmed to die. Sampling problems inherent to electron microscopy make it exceedingly difficult to rule out this possibility in normal embryos using ultrastructural criteria alone. And, of course, if neurons are predestined to die, they may reflect this fate in early biochemical or physiological differences that are not (or not easily) revealed in the electron microscope. Obviously, a new or different approach is needed if this question is to be resolved. In fact, two different procedures have been used recently to attack this issue. The first approach utilizes an old technique involving the early surgical removal of the target tissue (e.g., the limb-bud) of the specific neurons being studied. The second approach requires surgical, pharmacological or physiological procedures in an attempt to alter (enhance and/or prevent or retard) naturally-occurring cell death. The evidence from these two approaches concerning spinal motoneurons will be dealt with separately.

3.1. Induced cell death

In their classic study of cell death in the chick spinal ganglia, Hamburger and Levi-Montalcini (1949) showed conclusively that there is an overproduction of spinal ganglion cells during normal development. They interpreted this observation as evidence for a competition among the cells, either for a limited supply of a hypothetical trophic agent or for a limited number of 'synaptic' targets (i.e., sensory end-organs) in the muscle, viscera and skin – in either case, a process in which the losers ultimately die. In support of this contention they showed that if

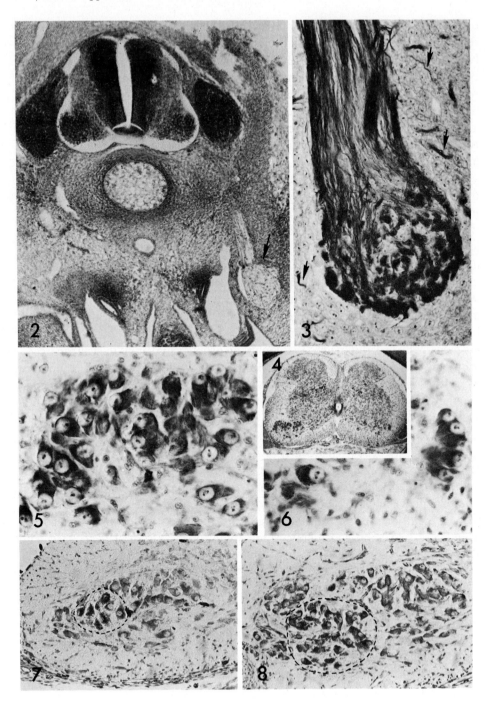

one removes the early prospective limb-bud at a stage prior to innervation – thus avoiding axotomy – the number of sensory neurons which die is greatly increased.* Later, Hamburger (1958) showed that the somatic motoneurons in the LMC respond in a similar fashion to limb-bud removal – indeed, in this situation over 90% of the LMC neurons die – and he also provided evidence that initially the deprived motoneurons undergo a normal differentiation despite their inability to innervate their usual muscular targets in the limb.

On the basis of these early findings, it seemed reasonable to us to use the limb-bud removal situation to investigate the question of whether neurons that will ultimately die exhibit histological, histochemical or biochemical differences even before frank signs of cytological degeneration are detectable in the light microscope. The rationale behind this approach was that if induced cell death is merely a quantitative exaggeration of naturally-occurring cell death, then we would be dealing with a population of neurons of which most, if not all, would ultimately die.

As had been reported earlier by Hamburger and others, we also found that the early removal of the leg-bud resulted in an almost total depletion of motoneurons (Figs. 1, 3 and 6). Observations at the light microscopic level indicated that the deprived cells initially differentiated normally and appeared indistinguishable from

---

*It was chiefly this experimental evidence which led Hamburger and Levi-Montalcini to propose the competition hypothesis. They argued that if normally, cells die due to a naturally-imposed limitation in target size, then an *experimentally* increased reduction in target size should lead to even more cell loss, a prediction borne out by their findings.

Fig. 2. Photomicrograph of a cross-section through the mid-lumbar region of a 5-day embryo showing the formation of the neuroma (arrow) on the limb-bud removal (right) side. Thionin stain. ×120.

Fig. 3. Neuroma from a 5-day embryo stained with the Cajal–de Castro silver technique. Note that a few nerve fibers (arrows) have escaped from the neuroma. ×440.

Fig. 4. Photomicrograph of a cross-section through the mid-lumbar region of a 12-day embryo. Only a few healthy looking cells still survive in the deprived LMC (right side). Thionin stain. ×29.

Fig. 5. and 6. Higher magnification of the control (Fig. 5) and deprived (Fig. 6) LMC from Fig. 4. ×400.

Fig. 7. Photomicrograph of the LMC of lumbar segment 27 of a 10-day control embryo following HRP injection into the peroneus muscle. Cluster of heavily and moderately labelled cells are enclosed by dotted line. Cresyl Violet stain. ×130. (See Oppenheim, 1981b, for deatils.)

Fig. 8. Photomicrograph of the LMC of lumbar segment of an α-BTX-treated 10-day embryo following HRP injection into the peroneus muscle. Cluster of heavily and moderately labeled cells are enclosed by dotted line. Note the increased number of motoneurons and the increased number of labeled cells relative to Fig. 7. Cresyl Violet stain. ×130.

control cells located on the contralateral side of the spinal cord (Oppenheim et al., 1978) (Fig. 2). Moreover, data on cell size (Fig. 9), the ability to transport

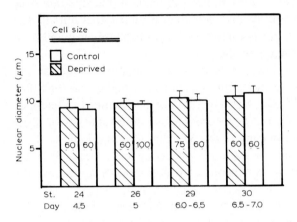

Fig. 9. Nuclear diameter of motoneurons (mean ± S.D.) in the LMC following limb-bud removal on day 2. Cells were drawn using a drawing tube and a 100 × oil objective (× 1250). Measurements were made of the greatest diameter of the nucleus. The sample size is indicated in the bars.

horseradish peroxidase (HRP) retrogradely, dendrite growth, the histochemically defined presence of acetylcholinesterase (AChE), biochemical measures of choline acetyltransferase (CAT) activity, and quantitative electron microscopy (synaptogenesis) all appeared normal in the deprived motoneurons up to the time when massive cell loss began. Ultrastructural comparisons of the cyto-differentiation of the deprived and control motoneurons also revealed no consistent differences, and axon counts in the ventral roots of stage 28 (5 day) embryos were similar for the control and deprived side, indicating that all the deprived cells had sent an axon as far as the ventral root.

Although it wasn't possible in these experiments to determine quantitatively whether all of the deprived neurons also contributed axons to the mixed *peripheral* nerve, measurements of the longitudinal thickness of the peripheral nerve taken roughly midway between the spinal cord and the body-wall, proximal to the thigh region, at stage 26, were similar on the two sides (control = 38 ± 5 μm, $n = 10$; deprived = 40 ± 4.1 μm, $n = 8$). In summary, based on a number of different criteria involving several different procedures, the target deprived somatic spinal motoneurons in these experiments appear indistinguishable from control neurons up to the time of onset of massive cell death.

A notable exception to this conclusion concerns the response of the axons of deprived motoneurons once they reach the vicinity of the trunk where the limb-bud attaches. Whereas the axons on the control side innervate the leg in an entirely typical fashion, the deprived axons end in a tight mesh or tangle of nerve endings (neuroma) some distance from the body wall at the approximate site of the lumbar plexus (Figs. 2, 3). Most axon profiles appear normal within the neuroma but no synapses, or even primitive contacts, have been observed. The fact that the deprived axons usually stop growing and form a neuroma before reaching the site of attachment of the limb – an observation also noted by Hamburger (1958) – may be significant in that it implies that cues normally supplied by the limb may exert a *tropic* influence on axonal growth even before the axons are in physical proximity with the limb. In addition to the implications this observation has for understanding the growth of specific axonal pathways, it also suggests that putative *trophic* factors supplied by the limb that are required for the survival and maintenance of motoneurons may be able to act at a distance. Thus, it is conceivable that the regulatory events involved in the control of cell survival may begin even before the axons have reached their targets; the sporadic, but consistent, early cell death that we and others have observed among this population of motoneurons may reflect the results of an early competitive process. A similar situation may exist in the *spinal ganglion* cells of the chick since it has been reported that some cells in this population degenerate very shortly after they become postmitotic and thus probably before their axons have reached peripheral targets (see below).

Because early limb-bud removal also *deprives* sensory neurons of their synaptic targets resulting in an induced cell loss in spinal ganglia (Hamburger, 1934; Hamburger and Levi-Montalcini, 1949), and since *naturally-occurring* cell death in the spinal ganglia extends throughout much of the period of motoneuron cell death, the question arises whether the natural death of limb motoneurons is somehow related to the development of extrinsic primary afferents. Although some evidence now exists on this point, the issue is still in dispute (Oppenheim, 1981a).

In collaboration with Dr. N. Okado, we have recently begun re-examining this problem using preparations in which the lumbar spinal ganglia are prevented from developing by surgical removal of the neural crest (Okado and Oppenheim, 1981). Following removal of the lumbar crest at stage 16, cell number in the LMC doses does not appear to be significantly altered, at least up to day 10, which is after the lion's share of natural cell death has occurred. Cell size is also unchanged under these conditions. However, a significant number of cells are lost after day 10 compared to controls. This later cell loss probably reflects a secondary transneuronal effect of deafferentation which is different from or unrelated to events regulating natural cell death. Preliminary experiments by Okado also show that following the removal of *intrinsic* spinal cord afferents, produced by early spinal cord transection, there appears to be no noticeable difference in cell number or cell size in the lumbar LMC up to days 10 to 12, after which, however, a secondary cell loss occurs. To summarize, the absence of either intrinsic or extrinsic afferents to somatic motoneurons during the period of *natural* cell death does not appear to alter the magnitude or the rate of cell loss.

3.2. Peripheral or target innervation prior to the onset of cell death

From the studies involving early hind limb removal, it is clear that motoneurons later destined to die can begin differentiation including axonal outgrowth. In this situation virtually all peripherally deprived cells send axons into the ventral root and most, if not all, appear to grow as far as the neuroma. It isn't clear from these data, however, whether this is also true of cells destined to die in normal embryos or, more importantly, whether in normal embryos *all* axons actually innervate the leg before the cell death period begins. For instance, it is conceivable that only a certain proportion of the total number of neurons ever gain access to the limb and that the remainder die by virtue of this failure. If this were the case it would require a rather serious re-thinking of the original competition hypothesis which tacitly assumes that initially the axons of all (or most) neurons in a given population gain access to their target.

Although as mentioned above, there may be a quantitatively small amount of *sporadic*, natural cell loss of neuroblasts or young motoneurons prior to limb innervation, recent evidence strongly suggests that the vast majority of lumbar motoneurons in the chick innervate the leg prior to the onset of massive cell death. Comparisons of motoneuron cell-body counts with the number of ventral root axons indicate that there is a virtual one-to-one relationship between the two (Chu-Wang and Oppenheim, 1978b). In both the brachial and lumbar region limb innervation begins between 4 and 5 days of incubation (stages 23 to 25). Once the motor axons leave the ventral root region, they are joined by sensory and sympathetic axons from which they are indistinguishable ultrastructurally at these early stages. And because both sensory and autonomic ganglionic neurons also exhibit natural cell death, it isn't possible to determine the number of motor axons (vs. the number of sensory or sympathetic axons) entering the limb at these early stages. Consequently, a different approach is needed to answer this question.

With the recent advance of retrograde labelling techniques using horseradish peroxidase (HRP) in embryonic systems, it is now possible to backfill all neurons innervating a target and, by counting the labelled cells, to establish what proportion of the total population innervates a target prior to cell death. By using this approach we have been able to show that virtually all motoneurons in at least some regions of the lumbar spinal cord innervate the leg by embryonic day 5 (stage 27+) (Oppenheim and Chu-Wang, 1977). Moreover, by examining these preparations in the electron microscope we have found that degenerating motoneurons contain HRP, indicating that these cells had axons in the limb at the time of injection, several hours earlier, after which they apparently underwent natural cell death. Thus it seems rather clear that prior to the massive loss of motoneurons all of these cells had sent axons into the limb.

At the time of the initial innervation of the limbs, the presumptive limb muscles in the wing and leg are represented by a dorsal and a ventral muscle mass. Only at about stages 29 to 30 or $6\frac{1}{2}$ days of incubation do these early muscle masses first begin to cleave into the individually recognizable muscles of the adult, a process which then continues until about stage 34 (day 8). In light of this temporal sequence, it becomes of interest to determine whether most of the axons from the motoneurons comprising specific motor pools innervate their respective muscles (or presumptive muscle regions) prior to the onset of the major period of cell death. Whereas the evidence from the HRP labelling experiment implied that the axons of all of the motoneurons have entered the limb by stage 28, these data do not answer the question of whether each muscle anlage has received its full complement of innervation at this time. Innervation may be a dynamic process in the sense that 'new' axons may project to a muscle at the same time that other earlier arrivals are degenerating.

Theoretically, it should be possible to completely fill individual muscles or presumptive muscle regions with HRP at early stages so as to establish the number of motoneurons projecting to each muscle. Since cell death in the entire lumbar region involves the loss of 40 to 50% of the neurons present on day 5, one would predict that roughly twice as many cells would project to a given muscle or muscle anlage at this time, compared to day 10 when cell death is almost complete. As Landmesser (1978) has pointed out, however, in actuality this approach has a number of shortcomings. She has shown that by adding together all of the cells labelled at stage 28 following HRP injections into each of the four major muscle masses of the leg (i.e., the dorsal and ventral thigh and shank) only about 30% of the total cells present at that stage appear labelled. Notwithstanding the fact that certain parts of the *limb* were inadvertently left uninjected in her experiments, the presence of 69% unlabelled cells leaves open the possibility that many axons may still be extending processes to their final targets at this time. The apparent discrepancy between the results of Landmesser and our own previous findings, in which virtually all motoneurons in the caudal lumbar region were labelled following HRP injections into the ventral muscle mass at Stages 27 and 28, may be more apparent than real. We used rather massive injection volumes (1 to 2 μl) and a considerably more concentrated HRP solution (50% vs. 5%), which could lead to the uptake of HRP by a larger number of neurons and/or to their increased visualization.

An alternative procedure for helping to resolve this question is to compare the number of axons entering a specific muscle at different stages before, during, and after the cell death period. Because we have been recently focusing on the relationship between cell death and neuromuscular differentiation in the brachial motoneurons which innervate the two wing muscles, anterior and posterior latissimus dorsi (ALD and PLD), data on axon numbers innervating these two muscles are presented in the subsequent section. One major reason for choosing these muscles is that they represent clear cut examples of slow tonic (ALD) and fast twitch (PLD) muscles in the chick. An additional consideration is that a great deal is known about the physiological, morphological and biochemical characteristics of these two muscles (Vrbová et al., 1978).

4. Normal development of ALD and PLD muscles and their innervation

4.1. Motoneurons

The motoneurons innervating ALD and PLD muscles are located primarily in the LMC aspect of segments 14 and 15 with a small contribution from motoneurons in segment 13 (DeSantis et al., 1977). At stage 25 (4.5 to 5 days) which is after the completion of proliferation and migration of prospective LMC neurons, there are approximately 7,000 motoneurons in segment 14 (Fig. 10). The major period

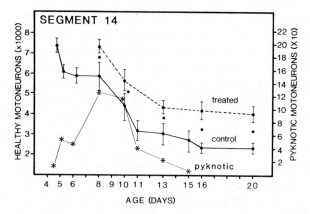

Fig. 10. The number of healthy and degenerating motoneurons in brachial segment 14. Treated animals received daily injections of curare or α-BTX on days 5 to 9 as previously described. The sample size = 5 to 7 embryos at each stage for *healthy* and 2 to 3 for *pyknotic* (for dosage see legend, Fig. 25). (✱) $P < 0.002$ between treated and control. Pyknotic data from control only.

of cell loss in segment 14 occurs between stages 25 and 37 (11 days) during which time the number of motoneurons is reduced by 58% to about 3,000 cells. Because a number of other wing muscles are also innervated by motoneurons in segment 14, and because it has not been possible to selectively identify the specific motor pools for the ALD and PLD at these early stages in segment 14, it isn't possible to determine the precise period of the death of ALD and PLD motoneurons; although we assume that they degenerate over the entire cell death period for the brachial region, this assumption may not be correct.

It is of interest that there are degenerating motoneurons in the LMC of segment 14 during the period between 5 and 8 days when the value for healthy cells does not change appreciably (Fig. 10). Laing (personal communication) has also observed dying motoneurons throughout the brachial region of the chick at these stages. Since cell division ceases on day 5 in this population, these observations probably indicate that there is a continued migration of motoneurons into the LMC, thereby offsetting the early loss of cells by degeneration.

4.2. Peripheral nerves

Because the muscle masses in the wing are still in the process of cleaving into individual muscles at earlier stages (Sullivan, 1962), day 8 (stage 34) is the first time that we have succeeded in reliably identifying ALD and PLD muscles and their nerve supply. The regions of the nerves that we have used for the quantitative and qualitative description below are depicted schematically in Fig. 11. In our material the nerve which supplies the primary innervation to the PLD typically sends a small branch to the ALD; thus, the data for PLD were obtained at region

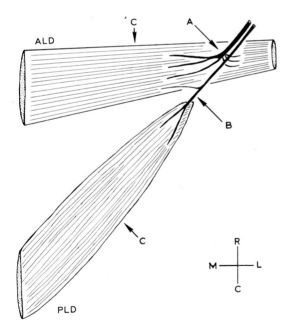

Fig. 11. Schematic illustration of ALD and PLD muscles and their innervation (left side of 1-day-old chicken) as viewed from a ventral perspective. (A) Area sampled for ALD axons; (B) area sampled for PLD axons; and (C) area sampled for muscle cell counts. Rostral (R), caudal (C), medial (M) and lateral (L) directions are indicated.

B in Fig. 11. Axon counts in the ALD nerve were obtained by subtracting the counts made at region B from region A★.

A summary of the data for axon counts is shown in Fig. 12A,B. Axon number

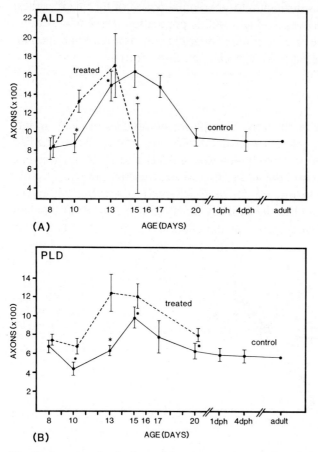

Fig. 12. Axon number in the peripheral nerves innervating ALD (A) and PLD (B) of control and treated animals. The treated animals received daily injections of curare or α-BTX as previously described. The sample size = 5 to 7 embryos at each stage. (✱) $P < 0.05$; (✲) $P < 0.001$.

★It is more important to note that these counts include *all* axons entering the respective muscles not just motor axons. Based on axon counts in the ventral and dorsal roots of 13-day-old control embryos (brachial segment 14), there are approximately 4 sensory axons for every 1 motor axon. On the assumption that this ratio is similar in the mixed peripheral nerve, then one can estimate that on day 8 (i.e., prior to massive cell death) there are approximately 130 somatic motoneurons supplying the PLD and 160 supplying the ALD. Furthermore, since we have not yet been successful in labelling the specific motor pools which supply these two muscles, the data on cell loss in Fig. 10 represent all motoneurons in segment 14 not just those which innervate the ALD and PLD muscles. A recent paper, published after the preparation of this chapter, includes a detailed report of the development of innervation of the ALD and PLD muscles (Bourgeois and Toutant, 1982).

increases until 15 days at which time there are 1,700 axons in the ALD and 950 in the PLD nerves. After day 15 axon number declines returning eventually to a level rather similar to the 8 day values. The large increase in axon number in the peripheral nerves at a time when massive cell death is occurring among the motoneurons in segment 14 (Fig. 10) may appear surprising. In a previous study (Chu-Wang and Oppenheim, 1978b), we reported that there is a virtual one-to-one relationship between motoneuron counts and *ventral root* axon counts in the *lumbar* region. If one takes into account the fact that brachial segment 14 contains about 1,500 medial motoneurons (MMC), whose axons also exit through the ventral root (Oppenheim, unpublished observations), then there is also a close to one-to-one correspondence between the number of motoneurons (LMC and MMC) and ventral root axons in segment 14. Consequently, we assume that the increase in axon counts up to day 15 in the ALD and PLD nerves reflects axonal branching rather than the initial outgrowth of new axons from the spinal cord.

Although it is also conceivable that the increased axon counts between day 8 and day 14 reflect the initial outgrowth of axons from the sensory and sympathetic ganglia, this appears unlikely. There is a naturally-occurring death of sensory and sympathetic cells during the period between 8 and 15 days (Hamburger et al., 1981; Oppenheim et al., 1982), so that one would actually expect a net *loss*, not an increase, of axons of sensory or sympathetic origin during this period.

The composition of the peripheral nerves on either day 13 or 15 compared to the period *after* day 15 also suggests that branching is occurring. At the earlier period more than 70% of the axons are small (<0.5 μm) and unmyelinated and they are almost always grouped in clusters. After day 15 when axon numbers decline, the larger unmyelinated axons (>0.5 μm) probably gradually become myelinated, whereas many of the small unmyelinated axons decline in number presumably due to degeneration and/or retraction.

Between day 13 and day 17 we have frequently observed a single Schwann cell forming myelin sheaths around each of 2 to 4 peripheral axons (Fig. 34). Usually only one of the axons in such a set appears cytologically normal, whereas the others appear to be degenerating. We previously reported similar 'supernumerary' myelin sheath profiles in the lumbar ventral roots (Chu-Wang and Oppenheim, 1978b). At that time we considered this to be a rare, but normally occurring phenomenon. Based on our present observations, however, it appears that this is a much more common phenomenon in developing *peripheral* nerves. Supernumerary myelination may represent a means whereby 'redundant' axonal branches are removed during normal ontogeny.

The phenomenon of transient polyneuronal innervation is known to occur in the avian ALD and PLD and its time course is rather closely correlated with the major changes in axonal counts observed by us in the peripheral nerves innervating these muscles (Atsumi, 1977; Bennett and Pettigrew, 1974; Srihari and Vrbová, 1978, 1980). Polyneuronal innervation, as defined electrophysiologically, represents the innervation of a single neuromuscular junction (NMJ) by more than one motoneuron. Since this phenomenon is usually transient, occurring only during a limited period in early development, and since in the chick its disappearance is not associated with cell death (Oppenheim and Majors-Willard, 1978), it must reflect increased axonal branching by developing motoneurons. Although it is not known where the axon branching occurs that is associated with polyneuronal innervation, our observations indicate that, at least in the ALD and PLD of the chick, part of this branching may originate from axons *outside* of the muscle, prior to target innervation, and part from branching *within* the muscle (see Section 5 and Table 1). Axonal branching *proximal* to the target has also been observed in other motoneuron populations during development (Oppenheim, 1981a). Already by day 6 (stage 30) there is a one-to-one relationship between the number of motoneuron cell bodies and the number of ventral root axons in segment 14 (unpublished observations). Thus, it seems reasonable to assume that by day 8 (stage 34) ALD and PLD muscles have received a full complement of motor (and probably sensory) innervation. If this assumption is correct, then much of the subsequent changes in axon number in the peripheral nerves observed by us must reflect changes in the number of axonal branches.

4.3. Neuromuscular junction

By 8 days (stage 34) neuromuscular contacts (NMJ) are already forming in both ALD and PLD muscles. Typically the NMJ is found on myotubes although occasionally we have found axon terminals containing synaptic vesicles in close apposition to myoblasts; however, postsynaptic membrane thickening has never been observed in these cases. At 8 and 10 days most of the NMJs in both muscles are contacted by only a single axon; rarely, there may be 2 axons contacting a single postsynaptic site (Figs. 13 and 14). One notable and consistent difference between the two muscles is that after 10 days there is often a bundle of axons situated at or adjacent to the NMJ in the ALD, whereas usually only 2 to 3 axons are seen at the NMJ in the PLD. Although NMJs are infrequent in the ALD and PLD at 8 to 10 days (relative to later stages), it is likely that even if functional contacts are formed on only a few myotubes at these early stages they could result in the activation of the entire muscle. Dennis (1981) has found that in the rat fetus the stimulation of only a few muscle cells in similar preparations quickly leads to the contraction of all myotubes in a muscle.

From 13 days on, the number of axons contacting a single NMJ increases in both muscles (Figs. 17,21). Based on serial cross and longitudinal sections of the endplate region, the increased axons at the NMJ in the ALD appear to arise from separate axons, whereas in the PLD the additional axon profiles at the NMJ appear to be due primarily to axonal branching close to the endplate. Moreover, in the ALD muscle axon profiles are usually in a cluster and are ensheathed by

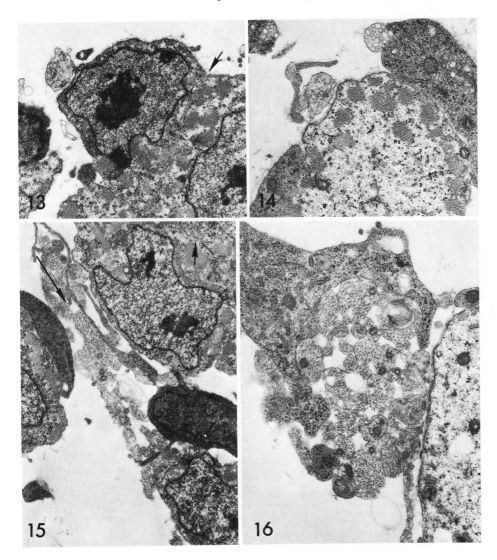

Fig. 13. Cross-section of a control ALD myotube from a 10-day embryo showing that only one axon terminal with synaptic vesicles is present at the endplate. Arrow indicates the narrow intercellular gap which separates the adjacent myotubes in a cluster. ×9,500.

Fig. 14. Cross-section of a control PLD myotube from a 10-day embryo showing one axon terminal at the endplate. ×11,850.

Fig. 15. Cross-section of an ALD myotube from an α-BTX-treated 10-day embryo showing that the NMJ receives several axon terminals. Long arrow indicates the source of axon terminals. Short arrow indicates the narrow intercellular gap. ×8,900.

Fig. 16. Cross-section of an α-BTX-treated PLD myotube from a 10-day embryo in which a NMJ receives many axon terminals from an axon bundle. ×12,250.

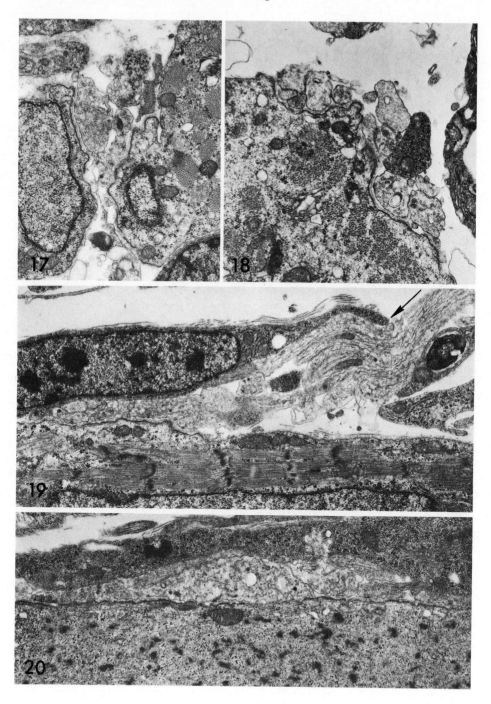

a single Schwann cell, whereas in the PLD muscle the individual axons are usually separated from one another by Schwann cell processes. Between day 15 and day 18 the number of axon profiles at the NMJ in the ALD muscle decreases from an average of 3–7 to 2–3, which number appears to be retained into maturity. In the PLD muscle the axon profiles are gradually reduced such that after days 18 and 19 the NMJ is usually always contacted by only a single axonal ending.

It is known that in the mature ALD, each muscle fiber has several NMJs located at regularly spaced intervals, whereas the PLD has only a single NMJ located along the length of each myofiber (Gordon et al., 1974, 1975). Although this pattern appears to be established rather early in both muscles, it isn't presently known whether there are more than the adult number of NMJs at the earliest stages of neuromuscular contact (i.e., before stage 36, 10 days). We have previously suggested that this might occur (Pittman and Oppenheim, 1979) and that the reduction of such 'ectopic' contacts may reflect, in part, the peripheral correlate of motoneuron cell death. Data presented below in Section 5 provide some indirect support for this hypothesis.

4.4. Muscle development

If the survival and maintenance of spinal motoneurons is dependent upon interactions with target muscles, then it is of obvious interest to understand as much as possible about the normal development of both motoneurons and muscle. It is clear from what follows below, however, that our present understanding of those aspects of muscle differentiation that are pertinent to the issue of motoneuron cell death is still far from complete. Nevertheless, the scanty information that is available is worth discussing since it may, if nothing else, suggest ideas for future research.

In many respects the development of the ALD and PLD muscles occurs in a fashion rather similar to that already described in detail for the neonatal rat (Ontell, 1977; Ontell and Dunn, 1978). In both the neonatal rat and the chick embryo immature muscle cells consist of a mixed population of primary fibers, satellite fibers and satellite myotubes. In the discussion that follows we include

---

Fig. 17. Cross-section of a control ALD myofiber from a 13-day embryo showing that an endplate receives 1 to 3 axon profiles. × 10,500.

Fig. 18. Cross section of an α-BTX-treated ALD myofiber from a 13-day embryo. Compared to the control (Fig. 17) the NMJ has many axon profiles. × 11,800.

Fig. 19. Longitudinal section of an α-BTX-treated ALD myofiber from a 13-day embryo in which a bundle of axons (arrow) make synaptic contact with a healthy myofiber. × 11,000.

Fig. 20. Longitudinal section of an α-BTX-treated ALD myofiber from a 13-day embryo showing axons still in synaptic contact with a degenerating myofiber. × 11,200.

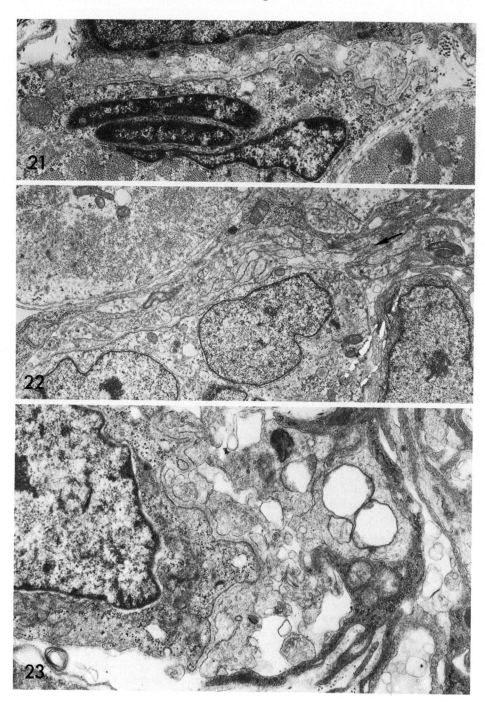

all of these myofilament-containing cell types under the common rubric of muscle cells.

Quantitative studies were carried out using cross-sections through the middle or girth region of each muscle (C, Fig. 11). Only after day 14 for the ALD and day 16 for the PLD do the respective muscles contain a majority of independent muscle fibers. Prior to this time the various myofilament-containing (muscle) cells occur in clusters enclosed by a common basement membrane. Depending on the developmental stage, each cluster may contain 2 to 14 muscle cells. Because the cells in the clusters are separated from one another by a gap of less than 20 nm, which is beyond the resolution of the light microscope, it is not possible to count the individual muscle cells using the light microscope alone. After a detailed examination of the clusters in the electron microscope, we discovered that clusters which are similar in size and shape typically contain similar numbers of muscle cells. Thus, we have proceeded by first examining ultra-thin sections with the electron microscope and counting the number of individual muscle cells contained within the different sized clusters in each muscle and at each developmental stage. The total number of clusters was then counted in adjacent thick epon sections with the light microscope. By multiplying the number of muscle cells per cluster by the number of clusters, and by including the number of independent muscle cells, it was possible to estimate rather accurately ($\pm 5\%$) the total number of muscle cells contained in an entire cross-section of the muscle. Although it has been reported (Kikuchi and Ashmore, 1976) that at early stages neuromuscular contacts are formed on only one muscle cell per cluster (usually on the so-called primary myotube), we have not found this to be necessarily true of the ALD or PLD. Consequently, we felt that in order to quantitatively compare axons and motoneurons with muscle cells at each stage, it was imperative to have relatively precise data on muscle cell number; by merely counting clusters one could possibly arrive at an erroneous conclusion.

At 8 days, the ALD contains clusters of muscle cells, myoblasts and other undifferentiated cells. Each cluster contains 1 to 3 muscle cells. At 10 days, more than 90% of the mature number of muscles cells are present (Fig. 24). Most of these are still in the myotube stage and each cluster contains 3 to 14 individual muscle cells. By 13 days, the ALD contains the full mature complement of muscle fibers; a majority of these are independent (i.e., non-clustered) immature myo-

---

Fig. 21. Cross-section of a control PLD myofiber from a 17-day embryo showing only a few axon profiles at endplate. × 17,000.

Fig. 22. Cross-section of a curare-treated PLD myofiber from a 17-day embryo. Note the numerous axon profiles at the endplate. Arrow indicates the source of axons. × 13,500.

Fig. 23. Cross-section of a curare-treated PLD myofiber from a 17-day embryo showing the presence of numerous axon profiles at endplate, × 16,500.

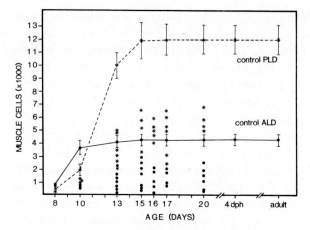

Fig. 24. Muscle cell number in control and treated ALD and PLD. Treated animals received daily injections of curare or α-BTX as previously described. Sample size = 5 to 7 animals at each stage for control groups. The raw data for each treated animal is indicated separately. (●) ALD; (✱) PLD. Both treated groups on day 8 are significantly different from control at $P < 0.05$. At all other ages, $P < 0.001$.

fibers with a peripherally placed nuclei. By 15 days, the ALD muscle consists almost entirely of independent myofibers.

The development of the PLD muscle is retarded by about 2 days relative to the ALD muscle. For instance, it is only at 13 days that the PLD attains 90% of the total adult complement of muscle cells (Fig. 24). Prior to this, all of the muscle cells occur in clusters and most are still in the myotube stage; by 16 days most of the muscle cells in the PLD consist of independent immature myofibers.

The mature ALD contains about 4,200 muscle fibers or myofibers whereas the PLD has approximately 12,000 muscle fibers. As mentioned, in both muscles these values are attained already 5 to 6 days before hatching. Since no new muscle fibers appear to be added after this, the subsequent increase in size and weight that continues until maturity is probably due entirely to the addition of myofibrils and new sacromeres. Finally, an interesting point concerning these data, and one which we shall discuss in greater detail below, is the apparent temporal mismatch between motoneurons and muscle differentiation; i.e., the total number of motoneurons is present and innervation of the limb begins long before the total number of muscle cells is attained.

The preceding summary of the normal development of motoneurons, muscles and their interrelationship, provides a framework within which to evaluate experimental perturbations of these processes which may lead to alterations of natural cell death. In the following section, we examine several different approaches to this problem.

5. Prevention of the naturally-occurring loss of spinal motoneurons

5.1. Increased target size

According to the competition hypothesis, a key element in the embryonic regulation of neuronal survival is the successful competition of some cells for a limited amount of trophic substance and/or a limited number of synaptic targets (and the death of the others). Consequently, it should be possible to modify cell survival by altering the size of the target (e.g., the limb-bud). Whereas the results following early limb removal are entirely consistent with this prediction – especially the evidence showing that motoneuron survival is proportional to the amount of remaining limb tissue (Hamburger, 1934) – the prevention or amelioration of cell death by an *increased* target size would be considerably more persuasive evidence for this aspect of the competition hypothesis. The creation of an enlarged target size by the transplantation of supernumerary limb buds is an old procedure in neuroembryology. Only recently, however, has this old, but powerful, approach been utilized in the study of naturally-occurring cell death.

In a recent replication of the early 'hyperplasia' experiments by Hamburger (1939), Hollyday and Hamburger (1976) transplanted a supernumerary leg bud in the thoracic region just rostral to the normal leg. The surgery was carried out at stages 17 to 18, before any motor axons have left the spinal cord in this region. Although no attempt was made to determine quantitatively the volumetric increase of total peripheral limb tissue, since only embryos with large, well-formed limbs (normal and supernumerary) were included in the data analysis, it seems likely that there was an effective doubling of target tissue in these preparations.

Cell counts of the LMC motoneurons in lumbar segments 23 to 30 at 12 and 18 days (after the cell death period) showed that there was an average of 14 to 15% more neurons present on the experimental side of the spinal cord compared to the contralateral, control, side. Since there were no differences between the two sides at 6 days, (i.e., prior to the massive cell death) one can infer that the addition of supernumerary limb does not affect the proliferation of motor neuroblasts.

Thus, in this experiment approximately 1/3 of the cells that normally die were rescued by increasing the size of the target region. In light of the fact that the target size was effectively doubled in these preparations, it may appear surprising that cell death wasn't prevented entirely – or at least to a greater extent – under these conditions. While the reason for this apparent discrepancy is presently unknown, it seems most likely that the physical location of the supernumerary limb in an ectopic (thoracic) position may have prevented many of the axons of lumbar motoneurons from gaining access to the additional target sites (or to additional trophic substance) provided by the extra limb. Retrograde labelling experiments with these preparations has, in fact, shown that at 12 days the axons supplying the extra limb are exclusively derived from motoneurons in the rostral lumbar segments 23 to 25. More convincing support for this explanation, however, requires that labelling experiments be done prior to the onset of cell death in order to determine the proportion of cells throughout the lumbar region which *initially* innervate the transplant and control limbs. Since other studies, some of which involved more favorable physical conditions for innervation of the supernumerary limb, have also failed to prevent a considerable proportion of naturally-occurring motoneuron loss (Oppenheim, 1981a) it remains an open issue whether the presence of an enlarged periphery is sufficient for the prevention of *all* naturally-occurring motoneuron death.

An alternative procedure for creating an increase in the size of the periphery, and one which lessens considerably the problem of all neurons having equal access to the additional targets, is the surgical elimination of some proportion of the competing neurons prior to the onset of cell death. For instance, in the frog, *Xenopus laevis*, cutting one of the three spinal nerves supplying the hindlimb reduces cell death in the remaining lumbar motoneurons from 75 to 40% (Olek and Edwards, 1977). Similarly, in the chick, Lance-Jones and Landmesser (1980) have found that surgical depletion of a portion of the motor pool for a specific muscle reduces cell death in the remaining motoneurons of that pool, presumably by reducing competition.

A particularly interesting study utilizing this approach involves neurons in the chick ciliary ganglion, a parasympathetic *visceromotor* ganglion supplying the iris and ciliary muscles in the eye. Pilar et al. (1980) have shown that by cutting two of the three major postganglionic branches of the ciliary ganglion just prior to the cell death period – thereby reducing the number of cells competing for innervation of the eye – cell death among the remaining cells supplying the third branch can be reduced by about 40%. An additional, and novel, outcome of this manipulation was that the surviving cells supplying the remaining nerve branch exhibited an accelerated maturation in that axon size, conduction velocity and myelination were all significantly increased at stage 40 compared to control axons; this latter effect was transient, however, since by stage 42 the differences had disappeared. The authors concluded that competition in the normal ganglion has a retarding influence on neuronal maturation. (Additional evidence for this proposition is provided by studies discussed below in which the cell death of spinal motoneurons – and presumably competition – has been reduced or eliminated by chronic treatment with neuromuscular blocking agents.)

Finally, by increasing the size of the sympathetic ganglia of the chick by Nerve Growth Factor (NGF) treatment, naturally-occurring neuronal death in the preganglionic visceral efferent (sympathetic) neurons in the thoracolumbar cord is reduced (see Section 7), thereby providing evidence in still another system that neuronal death is regulated by the size of the postsynaptic target.

5.2. Modification of motoneuron death by chronic treatment with pharmacological agents that affect neuromuscular activity

It is becoming increasingly recognized in neuroembryology that neural activity may play a significant role in at least certain steps in the ontogenetic organization of the nervous system (Harris, 1981). Several years ago, in an early phase of our investigation of the natural death of spinal motoneurons, it occurred to us that perhaps the massive loss of cells in this system might also be regulated, at least in part, by synaptic activity. Although it was already known at the time that neuromuscular contacts could develop in the presence of neuromuscular blocking agents (Freeman et al., 1976), the salient fact that the massive death of motoneurons in the chick spinal cord coincides with the structural and functional innervation of the limbs provided the impetus to explore this question more deeply.

Based on the assumption that only those cells that established a sufficiently large number of functionally stable neuromuscular contacts would survive, we expected that by chronically blocking neuromuscular activity during the normal period of cell death (i.e., from day 5 or 6 to day 10) cell loss might be exacerbated. Consequently, it came as a surprise when we discovered that the spinal cord of treated embryos sacrificed on day 10 contained several thousand more motoneurons than controls (Figs. 7 and 8) (Pittman and Oppenheim, 1978, 1979). Subsequent studies have confirmed these results (Laing and Prestige, 1978) and extended them to other species (Olek and Edwards, 1978) and neuronal systems (Creazzo and Sohal, 1979). In the most favorable cases virtually all cell death can be prevented. Moreover, a wide variety of pharmacological agents and neurotoxins which act both pre- and postsynaptically and which include depolarizing and competitive neuromuscular blockers are equally effective in this regard (Fig. 25). The increased number of motoneurons in these experiments is the result of reduced cell death in that there are significantly fewer degenerating cells in the LMC of treated embryos. Although at first we believed that the prevention of neuromuscular activity (i.e., behavioral paralysis) was the critical factor in producing this effect, it now appears that behavioral inactivity is a necessary but not a sufficient condition for the prevention of motoneuron death.

As summarized in Figs. 26 to 29, several additional agents that act at the neuromuscular junction (eserine, carbachol, nicotine and choline) produced a behavioral paralysis at least as great as curare or α-bungarotoxin (α-BTX) without preventing the death of motoneurons (Oppenheim and Maderdrut, 1981). Furthermore, related experiments with the frog, *Xenopus laevis*, in which neural activity was suppressed by the use of anesthetics such as chlorobutanol or procaine, have also failed to prevent cell death, whereas treatment with α-BTX, a neuromuscular blocker, was effective in preventing cell death in these same populations (Olek

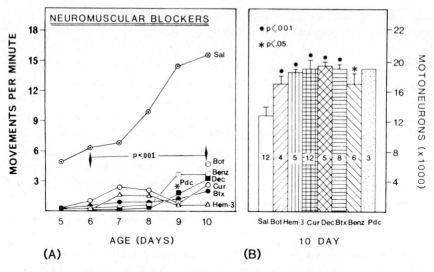

Fig. 25. The effects of treatment from day 5 or 6 to day 9 with a variety of neuromuscular blocking agents on embryonic motility (A), and motoneuron number in the lumbar LMC (B). *Sal*, 0.9% saline; *Bot*, botulinum toxin, 1200 LD50 on day 6; *Hem-3*, hemicholinium-3, 2 mg day 6, 4 mg days 7 and 8, 6 mg day 9; *Cur*, curare, 2 to 2.5 mg per day; *Dec*, decamethonium, 10 μg per day; *BTX*, α-bungarotoxin (or cobra-toxin), 100 μg per day; *Benz*, benzoquinonium, 0.5 mg day 5 and day 8; *PDC*, a derivative of phenyldiacetylcholine, is a novel competitive blocker with a shorter duration of action (i.e., it is metabolized more rapidly) than the other blockers used. *PDC* was synthesized for us by Dr. J. Cacolas, Department of Medicinal Chemistry, University of North Carolina, Chapel Hill. Sample size is indicated in the bars. Movements were significantly reduced at all ages by all of the blockers. Motoneuron values are compared statistically to the saline control. *Benz* was also significantly differently from curare at $P < 0.05$. The statistical comparisons in Figs. 25 to 29 were calculated using the Mann–Whitney test and the probability values are relative to either saline controls or curare-treated embryos.

and Edwards, 1978b)*. Motoneuron death in the chick can also be prevented by either botulinum toxin or hemicholinium-3 (Fig. 25), both of which effectively interfere with the release of acetylcholine (ACh), presynaptically. Since the ultimate effect of the general anesthetics used in the experiments with *Xenopus* embryos would also be the failure or diminution of ACh release it isn't clear why in this case such treatment failed to prevent cell death. It is conveivable, that in the frog sufficient ACh is released spontaneously under these conditions to ensure the normal loss of motoneurons. The failure of such agents as carbachol, eserine and nicotine to prevent cell death in the chick, despite their ability to suppress neuromuscular activity, may indicate that at the appropriate doses these drugs

---

*Regrettably, it has not been possible to use general anesthetics or other agents such as tetrodotoxin, with the chick embryo. Cardiac activity becomes sensitive to such treatment considerably earlier in the chick than in the frog. Consequently, doses sufficient to suppress neural activity are lethal.

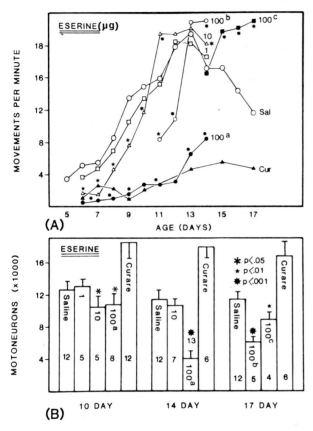

Fig. 26. The effects of chronic eserine treatment on embryonic motility (A) and motoneuron number in the lumbar LMC (B). Embryos were treated daily (once) with 1, 10 or 100 μg of eserine in 50 μl volumes. *Group 100^a*, treatment from days 5 to 13; *Group 100^b*, treatment from days 10 to 13; *Group 100^c*, treatment from days 13 to 16. The sample size is indicated in the bars. Although not shown in (B) the *100^b* group (n = 3) treated from days 10 to 13 also had approximately 50% fewer motoneurons than controls on day 14 (Mean = 6,000 vs. 11,700).

actually somehow mimic the role of neuromuscular function in regulating the extent of cell death.

Indeed, our rationale in using eserine, nicotine, carbachol and choline in these experiments was that it might prove possible to enhance or hasten the death of motoneurons by increasing receptor stimulation and/or muscle activity. Although chronic treatment with eserine or carbachol between days 5 and 9 (the period of normal cell loss) did, in fact, produce a significantly greater loss of cells on days 8 and 10 (Figs. 26,27), treament *after* day 10 was considerably more effective in this regard. For instance, chronic eserine treatment between day 10 and day 13 or between day 13 and day 17 resulted in 47 and 34 percent fewer motoneurons

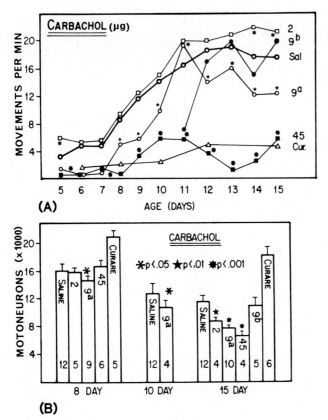

Fig. 27. The effects of chronic carbachol treatment on embryonic motility (A) and motoneuron number in the lumbar LMC (B). Embryos were treated twice daily with 2, 9 or 45 μg of carbachol in 50 μl volumes. *Group 9[a]*, treated from days 5 to 14; *Group 9[b]*, treatment from days 10 to 14; *Sal*, saline; *Cur*, curare (for comparison). At 8 and 10 days only the 9 μg group had significantly fewer motoneurons than the saline control. For illustrative clarity motoneuron data for the 2 and 45 μg groups are not depicted at day 10. The sample size is indicated in the bars.

on days 14 and 17, respectively. By contrast, eserine treatment between day 5 and day 9 resulted in only 17% fewer neurons on day 10 (Fig. 26).

Since the most marked effect of eserine on motoneuron survival occurred after the period of normal cell death it isn't clear what, if any, significance this result has for understanding naturally-occurring neuronal death. Eserine is known to produce myopathies in adult animals (Leonard and Salpeter, 1979) and we have observed similar ultrastructural abnormalities in some of the muscles of our eserine-treated preparations.* Despite our present uncertainty concerning the

---

*It is of some interest that a few embryos that were treated daily with eserine from day 5 to day 13 managed to hatch and appeared normal, behaviorally, after hatching despite having 50 to 60% fewer lumbar motoneurons than controls (Oppenheim and Maderdrut, 1981).

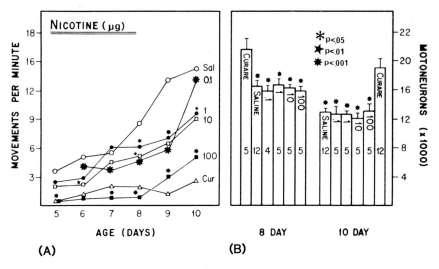

Fig. 28. The effects of chronic nicotine treatment on embryonic motility (A) and motoneuron number in the lumbar (LMC) (B). Embryos were treated twice daily with 0.1, 1,10, or 100 μg of nicotine in 50 μl volumes. Motoneuron number was not significantly different from the saline group at any dosage; however, all nicotine (and saline) groups were significantly different from curare. The sample size is indicated in the bars.

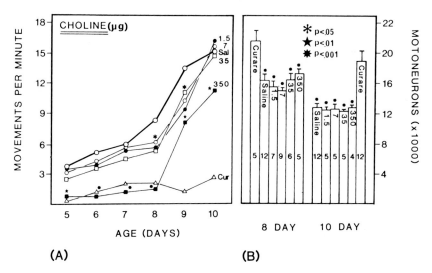

Fig. 29. The effects of chronic choline treatment on embryonic motility (A) and motoneuron number in the lumbar LMC (B). Embryos were treated twice daily with 1.5, 7, 35 or 350 μg of choline. Motoneuron number was not significantly different from saline at any dosage; however, all choline (and saline) groups were significantly different from curare. The sample size is indicated in the bars.

relevance of the eserine and carbachol data for understanding the mechanisms of cell death, it is conceivable that these findings do, in fact, reflect the disruption of functional neuromuscular interactions involved in the *normal* maintenance of cell survival. If this is the case, then the results (i.e., increased cell death) may, in fact, be consistent with our activity related model of cell death.

Finally, although the data are not presented here, we have found that chronic treatment from day 5 to day 9 with the potassium channel blocker, 4-aminopyridine (10, 50 or 100 μg/day), which was expected to increase synaptic activity, or receptor activation, also had no effect on either neuromuscular activity or motoneuron number in the lumbar LMC on day 8 or day 10. Without an independent measure of ACh release, however, we cannot conclude that these results are inconsistent with our model. It is conceivable that 4-aminopyridine does not produce the expected enhancement of ACh release in non-stimulated, but spontaneously active, embryonic motoneurons. Or perhaps the potassium channels in these immature neurons are not yet sensitive to the drug.

The reduction of degeneration among somatic motoneurons by neuromuscular blocking agents can be maintained indefinitely so long as the embryo remains immobile. Once neuromuscular activity is allowed to recover, however, the excess cells are lost.[†] In some cases, this delayed cell death results in an even greater loss of neurons than occurs during the normal cell death period. Why this excess cell loss occurs is not known. Furthermore, beginning the neuromuscular blockade at any time prior to the cessation of natural cell death on days 10 to 12 results in the rescue of a significant number of the neurons that have not yet degenerated (Oppenheim, unpublished observations); this is true for lumbar and brachial motoneurons. And, finally, initiating the neuromuscular blockade (curare) on either day 3 or 4, vs. day 5, and continuing treatment in all cases until day 9, results in an approximate 10% greater reduction in cell death, as determined from embryos sacrificed on day 10 (number of lumbar motoneurons on day 10 = $21,137 \pm 874$, $n = 4$ vs. $19,016 \pm 1,035$ $n = 7$). Thus, even the earliest dying motoneurons can be rescued by neuromuscular blockade.

Based on histological, morphometric (Fig. 30), and biochemical evidence (CAT) (Pittman and Oppenheim, 1979), the excess motoneurons resulting from the prevention of cell death appear to differentiate normally. They grow axons and establish contacts with the limb resulting in an apparent hyperinnervation of at least some muscles. In fact, as discussed below, there is some indication that the prevention of cell death may actually enhance some aspects of differentiation. As

---

[†]Since the chick embryo begins to depend upon pulmonary respiration shortly before hatching, virtually all paralyzed embryos die at this time. Recently, however, a few embryos treated with curare hatched and survived for 3 to 4 days. Despite the *absence* of a behavioral paralysis after hatching, these animals retained the excess motoneurons resulting from the earlier prevention of cell death (Oppenheim, 1982).

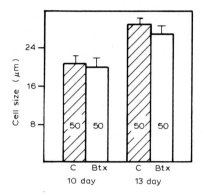

Fig. 30. Cell body size (diameter, mean ± SD) of LMC motoneurons in control and α-BTX-treated embryos. Samples were taken from lumbar segment 26. The outline of the entire cell body was drawn using a drawing tube and a 100× oil objective (×1,250). The sample size is indicated in the bars.

in the case of the ciliary ganglion noted above, this accelerated maturation may reflect the absence or lessening of competition between motoneurons resulting from the neuromuscular blockade.

Daily treatment of chick embryos with α-BTX or curare from day 5 to day 10 has no apparent effect on the survival or differentiation of neurons in either spinal

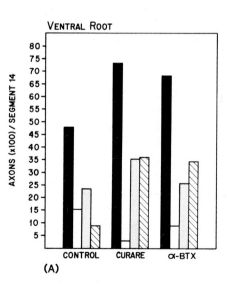
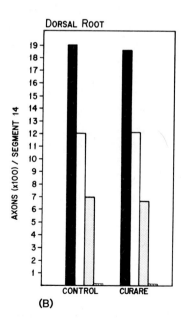

Fig. 31. Axon number and fiber composition in the ventral (A) and dorsal (B) root of brachial segment 14. The curare and BTX (α-bungarotoxin) groups were treated as previously described. (■) total; (□) small (< 0.5 μm) unmyelinated; (▨) large (> 0.5 μm) unmyelinated; (▧) myelinated. Note the marked increase in myelinated axons in the treated ventral roots.

sensory or sympathetic ganglia (Oppenheim, 1982). Axon counts in the dorsal root of brachial segment 14 (Fig. 31) show that the total number of axons as well as the proportion of the various classes of axons are also unchanged following such treatment.

In previous studies, we have focused primarily on the response of the *lumbar* motoneurons and their targets to chronic treatment with neuromuscular blocking agents. As discussed in Section 4, however, we have recently been studying the brachial motoneurons in segment 14 as well as two of the wing muscles innervated by these neurons, the slow tonic ALD and the fast twitch PLD. In the sections that follow we wish to describe in some detail the response of this system to chronic neuromuscular blockade.

5.3. Development of ALD and PLD muscles and their innervation following neuromuscular blockade

5.3.1. Motoneurons

Consistent with previous studies from this laboratory (see above), chronic treatment of chick embryos with neuromuscular blocking agents prevents a substantial amount of the naturally-occurring cell death in brachial segment 14 (Fig. 10). On day 8 cell death has been entirely prevented in this population and even as late as day 20, the treated animals have about 43% more motoneurons in segment 14 than controls. The increase in the number of motoneurons in segment 14 produced by curare treatment is the result of decreased cell death; on day 8 (stage 34) there are 70 to 80% *fewer* degenerating motoneurons in the embryos previously treated with curare.

5.3.2. Peripheral nerve

As summarized in Fig. 12, by day 10, the ALD nerve of treated animals has approximately 50% more axons than controls. All of the axons at this stage are small and unmyelinated. At day 13, there are slightly fewer axons in the treated ALD nerve and from day 13 on, there is a progressively greater decrease in axon numbers in the treated cases. For instance, on day 16, there are only about 800 axons in the treated nerve compared to about 1,500 in controls, a loss of about 45%. Proportionately, most of this loss involves the small, unmyelinated class of axons.

In contrast to the ALD, axon numbers in the treated PLD nerve appear to be greater, relative to the control, at all stages (Fig. 12). The largest increase in the number of PLD axons occurs between day 10 and day 13. By day 13 the treated PLD nerve has about twice as many axons as control animals. The majority of this increase involves the small, unmyelinated axons. Axonal differentiation also appears accelerated in the treated 13 day PLD nerve; whereas control animals have no myelinated axons in the peripheral nerve at this stage, the treated nerves always contain a few myelinated axons and the thickness of the myelin sheath resembles that of a normal 15 day embryo (Figs. 33 and 34). There are also approximately 70% more myelinated axons in the *ventral root* of 13-day curare or α-BTX treated embryos (Fig. 31A). Despite the fact that there is a loss of axons after day 13 in the treated PLD, even as late as day 20, just prior to hatching, there are still significantly more axons than in controls. The greatest loss of PLD axons in both treated and control animals involves the small, unmyelinated axons.

5.3.3. Muscle development
Embryos chronically immobilized with either α-BTX or curare show a marked decrease in the number of muscle cells in both ALD and PLD (Fig. 24). This reduction is already quite evident by day 8. From day 13 on, the ALD muscle has less than 30% of the normal number of muscle fibers and in some rare cases only a few muscle cells could be detected. The normal muscle mass is replaced by connective and fatty tissue (and the remaining ALD nerve). The PLD muscle responds less dramatically to the drug treatment in that approximately 50% of the normal (control) number of muscle fibers are still present after day 13. Although we have not made a detailed study of the cytological development of muscle cells following neuromuscular blockade, the surviving muscle fibers appear to have differentiated relatively normally. Degenerating muscle cells in various phases of break-down are frequently observed in the treated ALD and PLD (Fig. 20). Degenerating muscle cells are also occasionally seen in control animals, especially at 8 and 10 days of incubation.

The fact that the embryos in which cell death is prevented by neuromuscular blockade have considerably fewer muscle fibers (but *more* motoneurons), appears, on the face of it, to argue against any simple explanation of cell death in which target size, per se, is the limiting factor. Yet, if one assumes that embryonic muscle receives a full complement of innervation prior to, or concomitant with, the onset of massive cell death, then the fact that the control ALD and PLD muscles have only a small percentage of the total number of muscle cells at this stage may indicate that, in the *normal* situation, cell death is, in fact, related to the number of available muscle cells (McLennan, 1981). In the treated cases it is conceivable that neuromuscular blockade increases the amount or availability of whatever entity is required for cell survival despite a reduced muscle mass. By forcing early developing motoneurons to innervate developmentally older muscles, which have a full complement of muscle fibers, one may be able to experimentally test this notion.

5.3.4. Innervation and the neuromuscular junction
It has been previously noted by us as well as by other investigators (Gordon et al., 1974, 1975) that chronic paralysis results in a dramatic alteration of the innervation pattern of the ALD and PLD muscles. Both muscles become hyper-

innervated as indicated by the presence of an increased number of NMJs and an increased number of axonal endings at each NMJ. Both of these effects can be detected as early as 9 to 10 days of incubation in the ALD and PLD (Chu-Wang and Oppenheim, in preparation).

As described in Section 4, the *control* ALD and PLD muscles have only a single axon profile per NMJ at 8 and 10 days. In contrast, the treated muscles have 3 to 8 axon profiles per NMJ at this time and these values continue to increase after 10 days (Figs. 15, 16, 18–20, 22, 23). The increase is especially striking in the treated PLD, in which 7 to 14 axon profiles are found at each NMJ (vs. 2 to 3 in the control) between day 12 and day 17. By contrast, the treated ALD usually has only 2 to 3 more axon profiles per NMJ than the controls during these stages. Thus, although there are significant changes in the innervation of both muscles as a result of neuromuscular blockade, the effects are most striking in the PLD.

The treated ALD and PLD muscles show significant increases in both the number of NMJs and in polyneuronal innervation as early as day 8 to 9 whereas the greatest increase in axon number in the treated peripheral nerves (Fig.12) occurs somewhat later. Thus in addition to preventing cell death, the neuromuscular blockade may also induce branching or sprouting *within* the muscle. In order to assess this possibility, we have counted all of the axons in the mid-belly region (C, Fig. 11) of the control and treated ALD and PLD muscles at 8, 10 and

TABLE 1

The number of axons within the ALD and PLD muscle of control and treated[a] embryos

Age	ALD		PLD	
	Control	Treated	Control	Treated
8 d	1902 (2)	3145 (2)	592 (1)	1708 (1)
10 d	1023 (1)	1768 (1)	533 (2)	1392 (2)
13 d	551 (1)	1867 (1)	539 (1)	1025 (1)

[a] Embryos were treated with curare or α-BTX beginning on day 5 as previously described. The counts were made from electron micrographs of cross-sections through the region marked C in Fig. 11. Number of embryos indicated in parentheses.

Fig. 33. Cross-section of the PLD nerve from a 13-day control embryo showing different sized axons but no myelination. × 11,300.

Fig. 34. Cross-selection of the PLD nerve from a curare-treated 13-day embryo showing that the axons have apparently differentiated faster than the control as indicated by the precocious appearance of myelin sheaths. Note the presence of many small unmyelinated axons, which are usually in clusters. Arrows indicate the supernumerary myelin sheaths. × 11,300.

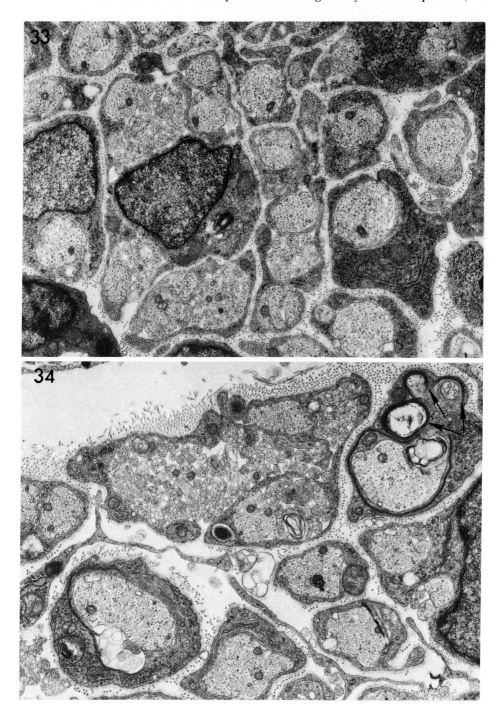

13 days of incubation. As shown in Table 1, there is a marked increase in the number of axons *within* the muscles of the treated animals. This difference is present as early as day 8, which is before the peripheral nerves *outside* of the muscle show a significant response to the treatment. Although there is currently some dispute over whether receptor blockade inhibits or induces sprouting within some adult mammalian muscles (Brown et al., 1981), our results clearly show that in the embryo such treatment can induce axonal sprouting and/or branching within the muscle and in the peripheral nerve proximal to the muscle.

The observation that axon number in the peripheral nerve innervating the treated ALD muscle falls markedly below control values by 16 days (and probably as early as 14 days) (Fig. 12), despite the presence of a greater number of motoneurons in segment 14, appears contradictory. As long as the motoneurons survive, one would have expected the treated animals to have increased numbers of peripheral axons at all stages. Although it is conceivable that some of the excess motoneurons can be maintained even in the absence of a contact with their normal target, this seems rather unlikely. Until it is possible to directly evaluate the number of cell bodies in the specific motor pool innervating the ALD muscle, however, one cannot rule out the alternative possibility that many of the excess motoneurons innervating the ALD muscle eventually die despite the treatment with neuromuscular blocking agents; the much greater loss of muscle fibers in the ALD versus the PLD muscle following such treatment may induce a secondary loss of motoneurons. It is well known that these two muscles respond differently in a number of other respects to the chronic blockade of neuromuscular activity (Vrbová, et al., 1978). Although it isn't clear what, if any, significance these differences may have for understanding the regulation of neuronal cell death, it does suggest that in the future more attention should be focused on a comparison of cell death in several specific motor pools serving muscles which differ in their developmental, physiological, biochemical and histological properties.

6. *How does chronic neuromuscular blockade prevent the death of spinal motoneurons? A Model*

We assume that the first step in the prevention of motoneuron death by treatment with either pre- or postsynaptic NMJ blocking agents involves the prevention of, or a reduction in, interactions between presynaptically released ACh and the ACh–R iontophore complex in developing muscle. Although it is conceivable that the variety of effective neuromuscular agents we have used may interfere with the release and/or the binding of a non-ACh anterograde trophic agent, this seems unlikely.

There is an increasing body of literature, which has been extensively reviewed recently (Edwards, 1978; Fambrough, 1978; Jansen et al., 1978), indicating that both the level and distribution of ACh–R in developing as well as adult muscle is controlled almost entirely by innervation and synaptic and/or muscle activity. Moreover, denervated or functionally inactivated muscle which contains high levels of extrajunctional ACh–R is also susceptible to hyperinnervation. Direct stimulation of denervated muscle can prevent or reverse the development of high levels of extrajunctional ACh–R and can inhibit the ability of denervated muscle to become hyperinnervated. Similarly, adult muscles exposed to neuromuscular blocking agents are capable of accepting additional innervation. Thus, the presence of high levels of extrajunctional ACh–R may allow hyperinnervation or may reflect a particular state of the muscle which allows multiple innervation. The release of ACh and its interaction with ACh–R in the muscle have been strongly implicated as playing a role in all of these events (Lomo, 1980); neurotrophic regulation of some of these phenomena have also been reported (e.g., Markelonis et al., 1982).

In an earlier publication we summarized our thoughts concerning a potential mechanism controlling the survival of embryonic motoneurons in the following way: 'A possible model for explaining both the permanent and transient hypothanasia (i.e., the prevention of cell death) seen following neuromuscular blockade would be that during early synapse formation an axon contacts a myotube and begins releasing small amounts of ACh which leads to a partial depolarization of the muscle membrane. Prior to and during this critical period of early membrane depolarization, more than one synapse could be maintained by each myotube. However, once the *frequency* and *extent* of membrane depolarization reached a certain level, the membrane properties would be altered such that only one synapse could be supported, and additional contacts would be lost. Following neuromuscular blockade, the amount of ACh interacting with receptors would decrease considerably, thereby decreasing membrane depolarization. This would result in the critical period being maintained and an increase in motoneuron survival. For muscle fibers which are normally multiply-innervated, similar events could also occur' (Pittman and Oppenheim, 1979, p. 444).

Although at present there is no direct evidence that the prevention of cell death in spinal motoneurons involves an increase in the number of ACh–R in limb muscle, pharmacological treatments similar to those used to prevent cell death result in increased extrajunctional ACh–R (Betz et al., 1980; Burden, 1977) and in an increased sensitivity of the muscle to ACh in the chick embryo (Gordon et al., 1975). Following the pharmacological prevention of cell death, at least some limb muscles become hyperinnervated from a greater than normal number of motoneurons. At present, however, it isn't clear to what extent the hyperinnervation may also reflect axonal branching or conversely to what extent the increased branching may contribute to the prevention of cell death. Independent evidence that many of the excess motoneurons resulting from the prevention of cell death do, in fact, innervate their target comes from the observation that in these preparations a greater than normal number of cells are labelled following HRP injection into specific limb muscles (Oppenheim, 1981b).

As implied in the above quote, in our original model we tacitly assumed that a significant reduction in muscle activity during the normal cell death period would be sufficient to prevent cell death. This now appears to be incorrect. Muscle activity (or at least a behavioral paralysis) induced by certain pharmacological agents (Figs. 26 to 29) does not prevent cell death. Although at present we have no explanation for these results, they should not necessarily be interpreted as inconsistent with our activity-related model of cell death. Since we do not yet know the mechanisms of action of these drugs in the embryo it is conceivable that they are acting to mimic the cellular or molecular effects of normal neuromuscular activity. A more valid and direct test of our hypothesis would involve the electrical stimulation of limb muscle and/or peripheral nerve during the course of naturally-occurring cell death. If the model is correct, we would predict that by using the appropriate parameters of electrical stimulation, cell death in normal embryos would be enhanced and the prevention of cell death in curare-treated embryos should be reduced.

We have been carrying out such electrical stimulation experiments and although they are technically rather difficult in that only about 10% of the embryos survive the experimental treatment we have been successful in obtaining data from several embryos in which one leg-bud (the right side) was chronically stimulated with either fine wire or suction electrodes for 5 to 23 h ($M = 12.3$ h) at various times between 6 and 8.5 days of incubation (Oppenheim and Núñez, 1982). Control embryos consisted of animals in which the electrodes were applied to the right leg but in which no current was passed. For some embryos curare or α-BTX was administered daily beginning on days 5 or 6 in order to prevent naturally-occurring cell death. (Controls received a similar regimen of saline treatment.) The numbers of both healthy and pyknotic cells in the LMC of both right and left sides of the spinal cord were counted in every 10th section through the entire lumbar region. The results are summarized in Fig. 32.

Because control embryos, in which electrodes were attached to the right leg, but in which no current was passed, were found to have similar numbers of both healthy and pyknotic LMC cells on the right and left sides of the spinal cord, we conclude that the procedures themselves have no apparent effect on cell death. Consequently, for the control data the right and left sides were combined for all further comparisons with the experimental or stimulated embryos. Irrespective of whether stimulation was begun at stages 31 to 33 or stages 34 to 35, the experimental embryos, as predicted, always had *fewer* healthy neurons and *more* pyknotic neurons in the LMC on the stimulated or experimental side of the spinal cord. This was true for both the saline and the curare/α-BTX treated embryos. The difference was especially clear in the case of pyknotic neurons, perhaps

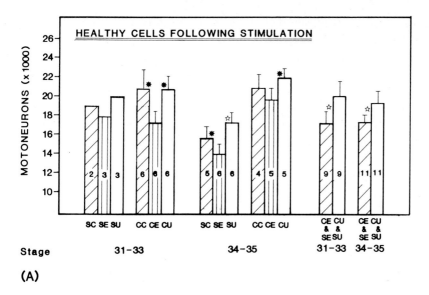

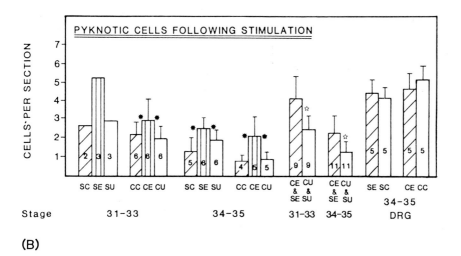

Fig. 32. The average number of healthy (A) and dying (pyknotic) (B) motor and sensory neurons (± S.D.) in the lumbar region following electrical stimulation of the right hind-limb at stages 31 to 33 or at stages 34 to 35. *SC*, saline control; *SE*, saline experimental (stimulated or right) LMC or DRG; *SU*, saline unstimulated (left) LMC or DRG; *CC*, curare/α-BTX control; *CE*, curare experimental (stimulated or right) LMC or DRG; *CU*, curare/α-BTX unstimulated (left) LMC or DRG. *CE & SE* and *SU & CU* represents pooled data from the saline plus curare/α-BTX embryos. Pyknotic sensory and motor neurons were counted in every 5th section of lumbar segments 23 to 30: (☆) $P < 0.01$; (★) $P < 0.05$ Mann–Whitney tests, one tailed. All statistical comparisons are relative to the experimental (stimulated) groups, SE or CE. The sample size is indicated in the bars.

because the relatively short duration of stimulation was not sufficient to allow the removal of large numbers of dying cells.*

Despite the fact that the mere placement of electrodes in the right leg of control embryos did not induce any changes in the ipsilateral LMC, it is still conceivable that the effects seen in the experimentally stimulated cases result from some non-specific pathological effect of the electrical current on immature neurons. This seems unlikely, however, since the number of degenerating *sensory* neurons in the lumbar dorsal root ganglia (DRG) of experimental embryos does not differ on the right and left sides (Fig. 32); similar to the motoneurons, sensory neurons in the lumbar DRG are still differentiating between stage 31 and stage 35 (although they already have axons in the limb) and they also exhibit natural cell death during this period. Thus it seems most likely that the electrical stimulation is specifically influencing mechanisms involved in the normal regulation of *motoneuron* survival. It is still conceivable, of course, that the spinal motoneurons are somehow especially susceptible to a pathological artifact associated with electrical stimulation. Against this, however, is the report that chronic electrical stimulation of the brachial spinal cord of the chick embryo for several days, at a time when the major period of cell death in the brachial region is over, does not reduce the number of axons in peripheral nerves (Toutant and LeDouarin, 1980). Thus, electrical stimulation, per se, does not appear to be pathological to embryonic motoneurons.

When considered together with previous evidence involving the use of pre- and postsynaptic pharmacological blockers in this system, these data support our working hypothesis that synaptic activity and/or muscle activity is directly involved in the regulation of motoneuron survival in the chick spinal cord. In view of the increasing possibility that target-derived trophic factors may regulate the survival of motoneurons (Bennett et al., 1980; Pollack, 1980), it seems plausible that synaptic transmission acts to regulate the availability (e.g., the synthesis, release, degradation, etc.) of this putative factor.

7. Cell death in a population of spinal visceral efferent neurons: the sympathetic preganglionic column of Terni

As noted in the introduction, naturally-occurring cell death is a common feature of the development of a wide variety of neurons in the central and peripheral nervous system of vertebrates. In the spinal cord of the chick embryo massive cell

---

*There is considerably more room for error in counting healthy neurons in that by using the presence or absence of the nucleolus as the major criterion for counting a cell, one may inadvertently include dying cells that are in the early stages of degeneration. In contrast, when focusing only on counts of frankly pyknotic cells it is much more unlikely that healthy cells will be included.

death has been reported in somatic motoneurons and in sensory neurons in the DRG. Among the many other types of spinal neurons present in the chick, two populations which have not previously been examined for the presence of cell death, but which are especially amenable to such an analysis, are the preganglionic sympathetic visceromotor neurons (column of Terni, CT) in the thoracic and sacral region and the sympathetic neurons in the paravertebral ganglia. Although Levi-Montalcini (1950) described the total disappearance of an 'abortive' or transient preganglionic visceromotor column in the cervical spinal cord of the chick, no information exists on cell death in either the permanent thoracic (sympathetic) or sacral (parasympathetic) CT or in the spinal sympathetic ganglia. Naturally-occurring cell death has, however, been reported in the *cranial* visceromotor nuclei and ganglia of the chick (Landmesser and Pilar, 1978; Wright, 1981). The thoracic CT cells offer an additional advantage, in that unlike the spinal sensory and motoneurons, these cells innervate other neurons (the sympathetic cells) whose size and number can be independently altered by the neurotrophic agent, Nerve Growth Factor. Therefore, it should be possible to produce graded increases in the size of the sympathetic ganglion targets and determine whether the neurons in the thoracic CT respond appropriately. Finally, since the CT neurons are cholinergic, they also offer the opportunity to explore whether pharmacological agents which block synaptic transmission at the cholinergic synapse in the sympathetic ganglion, can modify cell survival in either pre- or post-ganglionic cells.

The preganglionic CT in the chick embryo constitutes a discrete nucleus or column of cholinergic cells, which are distributed throughout the six thoracic and first two lumbar segments (thoracolumbar or sympathetic CT) and in the first 3 to 5 sacral segments (parasympathetic or sacral CT) (Fig. 35). The cells in the sympathetic CT are found next to the central canal in a dorsal position and can be readily backfilled with HRP following injections into the sympathetic paravertebral ganglia (Fig. 35).

The neurons comprising the thoracic and sacral CT begin their differentiation in the common motor column in the ventro-lateral region on day 4.5. Between day 4 and day 8 they migrate to their normal dorso-medial position adjacent to the central canal. Although individual cells can be recognized during their migration, the discrete preganglionic column or nucleus (CT) first becomes clearly discernible on day 8. Thus, it is only from day 8 on that reliable cell counts can be made in the CT.* Cell counts of both healthy and pyknotic neurons in the sympathetic CT were made from thionin-stained sections according to previously described criteria (see Oppenheim et al., 1982). In order to reduce the labor involved in

---

*We have never observed degenerating thoracic or sacral CT cells during their migration between 4.5 and 8 days of incubation.

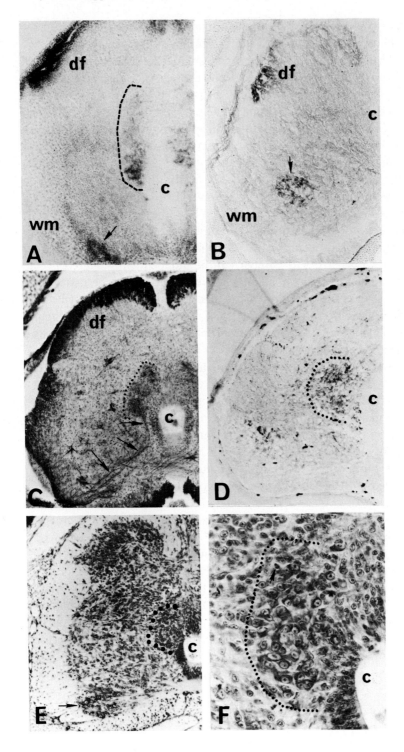

counting cells, we have restricted our analysis to the last two thoracic segments (T 5 to 6) and the first lumbar segment (L1); cells in the parasympathetic CT were counted in the first two sacral segments (S1–2).

Between day 8 and day 10, there is an approximate 25 to 30% reduction in cell number in the thoracic and lumbar CT (Table 2). It appears as if no significant loss occurs after day 10. Between day 8 and 15, there is about a 20% decline in cell number in the sacral CT (data not shown). Daily treatment with NGF from day 3 to day 9 reduced cell death in the thoracic and lumbar CT on day 10 in a dose dependent fashion but had no apparent effect on the cell loss in the sacral CT (Table 2 and Fig. 35). Daily treatment with NGF between day 10 and day 14 had no effect on cell number in the thoracic, lumbar or sacral CT on day 15. Daily treatment with NGF (20 μg) between day 3 and day 9 significantly reduced the number of pyknotic neurons in the thoracic CT on days 8 and 10 (Table 2). Similar treatment had no apparent effect on the number of pyknotic cells in the sacral CT.

The fact that NGF treatment failed to have an effect on cell number in the *sacral* (parasympathetic) CT but did reduce cell death in the *thoracolumbar* (sympathetic) CT argues against a direct or primary effect of NGF on preganglionic cells. Rather, in accord with the well-known specific effects of NGF on peripheral adrenergic neurons in the chick, these results imply that the effects of NGF on the thoracolumbar CT occur secondarily as the result of a direct influence on the sympathetic ganglia. Indeed, we have found that NGF treatment not only increased the size of individual sympathetic neurons but also increased the number of neurons in the ganglia by reducing cell death (Table 3). There is a dose-dependent relationship between sympathetic ganglion volume and the prevention of cell death in the CT. Since the validity of the competition hypothesis of neuronal death ultimately rests on the results of peripheral or target enlargement the present results provide striking confirmation of this notion in that an apparent graded enlargement of the *thoracolumbar* sympathetic ganglia produced a related increase in the survival of the preganglionic neurons.

Chronic treatment of chick embryos from day 5 or 6 with curare or α-BTX, both of which are nicotinic receptor antagonists that are much more effective in

---

Fig. 35. Ten-day control chick embryo spinal cord. (a) Cross-section of thoracic spinal cord (T5) stained for AChE. Broken line encloses the CT. Arrow indicates somatic motoneurons ($\times 265$). (b) Cross-section of lumbar (L4) spinal cord stained for AChE. Arrow indicates the somatic motoneurons. Note absence of the CT at this level ($\times 265$). (c) Cross-section of T5, reduced silver stain. Broken lines encloses the CT. Small arrows indicate the efferent fibers from the CT ($\times 265$). (d) Cross-section of T6 following HRP injection into sympathetic ganglion. Broken line encloses the CT containing HRP reaction product ($\times 265$). (e) Cross-section of T5 stained with thionin. Broken line encloses the CT. Arrow indicates the somatic motoneurons ($\times 265$). (f) Enlargement of CT from (e) ($\times 516$). Arrow indicates a pyknotic cell. Abbreviations: c, central canal; df, dorsal funiculus; wm, prospective white matter.

TABLE 2
Effect of nerve growth factor on the number of healthy and pyknotic neurons in the column of Terni

Stage (H/H)	Age (Days)	Control	Cytochrome c (20 µg)	NGF (20 µg)	NGF (10 µg)
Healthy neurons (T5–6)					
34	8	7,310 ± 761 n = 5	7,643 ± 594 n = 4	7,680 ± 856 n = 8	7,575 ± 831 n = 5
36	10	5,442 ± 497[c] n = 12	5,714 ± 804[c] n = 4	6,790 ± 727[b] n = 9	6,395 ± 645[b] n = 5
41	15	5,730 ± 833[c] n = 4		5,615 ± 563 n = 4	5,827 ± 711 n = 4
Healthy neurons (L1)					
34	8	2,010 ± 376 n = 12	1,979 ± 233 n = 4	1,865 ± 315 n = 4	
36	10	1,464 ± 298 n = 11	1,257 n = 3	2,175 ± 306[b] n = 7	1,742 ± 285[b] n = 5
Pyknotic neurons (T5–6)					
34	8	0.57 ± 0.21 n = 5		0.45 ± 0.18[a] n = 5	
36	10	1.66 ± 0.33[c] n = 9		1.03 ± 0.47[b] n = 11	
Pyknotic neurons (L1)					
34	8	0,71 ± 0.11 n = 4		0.33 ± 0.26[b] n = 4	
36	10	1.63 ± 0.50[c] n = 4		0.87 ± 0.32[c] n = 4	

Values are the mean ± S.D.
[a]Significantly different from the control group at $P < 0.05$.
[b]Significantly different from the control group at $P < 0.02$.
[c]Significantly different from the 8-day group at $P < 0.02$. The 10-day, 10-µg vs. 20-µg, groups are significantly different at $P < 0.10$ (T5–6) and $P < 0.05$ (L1). All P-values in Tables 1 to 3 were calculated using the Mann–Whitney U-test, two-tailed. As a labor saving device neuron counts were restricted to T5–6 and L1. Healthy neurons were counted in every 3rd section through these segments at 400×, 250×, or 160× depending on the stage. Degenerating neurons were also counted in every 3rd section under oil immersion (1,250×). (See Oppenheim et al., 1982 for details.)

blocking neuromuscular than ganglionic transmission, had no apparent effect on naturally-occurring cell death in the thoracolumbar CT at doses that significantly reduced cell death in the somatic spinal motoneurons in the LMC (Table 4). In contrast, hemicholinium-3, a potent inhibitor of high-affinity choline uptake that is equally efficient in blocking neuromuscular *and* ganglionic transmission, substantially reduced cell death in the thoracolumbar CT, the lumbar LMC and the sacral CT. Although more extensive studies, using additional pharmacological

TABLE 3
Effect of nerve growth factor on the number of healthy neurons in the sympathetic ganglia

Stage (H/H)	Age (Days)	Control	NGF (20 μg)	NGF (10 μg)
T5–6				
36	10	5,296 ± 698 n = 4	9,800 ± 1,073[c] n = 4	8,376 ± 801[c] n = 4
41	15	6,285 ± 421 n = 4	7,550 ± 544[a] n = 4	
L1				
36	10	11,312 ± 735 n = 4	15,688 ± 907[c] n = 4	

Values are the mean ± S.D.
[a]Significantly different from the control group at $P < 0.05$.
[b]Significantly different from the control group at $P < 0.02$.
[c]Significantly different from the control group at $P < 0.002$. The 10 μg group (T5–6) at 10 days is also significantly different from the 20 μg group at $P < 0.05$. Ganglion neurons were counted in every 5th section at 250×.

TABLE 4
Effect of hemicholinium-3 on the number of healthy and pyknotic neurons in the L1 column of Terni

Stage (H/H)	Age (Days)	Control	Curare- α-bungarotoxin	Hemicholinium-3
Healthy neurons				
34	8	2,010 ± 376 n = 12	1,973 ± 252 n = 4	
36	10	1,464 ± 298[a] n = 11	1,498 ± 208 n = 4	2,006 ± 327[c] n = 4
Pyknotic neurons				
34	8	0.71 ± 0.11 n = 4		
36	10	1.63 ± 0.50[a] n = 4		0.95 ± 0.57[b] n = 4

HC-3 was given twice a day using the following schedule: day 6, 1 mg; days 7 and 8, 2 mg; day 9, 3 mg. α-Bungarotoxin was given daily using the following schedule: day 6, 100 μg, days 7 and 8, 80 μg; day 9, 75 μg. Curare (2 to 2.5 mg) was administered daily on days 5 through 9. Values are the mean ± S.D.
[a]Significantly different from the 8-day control at $P < 0.02$.
[b]Significantly different from the 10-day control at $P < 0.05$.
[c]Significantly different from the 10-day control at $P < 0.02$.

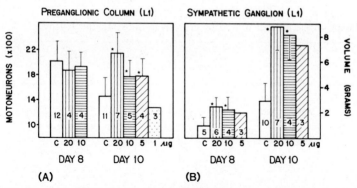

Fig. 36. Cell number in the preganglionic visceromotor column (A) and sympathetic ganglion volume (B) in lumbar segment 23 (L1) following daily treatment with nerve growth factor (1, 5, 10, 20 μg) from day 3 to day 7 or day 9. In (A) the 10-day, 20, 10 and 5 μg groups are significantly different from control at $P < 0.02$. In (B) the 8 and 10 day, 20 and 10 μg groups are significantly different from control at $P < 0.02$) (★) or $P < 0.002$ (✱). The sample size is indicated in the bars.

agents, are presently under way (Maderdrut and Oppenheim, 1982) the present preliminary data imply that naturally-occurring cell death of preganglionic neurons is also regulated by the level of synaptic activity at the ganglionic junction. Thus, both the size of their innervation field and the level of synaptic activity at the ganglionic junction may regulate the survival of sympathetic preganglionic neurons.

8. Function of naturally-occurring neuronal death

In accord with traditional neo-Darwinian theory, which holds that virtually all the traits and characters of an organism reflect adaptations that evolved by natural selection, it has been tacitly assumed that naturally-occurring neuronal death represents an adaptive feature of neurogenesis, in that an initial overproduction and subsequent reduction in neuronal number is believed to confer a selective advantage to those animals which exhibit this phenomenon. Although this would appear to be a plausible argument, it nonetheless needs to be emphasized that in the absence of direct proof it remains a hypothesis to be tested experimentally; it should not be forgotten that there are alternative evolutionary forces, besides those involving direct heritable adaptations for the acquisition of both developmental and adult traits (Gould, 1977; Lewontin, 1979).*

---

*By stating this, we do not wish to give the impression that we believe that natural selection is of minor significance in neurogenesis. On the contrary, there is every reason to believe that natural selection has been as strong a force in the attainment of each step in neurogenesis as in the final

If embryonic cell death does contribute to the normal development of an animal, then one would expect that the prevention of cell death would lead to abnormal development and reduced chances of survival. An instructive example of such a manipulation involves mesenchymal cell death in the chick limb-bud. The posterior necrotic zone (PNZ) of the chick wing, as discussed earlier, represents a region of massive mesenchymal cell death which is thought to be involved in the developing morphology of the upper forelimb. Surprisingly, however, Saunders (1966) has shown that the prevention of cell death in the PNZ doesn't seem to impair normal development of the wing (but see Hinchcliffe and Johnson, 1980). In the light of such negative data, one should perhaps exercise some caution in assuming a priori that neuronal death is adaptive. Until it can be shown that the prevention of neuronal death results in abnormal physiological or behavioral function, proposals concerning the specific biological role of this phenomenon, however plausible, will remain inferential.

Despite the absence of direct proof for the biological adaptiveness of naturally-occurring neuron death, several specific proposals have been suggested for explaining the adaptive role of cell death in neurogenesis. Included among these are: (a) that more neurons are transiently needed during neurogenesis in order to ensure the normal survival and development of a certain sub-population of cells in either the pre- or postsynaptic components of an interacting system. For instance, it is possible that normal muscle differentiation in the embryo requires the presence of considerably more motoneurons than are present in the adult. This we will call the *transient* or *provisional hypothesis* of cell death; (b) owing to either heritably or congenitally derived variations in the number of available postsynaptic targets, the nervous system may have evolved a mechanism of cellular redundancy or overproduction in order to ensure that all available targets are adequately innervated. This proposal, the *redundancy hypothesis*, assumes that a principal tenet of neuronal organization involves some sort of precise matching between two populations of cells (i.e., between neurons and neurons, neurons and muscle, or neurons and sensory receptors) (Katz and Lasek, 1978). In the case of the vertebrate limb, it is known from the study of genetic mutants that there is considerable genetic variation for limb size and structure. An initial overproduction of motoneurons followed by the regression of those which fail to establish synaptic connections (or fail to obtain sufficient trophic substance) could be envisioned as a mechanism for dealing with this variation. Available evidence supports this notion in that animals with genetic or congenital defects, in which parts of, or even the entire, limb may be absent, duplicated or otherwise

---

product, the mature nervous system. We only wish to draw attention to the fact that some developmental and adult traits may have arisen as a result of genetic drift, pleiotropic gene action, allometry or developmental noise, as well as by the more common direct selection of specific genes.

modified in size, have been shown to exhibit concomitant changes in spinal neurons and in limb innervation. Similarly, *experimentally* produced alterations in limb size also results in quantitative changes in the number of spinal motoneurons (Hamburger, 1934)*; and (c) an off-shoot of the redundancy hypothesis, and one which has practically acquired independent status as an explanation of cell death, is the proposal that the ability of neurons to survive depends on their innervating the *correct* postsynaptic target. According to this *error-correction hypothesis*, motoneurons within a specific spinal motor pool (e.g., the gastrocnemius) can only survive, or at least they have a greater probability of surviving, if they innervate the gastrocnemius muscle vs. another limb muscle.

In its most extreme form, the error correction hypothesis assumes that all of the cells that die in a population do so owing to the innervation of grossly incorrect targets. In other words, in the case of the chick hindlimb, 40 to 50% of the motoneurons – the extent of natural cell death – would be expected to innervate the wrong muscles. Although a small proportion ($< 10\%$) of motoneurons may die owing to the formation of such inappropriate neuromuscular connections (Oppenheim, 1981a), the lions share appear to die despite having innervated their grossly correct targets from the very earliest stages of limb innervation. Moreover the prevention of cell death by neuromuscular blockade in the chick does not alter the location of motor pools in either the lumbar or the brachial spinal cord on embryonic day 10 (stage 36) (Oppenheim, 1981b). Since the drug treatment in these experiments was begun at the earliest stages of limb innervation, the results strongly imply that virtually all motoneurons, even those that would normally have died in the absence of the blockade, had innervated their correct muscles. Finally, forcing motoneurons to innervate the wrong muscle during development also does not lead to enhanced cell death (Bennett et al., 1979; Hollyday, 1980; Lance-Jones and Landmesser, 1980; Summerbell and Stirling, 1981).

Despite the apparent strength of evidence against the error correction hypothesis of cell death, there remain some contradictory results (Lamb, 1977, 1979). Nonetheless, although the issue is by no means entirely settled, at present the

---

*Since the deserved demise of the biogenetic law of recapitulation, there have been few attempts to study or understand the relationship between ontogeny and phylogeny. This appears to be changing. It is now widely recognized that ontogeny is the cause of phylogeny (and not vice versa), in that it is heritable modifications in individual ontogenies which give rise to a particular phylogenetic series. An individual's present ontogeny poses severe constraints on whether, as well as how, mutations are successfully incorporated into the genome and expressed in the phenotype. Attempts to understand this relationship with regard to the vertebrate nervous system have been discussed in several recent papers (Bonner, 1982; Ebbesson, 1980; Gould, 1977; Katz and Lasek, 1978; Katz et al., 1981; Oppenheim, 1981c) and thus it is a subject that promises to be of considerable interest and significance to the field of neuroembryology.

consensus appears to be that the attainment of a specific neuronal connectivity has been of only minor significance in the evolution of naturally-occurring death of motoneurons. We hasten to add, however, that the situation may be entirely different for other kinds of neurons.

Regrettably, there has been little effort devoted to experimentally testing the *transient hypothesis* of cell death. Yet, there is increasing interest in the notion that a variety of different neuron types may exert transient trophic or inductive-like influences on the early differentiation of other neurons or targets (Lauder et al., 1981; Oppenheim, 1981c). It is conceivable, though entirely unsupported at present, that certain transient inductive-like influences may require an excess number of 'presynaptic' neurons some of which are then lost following their provisional role as morphogenetic 'inductive' or trophic agents. It is known, for instance, that embryonic motoneurons regulate the differentiation of various properties of their target muscles (see above). Thus, it doesn't seem unreasonable that the production by the muscle of a trophic agent necessary for motoneuron survival is also regulated by the motoneurons themselves. Since in most cases, approximately one-half of the neurons in a population undergo cell death, one only needs to make the assumption that it takes two embryonic motoneurons to 'induce' enough trophic substance to maintain one of them. The recent experiments of Lamb (1980), in fact, provide support for this idea.

Finally, in light of our observation that there appears to be a temporal mismatch between motoneurons and muscle development (see above), the possibility cannot be excluded that the death of motoneurons is an inevitable, non-adaptive result of this mismatch. That is, the temporal differences in the development of motoneurons and muscles may be constrained for reasons that are not yet clear and the resulting death of motoneurons may simply be an inevitable pleiotropic effect of this constraint.

In summary, the adaptive significance of naturally-occurring neuronal death remains largely a mystery. Although a number of interesting and plausible hypotheses are presently being considered, it remains to be seen which, if any, of these are most relevant for the naturally-occurring loss of spinal motoneurons.

Acknowledgements

The research of our own described here was supported by National Science Foundation grant, number 81-40397, by a grant from the Amyotrophic Lateral Sclerosis Society and by funds provided by the North Carolina Division of Mental Health. We wish to express our gratitude to R. Jones, R. Worsham and C. Suggs for technical assistance and to R. Daniels for typing, editing and remaining cheerful.

References

Atsumi, S. (1977) J. Neurocytol 6, 691–709.
Bennett, M.R., Lindeman, R. and Pettigrew, A.G. (1979) J. Embryol. Exp. Morphol. 54, 141–154.
Bennett, M.R. and Pettigrew, A.G. (1974) J. Physiol. (London) 241, 515–545.
Bennett, M.R., Lai, K. and Nurcombe, V. (1980) Brain Res. 190, 537–542.
Betz, H., Bourgeois, J.-P. and Changeux, J.-P. (1980) J. Physiol. (London) 302, 197–218.
Bonner, J.T. (Ed.) (1982) in Evolution and Development, Springer–Verlag, Berlin.
Bourgeois, J.-P. and Toutant, M. (1982) J. Comp. Neurol. 208, 1–15.
Brown, M.C., Holland, R.L. and Hopkins, W.G. (1981) Ann. Rev. Neurosci. 4, 17–42.
Burden, S. (1977) Dev. Biol. 56, 317–329.
Carr, V.M. and Simpson, S.B. (1978) J. Comp. Neurol. 182, 727–740.
Chu-Wang, I.-W. and Oppenheim, R.W. (1978a) J. Comp. Neurol. 177, 33–58.
Chu-Wang, I.-W. and Oppenheim, R.W. (1978b) J. Comp. Neurol. 177, 59–86.
Cowan, W.M. (1973) in Development and Aging in the Nervous System (Rockstein, M., ed.) pp. 19–41, Academic, New York.
Creazzo, T.L. and Sohal, G.S. (1979) Exp. Neurol. 66, 135–145.
Dennis, M.J. (1981) Ann. Rev. Neurosci. 4, 43–68.
DeSantis, M., Hoekman, T. and Limwongse, V. (1977) Brain Res. 119, 454–458.
Ebbesson, S.O.E. (1980) Cell Tissue Res. 213, 179–212.
Edwards, C. (1979) Neuroscience 4, 565–584.
Fambrough, D.M. (1979) Physiol. Rev. 59, 165–227.
Freeman, S.S., Engel, A.G. and Drachman, D.B. (1976) Ann. N.Y. Acad. Sci. 274, 46–59.
Glücksmann, A. (1951) Biol. Rev. 26, 59–86.
Gordon, T., Perry, R., Tuffery, A.R. and Vrbová, G. (1974) Cell Tissue Res. 155, 13–25.
Gordon, T., Tuffery, A.R. and Vrbová, G. (1975) in Recent Advances in Myology (Bradley, W.G., Gardner-Medwin, D. and Walton, J.N., eds.) pp. 22–26, Elsevier, New York.
Gould, S.J. (1977) Ontogeny and Phylogeny, Cambridge, Mass., Belknap.
Hamburger, V. (1934) J. Exp. Zool. 68, 449–494.
Hamburger, V. (1939) Physiol. Zool. 12, 268–284.
Hamburger, V. (1958) Am. J. Anat. 102, 365–410.
Hamburger, V. (1975) J. Comp. Neurol. 160, 535–546.
Hamburger, V. Brunso-Bechtold, J.K. and Yip, J. (1981) J. Neurosci. 1, 60–71.
Hamburger, V. and Levi-Montalcini, R. (1949) J. Exp. Zool. 111, 457–202.
Harris, W.A. (1981) Ann. Rev. Physiol. 43, 689–710.
Hinchcliffe, J.R. and Johnson, D.R. (1980) The Development of the Vertebrate Limb: An Approach Through Experiment, Genetics and Evolution. Claredon, Oxford.
Hollyday, M. (1980) in Current Topics in Developmental Biology (Hunt, R.K., ed.) Vol. 15, pp. 181–215, Academic Press, New York.
Hollyday, M. and Hamburger, V. (1976) J. Comp. Neurol. 170, 311–320.
Jacobson, M. (1978) Developmental Neurobiology, New York, Plenum.
Jansen, J.K.S., Thompson, W. and Kuffler, D.P. (1978) Prog. Brain Res. 48, 3–18.
Katz, M.J. and Lasek, R.J. (1978) Proc. Natl. Acad. Sci. USA 75, 1349–1352.
Katz, M.J., Lasek, R.J. and Kaiserman-Abramof, I.R. (1981) Proc. Natl. Acad. Sci. USA 78, 397–401.
Kikuchi, T. and Ashmore, C.R. (1976) Cell Tissue Res. 171, 233–251.
Laing, N. and Prestige, M. (1978) J. Physiol. (London) 282, 33–34P.
Lamb, A.H. (1977) Brain Res. 134, 145–150.
Lamb, A.H. (1979) Dev. Biol. 71, 8–21.
Lamb, A.H. (1980) Nature (London) 284, 347–350.

Lance-Jones, C. and Landmesser, L. (1980) J. Physiol. (London) *302*, 559–580.
Landmesser, L. (1978) J. Physiol. (London) *284*, 391–416.
Landmesser, L. and Pilar, G. (1978) Fed. Proc. *37*, 2016–2022.
Lauder, J., Wallace, J. and Krebs, H. (1981) Serotonin: Current Aspects of Neurochemistry and Function (Haber, B., Gabay, S., Issidoridis, M. and Alivisados, S.G., eds.) pp. 477–506, Plenum, New York.
Leonard, J.P. and Salpeter, M.M. (1979) J. Cell Biol. *82*, 811–819.
Levi-Montalcini, R. (1950) J. Morphol. *86*, 811–819.
Lewontin, R.C. (1979) Behav. Sci., *24*, 5–14.
Lømo, T. (1980) Trends Neurosci. *3*, 126–129.
Maderdrut, J.L. and Oppenheim, R.W. (1982) Soc. Neurosci. Abst. *8*, 638.
Markelonis, G.J., Oh, T.H., Elderfrawi, M.E. and Guth, L. (1982) Dev. Biol. *89*, 353–361.
McLennan, I. (1981) Soc. Neurosci. Abst. *7*, 291.
Okado, N. and Oppenheim, R.W. (1981) Soc. Neurosci. Abst. *7*, 291.
Olek, A.J. and Edwards, C. (1977) Soc. Neurosci. Abst. *3*, 115.
Olek, A.J. and Edwards, C. (1978a) Soc. Neurosci. Abst. *4*, 122.
Olek, A.J. and Edwards, C. (1978b) Brain Res. *191*, 483–488.
Ontell, M. (1977) Anat. Rec. *189*, 669–690.
Ontell, M. and Dunn, R.F. (1978) Anat. Rec. *152*, 539–556.
Oppenheim, R.W. (1982) Soc. Neurosci. Abst. *8*, 708.
Oppenheim, R.W. (1981a) in Studies in Developmental Neurobiology: Essays in Honor of Viktor Hamburger (Cowan, W.M., ed.) pp. 74–133, New York, Oxford.
Oppenheim, R.W. (1981b) J. Neurosci. *1*, 141–151.
Oppenheim, R.W. (1981c) in Maturation and Behavior (Connolly, K. and Prechtl, H.F. eds.) pp. 73–109, Philadelphia, Lippincott.
Oppenheim, R.W. and Chu-Wang, I.-W. (1977) Brain Res. *125*, 154–160.
Oppenheim, R.W., Chu-Wang, I.-W. and Maderdrut, J.L. (1978) J. Comp. Neurol. *177*, 87–112.
Oppenheim, R.W. and Maderdrut, J.L. (1981) Soc. Neurosci. Abst. 7, 291.
Oppenheim, R.W., Maderdrut, J.L. and Wells, D. (1982) J. Comp. Neurol. *210*, 174–189, 1982.
Oppenheim, R.W. and Majors-Willard, C. (1978) Brain Res. *154*, 148–152.
Oppenheim, R.W. and Núñez, R. (1982) Nature (London) *295*, 57–59.
Pilar, G., Landmesser, L. and Burstein, L. (1980) J. Neurophysiol. *43*, 233–254.
Pittman, R. and Oppenheim, R.W. (1978) Nature (London) *271*, 364–366.
Pittman, R. and Oppenheim, R.W. (1979) J. Comp. Neurol. *187*, 425–446.
Pollack, E.D. (1980) Neurosci. Lett. *16*, 269–274.
Saunders, J.W. (1966) Science *154*, 604–612.
Srihari, T. and Vrbová, G. (1978) J. Neurocytol. *7*, 529–540.
Srihari, T. and Vrbová, G. (1980) Develop. Growth Diff. *22*, 645–657.
Sullivan, G.E. (1962) Aust. J. Zool. *10*, 458–518.
Summerbell, D. and Stirling, V. (1981) J. Embryol. Exp. Morph. *61*, 233–247.
Toutant, M. and LeDouarin, G. (1980) IRCS Med. Sci. *8*, 408–409.
Vrbová, G., Gordon, T. and Jones, R. (1978) Nerve–Muscle Interaction, Chapman and Hall, London.
Wright, L.L. (1981) J. Comp. Neurol. *199*, 125–132.

CHAPTER 4

Neuromuscular development in tissue culture

L.L. RUBIN and K.F. BARALD

Department of Neurobiology, Rockefeller University, 1230 York Avenue, New York NY 10021, USA and Department of Anatomy, University of Michigan, School of Medicine, Ann Arbor MI 48104, USA

1. Introduction

This chapter is divided into two main parts. The first describes recent studies on the fusion of mononucleated myoblasts to form multinucleated myotubes in tissue culture. It concentrates primarily on approaches which are currently being used to explore the fusion process and to examine the influence fusion exerts on the synthesis of muscle-specific proteins. The second part is a detailed examination of events occurring during nerve–muscle synapse formation in culture. The sequence of events involved in synaptogenesis is described, and studies aimed at specifying the types of cellular and molecular interactions undergone by neurons and muscle cells are presented.

2. Myogenesis in vitro

2.1. Overview

This review will discuss only selected aspects of myogenesis in vitro. Applications of new techniques such as gene cloning (Yablonka and Yaffe, 1977; Merlie et al., 1981; Merlie and Sebbane, 1981; Shani et al., 1981), the cloning of myogenic cells (Hauschka et al., 1977), monoclonal antibodies (Fambrough et al., 1982a,b; Fischman and Masaki, 1982; Horwitz et al., 1982; Grove et al., 1981) and somatic cell hybridization (Blau and Webster, 1981; Wright and Gros, 1981) have begun to be applied to the study of myogenic events with the result that a body of new information, some of it in direct contrast to previous studies, is beginning to emerge.

In discussing studies of myogenesis in vitro, the following topics will be addressed: cell–cell interactions among prefusion myoblasts that result in fusion of mononucleated myoblasts into multinucleated myotubes; molecular events that occur just prior to fusion of myoblasts in vitro; and the fusion event itself. Fusion will be examined from the perspectives of cell–cell interactions and molecular synthesis resulting in the subsequent production of muscle- (i.e., myotube) specific molecules such as myosin, actin, tubulin, etc., and the changes in cell-surface components that accompany synthesis of some of these molecules. In addition, we will comment briefly on the problems raised by species differences in attempts to generalize about certain events in myogenesis. Since several excellent reviews of various aspects of myogenesis are available that summarize the information prior to 1977 (Murray, 1972; Nelson, 1975; Merlie et al., 1977) studies subsequent to 1977 will be highlighted.

2.2. Primary cultures and cell lines

Mononucleated myoblasts can be isolated from a number of primary embryonic muscle sources. Those of chick (Shimada et al., 1969; Fischbach, 1972; Patterson and Strohman, 1972; O'Neill and Stockdale, 1972a,b; 1974; Shimada and Fischman, 1973; Doering and Fischman, 1974; Prives and Patterson, 1974; Buckley and Konigsberg, 1977; Gardner and Fambrough, 1982; Horwitz et al., 1982), rat (Burstein and Shainberg, 1979; Stygall et al., 1979; Shani et al., 1981), mouse (Hauschka and Konigsberg, 1966; Richler and Yaffe, 1970; Robbins and Yonezawa, 1971; Giller et al., 1973; Powell and Fambrough, 1973; Giller et al., 1977), bovine (Whalen et al., 1976; Gospodarowicz et al., 1976), and human origin (Blau and Webster, 1981; Walsh and Ritter, 1981) have been grown in culture. In addition, a number of cell lines derived from rodent muscle have been isolated, cloned and characterized (Yaffe, 1968, 1971; Podleski et al., 1979; Linkhart et al., 1981; Ewton and Florini, 1981). These cells undergo the process of myogenesis in the culture environment; they attach to the substratum, migrate upon it, undergo cell division, adhere to one another in regular arrays, establish cell–cell communication, and fuse to form multinucleated myotubes. The syncytia that result after fusion of myoblasts from primary cells or cell lines are capable of synthesizing specific components that are peculiar to muscle, but there is an ongoing controversy over whether some of the characteristic enzymes (Easton and Reich, 1972; Delain and Wahrmann, 1975) and other muscle-specific proteins such as myosin and tropomyosin (Emerson and Beckner, 1975; Carmon et al., 1978) are restricted to, or uniquely synthesized in post-fusion muscle cells, or whether they may be produced by myoblasts prior to the fusion event (Trotter and Nameroff, 1976; Holtzer et al., 1972; Shainberg et al., 1971; Patterson and Strohman, 1972; Shani et al., 1981). This raises questions about the relationship of fusion and terminal muscle cell differentiation. Are they related in a cause and effect manner or are they independent events?

Stockdale (1982) has recently reviewed evidence that all myoblastic cells are not equivalent. There is recent evidence to support the theory that clonal lines of myoblasts may in fact follow different courses of myogenesis. Hauschka et al. (1977) have cloned myoblasts from limb buds of early human fetuses and from avian embryonic sources and find, in both cases, that two basic types of myoblasts exist within the developing lines. 'Early myoblasts' have been characterized as requiring factors in conditioned medium for differentiation; they form short myotubes that have few nuclei. The 'late' type, on the other hand, do not require conditioned medium factors and form large, branched fibers with many nuclei (Haushka et al., 1977). Bonner (1978) has also provided some experimental support for this theory.

2.3. Cell cycle and fusion

Holtzer's theory (reviewed in Holtzer et al., 1972, 1975b; Yeoh et al., 1978) that myoblasts 'withdraw from the cell cycle' before fusing still remains controversial. In Holtzer's theory the myoblasts' last mitosis, called a 'quantal mitosis' was thought to precede withdrawal into a G_0 state and was hypothesized as a necessary prerequisite for fusion. According to Holtzer's theory (Holtzer et al., 1972) the cell cycle can result in two daughter cells with 'synthetic pathways' identical to that of the mother cell; such a cell cycle is called a 'proliferative' cell cycle. Alternatively, it can result in the production of two daughter cells with pathways different from the mother cell. The latter is termed a 'quantal' cell cycle and is seen as a means of introducing 'genetic diversity' into the replication process. In establishment of a myogenic lineage from cells of the blastula, a number of divergent events must necessarily occur during subsequent generations of cells until there emerge cells that possess all of the capabilities of synthesizing molecules peculiar to differentiated muscle.

Holtzer and his colleagues developed the hypothesis that DNA synthesis with subsequent division was an 'obligatory condition' for the terminal events in myogenesis to occur (Holtzer et al., 1972). The 'quantal mitosis' is thought to serve two functions: it serves as a preparative transition step for the fusion event and it 'synchronizes' myoblasts in such a way that leaves them at the same state to undergo fusion. Just prior to fusion, myoblasts cease to synthesize DNA (Stockdale and Holtzer, 1961) and to divide.

Holtzer's underlying hypothesis is that when post-mitotic myoblasts actually begin the synthesis of muscle-specific proteins such as myosin and actin, the cells have been determined along myogenic lines for many generations. Support for the 'quantal mitosis' theory has come from other researchers in the field (Brunk, 1979; Blau and Epstein, 1979). However, this theory has been criticized experimentally and conceptually by Buckley and Konigsberg (1977) and Konigsberg et al. (1978) whose experiments on differentiating chick wing muscle cells in vivo have suggested that 92% of the cells that have not already fused retain their proliferative capacity, although their doubling times may be considerably lengthened. This was also found to be the case in vitro (Buckley and Konigsberg, 1974). O'Neill and Stockdale (1972) also demonstrated that in vitro, changing the culture medium produced an additional mitotic phase in cells that would have fused if the medium had not been replaced. However, much of the current literature in the field has centered around the quantal cell-cycle hypothesis and many interpretations of events in myogenesis have been based on this framework.

2.4. Advantages of culture systems for studies of myogenesis

The culture systems that have been used for examination of molecular, physical and cell-biological aspects of muscle fusion range from those involving primary cultures of embryonic myoblasts from a number of species (enumerated above) to myogenic cell lines such as those derived by Yaffe (see page 27 of Linkhart et al., 1981 for a list). Certain molecular aspects of myogenesis are therefore relatively accessible and can be studied with a number of biochemical techniques. The environment can be manipulated so that the fusion process can be synchronized (Shainberg et al., 1969, 1971; Partridge and Jones, 1977; Linkhart et al., 1981), delayed for example by transformation with tumor viruses (Fiszman and Fuchs, 1975), or by treatment with agents such as EGTA, bromodeoxyuridine (BudR) or cyclohemixide (Kalderon et al., 1977) or prevented entirely, for example by drugs such as diazepam (Bandman et al., 1978). This allows one to investigate a great many different aspects of the fusion process and to try to determine cause and effect. In the case of inhibition by diazepam, chick embryo myoblasts exposed to high concentrations (100 μM) of the drug do not fuse; lower concentrations allow fusion to occur but differentiative events, such as synthesis of myosin heavy chain, are blocked. Both effects are reversible on removal of the drug (Bandman et al., 1978). The ability to block fusion but not differentiation also occurs with a number of other reversible inhibitors as well, such as Ca^{2+} (Shainberg et al., 1969) where at concentrations blocking fusion, synthesis of myosin, myokinase, and other products of muscle cell differentiation still proceed. The problems presented by such approaches to the study of fusion and differentiation are that, in general, the mechanisms of action of the drugs used to delay or inhibit fusion are unknown, the concentrations of drug necessary to inhibit fusion are often extremely high, and application of the drug results in numerous side effects besides the desired inhibitory effect on fusion. In addition, effects of certain drugs at a given concentration on one species may be ineffectual or cause unwanted side effects or death in another. Only recently has it been possible to determine if drugs affect gene expression, transcription, translation or processing of various muscle-specific proteins. With increased use of cDNA probes and other tools of molecular genetics, the sites of action of various inhibitors may become clear (Shani et al., 1981).

It has long been known that the fusion process is Ca^{2+}-dependent (Shainberg, et al., 1969). Such information has been useful in culture studies where such agents as EGTA (Kalderon et al., 1977) or Ca^{2+}-medium depletion (Shainberg et al., 1969) have been used to synchronize the fusion event. The actual site of action of Ca^{2+} in triggering fusion after depletion is still unknown.

2.5. Cell–cell interactions among prefusion myoblasts

Knudsen and Horwitz (1978), in attempting to define stages in the process of myoblast fusion, have divided the process into the following recognizable sequential events: cell–cell recognition between myoblasts, adhesion, membrane union, and subsequent morphological changes. The first three stages involve myoblasts alone or myoblast–myotube interactions; the last the myotube alone. The stages have been defined and characterized by the use of different agents and 'manipulations' that affect each step in different ways.

Horwitz et al. (1982) have characterized the recognition step as being protein-mediated, Ca^{2+}-dependent and involving cytoskeletal elements. The adhesive step is presumed to be mediated through gap-junctions or gap-junction-like structures (see Kalderon and Gilula, 1979), that allow direct communication among fusing cells. The plasma membrane is presumed to play a large role in these early stages in fusion, particularly the membrane lipids.

2.6. Early cell–cell interactions in the fusion process

In Fear's cinematographic study of myoblast fusion (1977), he found that cells aligned in a side by side orientation did not fuse but eventually migrated past each other to assume an end-to-end configuration. Ruffling membranes were seen to be in contact with one another and subsequent fusion occurred about 6 h after the initial contact of the ruffling membranes. This finding did not support earlier studies of Fischman et al. (1967) and Shimada (1971) that suggested that myogenic cells were aligned in a side-to-side configuration prior to fusion. These earlier studies, however, did not follow the whole course of the interaction by time lapse cinematography, so that although cells do align themselves in a side-to-side configuration initially, the subsequent movement to an end-to-end position may have been missed. After alignment, the myoblasts presumably develop gap junctions with adjacent cells.

2.7. Gap junctions and cell–cell communication

Kalderon et al. (1977), in an electron microscope study, found gap junctions in prefusion embryonic chick thigh muscle myoblasts. Several small (20–300 nm) gap junctions were found in myoblasts aligned side-by-side and were characterized by close apposition of the two plasma membranes of the adjacent cells. The size of these junctions increased with duration of cell contact. In some regions of junctional contact, the authors observed microfilament accumulations. These gap junctions were present 2 h *before* fusion took place in these cells. In addition, the cells were found by electrophysiological assays to be ionically and metabolically coupled. When the fusion event was arrested by blocks with EGTA, BudR or cycloheximide, both gap junctions and ionic coupling were still detected. Therefore they concluded that gap junctional communication alone is not sufficient to trigger myoblast fusion.

In a subsequent study, Kalderon and Gilula (1979) followed the process of myoblast fusion by electron microscopy and freeze fracture. They found that the generation of multinucleated myotubes was accomplished by fusion of two monolayered plasma membranes (one from each of the fusing cells) into a single bilayer that appears free of particles in freeze-fracture replicas. The single bilayer subsequently disappears with the result that the cytoplasm of the two cells is contiguous. They reported that fusion of the two plasma membranes takes place in areas of the membrane that do not contain intermembranous particles. In addition, cytoplasmic unilamellar, particle-free vesicles have occasionally been seen associated with these regions. Such vesicles are present in normal myoblasts but absent in fusion-arrested myoblasts. The authors have proposed a model for myoblast fusion that suggests that the cytoplasmic vesicles actually generate a particle-free fusion-competent area of the membrane by fusing with the bilayer.

The studies and conclusions of Kalderon and Gilula have led to a number of experiments directed at the state of the components of the plasma membrane before, during and after fusion, some of which have focused on specific transitions in membrane lipids.

2.8. Involvement of membrane lipids in the fusion event

A number of studies have focused on lipid interactions and transitions prior to and during the fusion event. Prives and Shinitzky (1977) demonstrated that, prior to fusion, there is an increase in membrane fluidity in the plasma membrane of muscle cells. VanderBosch et al. (1973) looked at the effects of alterations in temperature and lipid composition on membrane fusion and found that the rate of fusion was affected by such changes and that Ca^{2+} was also involved.

Horwitz et al. (1978) have found that the inclusion of 25-OH cholesterol (an inhibitor of cholesterol synthesis or availability) in cultures of prefusion myoblasts of either primary or cell-line origin inhibits fusion of these cells. In addition, if the fatty acyl chains are enriched with eliadate or the polar head groups with phosphatidylethanolamine (PE), fusion is also inhibited but cells align and interact to a significant extent, which they do not do in the presence of 25-OH cholesterol. Their conclusion is that the lipid alterations influence the fusion process in ways specific to a given lipid alteration. They believe that the different lipid alterations act at different stages of the fusion process (cited in their model above). Fatty acyl and PE-enrichment are postulated to act at the recognition stage but prior to membrane union. If the sterol levels are lowered by the addition of 25-OH cholesterol, they postulate that this inhibits either the recognition process or some event just prior to it.

2.9. Lipid interactions in fusion

The role of lipids in early events in membrane fusion has also received circumstantial support from the work of Trotter and Nameroff, 1976; Horwitz et al., 1978; and Horwitz et al., 1979) who were able to demonstrate that phospholipase C inhibits fusion without affecting the recognition events or cell cycle parameters. In addition, a report by Poole et al. (1970) had previously implicated lipophilic agents as promoters of the fusion process. Dahl et al. (1978) prepared plasma membrane vesicles from cultured myoblasts at different stages in the fusion process. They reported that fusion competency of these vesicles was Ca^{2+}-dependent and also depended on the 'state of maturation' of the myoblast membranes. In part, this was related to the presence of protein components (revealed by freeze-fracture) in the plasma membrane vesicles that were suggested to render the vesicles 'fusion competent'. If artificially-produced vesicles are constructed of suitable lipids (Poole et al., 1970; Kantor and Prestegard, 1975; Papahadjopoulos et al., 1976), they too will fuse and this fusion is Ca^{2+}-dependent (Papahadjopoulos et al., 1976; Razin and Ginsburg, 1980).

In an attempt to relate such biophysical data to events in myogenesis, Cornell and Horwitz (1980) and Cornell et al. (1980) have recently postulated that myogenesis may involve a depression in lipid synthesis as part of the sequence of events leading to the production of myotubes. The depression in lipid synthesis could also conceivably occur as a result of G_1 cell-cycle arrest as myoblasts differentiate and fuse. They have reported that limiting cholesterol (by restricting its availability with 25-OH cholesterol, for example) also affects DNA, RNA and protein synthesis in myoblasts. In addition, Horwitz et al. (1982) have recently reviewed evidence that phosphatidylserine (PS) and phosphatidylethanolamine (PE) are likely to be involved in the fusion process. It is still not clear what triggers the proposed decrease in lipid synthesis or whether this is a necessary prerequisite for fusion or only an incidental event associated with it. If changes in lipids are not causal in membrane fusion, other molecules must be involved in at least the recognition event. Lipids are good candidates for involvement in the adhesion process of fusion, but initial cell–cell interactions are probably mediated by proteins and glycoproteins as they are in many other cell–recognition events in other systems.

2.10. Effects of other membrane alterations or perturbations on the fusion process

2.10.1. Lectins as mediators of muscle cell fusion

One possible means of cell–cell recognition that has been postulated to be fusion-related is the involvement of cell-surface receptors. Previous reports by several groups (Teichberg et al, 1975; Nowak et al., 1976 and Gartner and Podleski, 1976) have implicated lectin-like cell-surface molecules as mediators in the fusion processes of both chick muscle and L6 myoblast cell lines. The presence of the cell-surface lectin(s) was reported to increase and become greatest at the period of differentiation and fusion. A more recent report (Kaufman and Lawless, 1980) has failed to confirm that a β-D-galactoside-binding lectin was implicated either in the fusion of rat L8 myoblast cell lines or in newborn rat myoblasts, although they confirmed its presence. They also found the 'lectin' present in non-fusing variants of the L8 cell line; most of it was detected *within* the cells rather than on the cell surface by immunofluorescent antibodies.

Lectins, such as concanavalin A (con A) and wheat germ agglutinin (WGA) among others, also inhibit fusion of myoblasts to some extent rather non-specifically (Den et al., 1975; Shainberg and Burstein, 1976; Sandra et al., 1977; Furcht and Wendelschafer-Crabb, 1978; Burstein and Shainberg, 1979).

Sandra et al. (1977) found that it was necessary first to perturb the membranes of myoblasts with trypsin for tetrameric con A to have its effect. Dimeric con A had no effect on the fusion process; therefore the inhibition was dependent on the ability of con A to cross link the receptors. They suggest that some component(s) involved in fusion that is present in the membrane may become actively involved in the fusion process by spatial redistribution. They believe this is probably a glycoprotein and that it acts during the fusion step per se since con A had no effect on cell–cell recognition steps, or the alignment of the myotubes.

In Furcht and Wendelschafer-Crabb's study, on L6 myoblasts, con A binding to both undifferentiated and differentiated *fixed* myoblasts revealed a uniform distribution of con A binding sites. However, if live cells were lectin-treated, a redistribution of binding occurred on the differentiated, but not the undifferentiated cells. They too proposed that con A receptors might have been interacting with a cytoskeletal protein. Con A also bound to fibronectin in undifferentiated myoblast cultures; but this fibronectin matrix was lost (or diminished) as differentiation into myotubes occurred. They postulated that less ordered plasma membranes (random arrays of plasma membrane intermembranous particles and unrestricted mobility of con A receptors) was 'required to accommodate' the cell fusion process.

Burstein and Shainberg (1979), however, pointed out that although con A inhibited fusion of myoblasts in both the rat and the chick, probably through a similar mechanism to that of Ca^{2+} deprivation, in the rat, the appearance of several muscle-specific proteins was tightly coupled to the fusion event. In the chick this did not appear to be wholly the case.

2.10.2. Other effects on cell surface-mediated events

Local anaesthetics and barbiturates have also been shown to affect the fusion of myoblasts reversibly (Stygall et al., 1979). Although this is probably mediated through the membrane, the mechanism of action is complex and unknown.

Significant changes occur in cell-surface and cell-surface-associated proteins, glycopeptides and glycoproteins during the fusion process (Chen, 1977; Wahrmann et al., 1980; Curtis et al., 1980; Grove et al., 1981; Fambrough et al., 1982a, 1982b). New approaches to the study of cell membrane and muscle specific protein changes during myogenesis, including the production of monoclonal antibodies to cell-surface molecules, somatic cell hybridization and lipid-probe studies are allowing approaches to some fundamental questions the answers to which have remained elusive. However, contradictory reports have appeared relating to the amounts and effects of certain cell-surface associated proteins such as fibronectin and their roles in fusion. Chen (1977) has reported that the amount of myotube associated fibronectin (LETS protein) decreases on fusion in L8 myoblast cell lines. He has postulated that fibronectin may be responsible for cell–cell adhesion in prefusion myoblasts. He has further postulated that cell-surface associated proteases may be responsible for the decrease in fibronectin and that such proteases may 'process' recognition molecules. Podleski et al. (1979) in reporting on another cell line, L6, have found that cell-surface associated fibronectin at low levels enhances fusion but at high levels (and if exogenous fibronectin is added) enhances cell division. Hynes et al. (1976) found an increase in fibronectin in post-fusion myotubes. Gardner and Fambrough (in press) have attempted to investigate the synthesis and involvement of fibronectin in myogenesis by producing monoclonal antibodies to mononucleated cells and using the monoclonal to effect complement-mediated lysis of mononucleated cells in the cultures. This leaves cultures of pure myotubes and eliminates both myoblasts and fibroblasts. In addition, they generated a monoclonal antibody to fibronectin and used it in immunofluorescence studies of fibronectin during myogenesis. They found that cell-associated fibronectin is extensive on the substratum in motile prefusion myoblasts; that a major change occurs in distribution between days 2–3 in vitro when fibronectin is present in discrete blocks on the cell surface. The focal accumulations are present both on abutting surfaces of closely apposed cells and free surfaces. This is a period of intense fusion activity. Newly formed myotubes have very little fibronectin on their cell surfaces, although cell substratum associated fibronectin continues to accumulate. Such monoclonal antibodies have an obvious advantage for unraveling complex problems such as appearance, localization and involvement in myogenesis of fibronectin.

2.10.3. Effects of viral transformation on muscle cell fusion

Another approach to the study of muscle cell fusion that has raised some interesting questions about cell membrane interactions in the fusion process has been the use of transforming RNA-tumor viruses to inhibit or slow down the fusion process while leaving the recognition and alignment sequences unaffected (Fiszman and Fuchs, 1975; Hynes et al., 1976). In Hynes et al.'s (1976) study, transformed lines of rat myoblasts did not show an increase in myosin synthesis usually associated with fusion. Fibronectin was also greatly reduced or absent on the surfaces of the transformed lines. Further, all of the cell lines from which the viral genome could be rescued showed a complete block of the fusion process as well as a 10-fold drop in fibronectin (LETS protein) synthesis (see discussion of fibronectin above).

Temperature sensitive Rous sarcoma viruses transform sensitive chick cells such that inhibition of fusion is seen at the permissive temperature for virus expression, 37 °C (Fiszman and Fuchs, 1975; Fiszman, 1978). In addition to the fusion block, the myotubes did not express certain muscle-specific proteins. When the virus infected cells were returned to the non-permissive temperature (41 °C) for virus expression, a normal program of fusion and differentiation ensued.

In Hynes et al.'s interpretation of the fusion block by transforming viruses (1976) they hypothesize that all of the synthetic and differentiative events following fusion are dependent on events in the cell cycle that occur in G_1 accompanied by a *withdrawal* from the cell cycle. The viral transformation essentially keeps these cells in a proliferative phase (but see Buckley and Konigsberg, 1977), thus preventing these events from occurring.

The studies that deal with the initial recognition events and the subsequent cell fusion events have served to point out areas for further investigation. It is still unclear what triggers the initial events in myogenesis. It is also unclear just how closely events in vitro mimic those in vivo.

2.11. Potential triggers of myogenesis

2.11.1. The involvement of the substratum
The work of Konigsberg and his group (Konigsberg, 1963, 1977; Hauschka and Konigsberg, 1966) first called attention to the fact that collagen substrata markedly enhanced myotube differentiation. Myogenic cell lines have been shown to synthesize several forms of collagen (Garrels, 1979; Sasse et al., 1981). Subsequent studies by several investigators have been directed to the involvement of fibronectin in cell-surface events in myogenesis (see discussion above). Some investigators have found increases in fibronectin levels during the fusion process (Hynes et al., 1976); Walsh and Phillips, 1981). Studies by Furcht et al. (1978), and Chen (1977) have shown that although fibronectin matrices are present on cultured myoblasts, these disappear when myotubes are formed. Chiquet et al. (1979) have found that serum fibronectin is a factor necessary for the attachment of myoblasts to collagen substrata. The involvement of this attachment can now be investigated with monoclonal antibodies such as those reported by Gardner and Fambrough (unpublished).

2.11.2. Growth factors
Stockdale (1982) has recently reviewed the influences of culture medium components on myoblast proliferation and fusion. Proliferation of both primary myoblast cultures and cell lines is dependent principally on serum and components of embryo extract [probably somatomedins and/or platelet derived growth factor(s)] (Stockdale, 1977; Yaffe, 1973; Konigsberg, 1977). Other growth factors such as fibroblast growth factor (FGF) have also been shown to affect myoblast proliferation (Gospodarowicz et al., 1976; Linkhart et al., 1980, 1981) and the appearance of muscle-specific proteins such as acetylcholine receptors (Gospodarowicz et al., 1976). However, these authors also found that FGF did *not* stimulate proliferation of chick myoblasts, although it affected bovine cells.

Cells that will differentiate into myotubes either initiate cell fusion along with the synthesis of muscle-specific proteins or remain in the cell cycle depending on the presence of specific medium components. While the presence of certain specific mitogens such as fibroblast growth factor and certain serum components keep cells in a proliferative state (Linkhart et al., 1981), other medium components or deprivation of mitogens cause cells to differentiate (Bischoff and Holtzer, 1967; O'Neill and Stockdale, 1972a; Konigsberg, 1971; Doering and Fischman, 1974; Nadal-Genard, 1978). Lim and Hauschka (1981) have begun to characterize a mitogen receptor for epidermal growth factor (EGF) on the cell surface of myogenic cells that disappears from the cell surface as differentiation occurs.

Insulin has been shown to affect myoblast differentiation (Rutter et al., 1973) and Ball and Sanwal (1980) have reported that high concentrations of insulin and glucocorticoids synergistically promote differentiation in some clones of the rat myoblast cell line L6. Glucocorticoids alone were found to increase cell adhesiveness. However, Gospodarowicz et al. (1976) found no effect of insulin on bovine myoblasts, another species difference that prevents one from drawing general conclusions about hormone effects.

Ewton and Florini (1981) have recently studied the effects of both low levels of insulin and somatomedins on the rat L6 myoblast cell line. Both caused an increase in the incidence of myoblast fusion. They concluded that this effect was not simply due to increased cell proliferation and subsequent fusion but hypothesized that the fusion promoting effects of insulin stemmed from the fact that it was a somatomedin analogue. They had shown previously that at high levels insulin interacted with the somatomedin receptor in both fat cells and fibroblasts (Florini et al., 1977; Ewton and Florini, 1980) and postulated that a similar interaction occurred in muscle.

Whether or not environmental cues trigger the fusion of myoblasts into myotubes or whether the fusion process is 'built in' to the developmental clock of myoblasts as maintained by Holtzer et al. (1975b) is still not clear. A more complete understanding of the mechanism of the fusion process per se may provide clues to the key events responsible for initiating the fusion event.

2.12. Cell fusion and production of muscle-specific components

As discussed briefly in the introduction to the myogenesis sequence, the differentiation of skeletal muscle in vitro has come under new scrutiny with the development of several new approaches; the foremost among these are, of course, gene cloning and the application of monoclonal antibody techniques.

Questions about the fusion trigger and whether fusion *itself* is a trigger for differentiation can now be investigated more minutely. From previous studies (reviewed in Trotter and Nameroff, 1976) it is possible that at least some cell-surface components of muscle differentiate independently of fusion, although some recent evidence may indicate that other components are inserted or synthesized in a fusion-dependent fashion (Shani et al., 1981). There is still controversy over whether muscle-specific macromolecules are synthesized at the same rate in unfused and fused cells.

Trotter and Nameroff's (1976) study contends that fusion is not necessary for myofilament assembly, T-tubule invagination, sarcoplasmic reticulum (SR) differentiation or formation of SR–T-tubule junctions. The authors did not address the possibility that they were examining the properties of a specific clone of myoblastic cells whose myogenic program is different from that of other clones.

Shani et al. (1981) have recently addressed the question of when mRNAs for myosin heavy chain, myosin light chain-2 and actin were present during myogenesis and when those proteins were synthesized during development. They found that accumulation of mRNAs for these proteins were detectable a few hours before the onset of cell fusion, and increased during the phase when cells were rapidly fusing. Synthesis of protein quickly followed elaboration of the mRNAs, therefore indicating that activation (translation) of stored message was not 'a major mechanism' for controlling the time of synthesis of the proteins and that synthesis and expression were fusion-related. Other reports support this general conclusion (Yablonka and Yaffe, 1977; Devlin and Emerson, 1979). However, in contrast, Robbins and Haywood, 1978 and Dym et al. (1978) found in their studies that most of the mRNA for these proteins was stored in an inactive form in prefusion myoblasts. Some of the differences in results may be ascribed to differences between the rat and chick systems, however, since these species differ in several other significant respects (see sections above). Some of these problems may be resolved with the use of specific probes for myosins such as the monoclonal antibodies recently reported by Fischman and Masaki (1982). These antibodies allow a finer tuning of mRNA probes in the chick that will permit similar studies to those done by Shani et al. (1981). In studies with these monoclonal antibodies on embryonic, neonatal and adult chicken muscle, Masaki et al. (1982) found that three MHC isoforms appear sequentially during myogenesis (see Whalen et al., 1981). These may be distinct gene products that accumulate sequentially during development. However an alternative explanation is that post-translational modifications may occur sequentially in a single gene product. Studies of cultured myoblasts and myotubes may be very valuable for these investigations. However, Devlin and Emerson's (1978) study on the quail has shown that fusion-triggered synthesis of MHC, two myosin light chains, two subunits of troponin and two of tropomyosin is identical to the kinetics of mRNA activation in these cells. They suggest that results of previous studies (Whalen et al., 1976; Patterson and Strohman, 1972) that purported to show that myoblasts of chick and calf were capable of synthesizing contractile proteins were biased by the presence of differentiated fibers (approximately 10%) even at very early stages of culture. In their studies (Devlin and Emerson, 1978) they say that virtually no ($< 0.5\%$) post-mitotic or fused muscle is present. Other proteins that are synthesized by myoblasts cease being synthesized after fusion as evidenced by 2-D gel analysis. These proteins have yet to be identified.

The ultimate solution to the question of whether and when transcriptional or translational controls of muscle-specific proteins operate in myogenesis will await sequence-specific assays for these proteins.

It is obvious, however, that application of new techniques (gene cloning, somatic cell fusion, and the application of monoclonal antibodies) will aid many investigations of myogenesis. Changes in cell specific proteins and timing of synthesis of RNAs are now monitorable with probes of great specificity in the form of cDNAs and monoclonal antibodies. Cultures of pure myotubes from primary sources have become easy to produce with the advent of monoclonal antibodies that eliminate all other cells (Bayne et al., 1980). The use of pure primary cultures may now alleviate problems posed by the use of cell lines some of which may behave very differently from the primary muscle from which they were derived. Somatic cell hybridizations of myogenic cells from different species allow new studies of gene depression to be conducted. The definition and analysis of various steps in myoblast fusion and differentiation are now accessible.

3. Synapse formation

3.1. Initial nerve–muscle contacts

3.1.1. Experimental systems

The use of dissociated nerve–muscle cultures to study initial events in the formation of nerve–muscle synapses is now quite widespread. In vitro systems offer clear advantages in the examination of cell–cell interactions in terms of both visualization of the cells and in biochemical and pharmacological manipulations. Thus far, much information has been gathered from in vitro studies, information which appears, for the most part, to be consistent with that obtained from in vivo studies. In this section, we highlight what has been learned about the central aspects of synapse formation, paying special attention to neuronal regulation of muscle acetylcholine receptors (AChRs) and acetylcholinesterase (AChE). We will focus on information obtained using primary cultures of neurons and muscle cells, but will mention results from experiments with cell lines where appropriate.

Most in vitro experimentation has involved adding neurons – either dissociated or in thin cross sections (explants) – to cultures of already fused muscle cells. While this does not totally reproduce the in vivo situation in which spinal cord neuronal processes enter the muscle mass prior to the main period of myoblast fusion (Landmesser, 1978b), it has proven to be most convenient. At the normal time of addition of neurons, myotubes have a rather high level of AChRs over their entire surface (compared to extrajunctional regions of innervated muscle) with occasional preexisting clusters (or 'hot spots') of receptor molecules (Fischbach and Cohen, 1973; Sytkowski et al., 1973) and some membrane-bound AChE probably present uniformly (Rotundo and Fambrough, 1980).

3.1.2. Growth of spinal cord neuronal processes
Very little is known about factors which determine the rate or direction of growth of motoneuron processes. Certainly, no study demonstrating an action comparable to that which NGF exerts on growing sensory ganglion neuronal processes is currently available. (It is not even clear whether these factors are bound to cell migration pathways or are soluble molecules secreted by particular groups of muscle.)

Letourneau (1975a,b) studied the growth of sensory ganglion neuronal processes on a variety of mostly artificial culture substrates of different adhesiveness. Although processes could elongate well on several of the substrates, they clearly preferred certain ones (polyornithine to collagen, for example) and would grow primarily on them. Thus, a hypothesis could be advanced which explains directionality of neuronal process growth based on differences in extracellular matrix material throughout the embryo. Of course, the manner in which 'tracks' are laid down and maintained would need to be explained.

There have also been numerous studies directed towards identifying soluble molecules which might influence the growth and differentiation of embryonic nerve processes. Dribin and Barrett (1980) found that growth of neuronal processes from explants of embryonic rat spinal cord could be enhanced by 7 days of treatment with conditioned medium obtained from primary muscle cultures, but also from fibroblast and lung cell cultures. In fact, the lung cell cultures were most effective. The active components were heat-resistant and had apparent molecular weights greater than 100,000. Henderson et al. (1981), using a somewhat similar assay which measured changes after approximately 24 h of treatment, however, found that growth of chick spinal cord neuronal processes was enhanced by a factor contained in serum-free medium conditioned by primary cultures of muscle cells. Substantial activity was also found in liver cell cultures and in skin cultures, but not in those of lung cells. The active components were trypsin-sensitive and had apparent molecular weights of 40,000, 500,000 and greater than 1,000,000. These studies emphasize that different results can be obtained even under mildly different assay conditions. It may be some time, therefore, before these studies can be reconciled and applied to the in vivo situation.

3.1.3. Specificity of synapse formation
Are the growth cone tips of spinal cord processes searching for particular muscle macromolecules prior to the onset of synapse formation in culture? Even the in vivo situation is not entirely clear. Landmesser and colleagues (Landmesser and Morris, 1975; Landmesser, 1978a,b; Lance-Jones and Landmesser, 1980a,b) demonstrated that particular groups of motoneurons in early chick embryos appear destined to innervate certain groups of limb muscles, but the generality of such extreme specificity is not known. Motoneurons reinnervating denervated adult muscle end preferentially at the original junctional sites (Letinsky et al., 1976); in this case, the growing nerve processes seem to recognize particular components of the muscle's basal lamina (Marshall et al., 1977; Sanes et al., 1978). Because of these two situations at least, there is much interest in delineating the molecular basis of any specificity in synapse formation.

Many different types of cholinergic neurons – spinal cord (Fischbach, 1972), chick ciliary ganglion (Hooisma et al., 1975; Betz, 1976), rat and chick sympathetic (in their induced cholinergic mode; Nurse and O'Lague, 1975; Rubin, unpublished observations) and retinal (Puro et al., 1977) – appear able to innervate cultured limb skeletal muscle. The situation is not totally random, however, since chick dorsal root ganglion neurons do not innervate muscle cells (Fishbach, 1972; Obata, 1977). What seems to be a minimal requirement is the ability of neurons to synthesize or be induced to synthesize ACh; this situation may not differ so much from that in vivo, since vagal afferents can innervate denervated adult muscle in frog (Landmesser, 1971).

Cross-species innervation is also not difficult to demonstrate. Mouse spinal cord can innervate rat and human muscle (Crain et al., 1970; Crain and Peterson, 1974), chick spinal cord will synapse on rat myotubes (Fambrough et al., 1974) and *Xenopus* neurons will innervate cells from a rat skeletal muscle line (Kidokoro and Yeh, 1981). Myotubes can be innervated by several different neurons and at numerous locations on the cell surface (Fischbach, 1972). Whatever molecules on the muscle cell surface do determine the location of synapses are still unidentified. If such molecules exist, however, they must be of a fairly general nature, presumably varying little from species to species and among certain types of target cells.

Since AChRs are present in clusters (Fischbach and Cohen, 1973; Sytkowski et al., 1973) on uninnervated muscle cells (even when myoblasts are removed from aneural limb buds, i.e., prior to any innervation; Bekoff and Betz, 1976), these clusters were once thought to be the sites of synapse formation. Thus, the idea was that growing nerve processes sought out and synapsed on regions of relatively high AChR density. The experiments of Frank and Fischbach (1979) and of Anderson et al. (1977), which will be described in detail in Section 3.2.1., showed that such processes do not seek out these regions, often (at least) forming synapses away from areas of clustered AChRs. Since there is certainly still a relatively high density of AChRs even away from clustered regions, these experiments should not be taken as absolute evidence that the AChR does not play a role as a recognition substrate during synapse formation. It is also true that AChRs can be blocked with small ligands – curare and α-bungarotoxin – during nerve–muscle interaction without inhibiting synapse formation (Cohen, 1972; Crain and Peterson, 1974; Rubin et al., 1980). Still, since these ligands are relatively small compared with the AChR itself, these experiments also do not rule out the involvement of AChRs in synapse formation. However, combined with those of Marshall et al. (1977) in vivo, they strongly suggest that this macromolecule is not the main determinant of the site of synapse formation.

A complete understanding of nerve–muscle recognition will almost certainly involve also an appreciation of the structure and chemistry of motoneuron growth cones. The morphology of growth cones of cultured neurons has been studied in some detail (Pfenninger and Rees, 1976), but their biochemical nature is largely unknown. Some studies of the carbohydrate composition of cell surface molecules have been carried out and suggest differences among the types of neuronal cells (Pfenninger and Maylié-Pfenninger, 1981).

In addition to more straightforward techniques of identifying the relationships between certain molecules and sites of synapse formation, some laboratories have turned to studies of cell–cell adhesion (Barbera et al., 1973; Gottlieb et al., 1976; Rutishauser et al., 1976). The hope of these experiments is that the same interactions which mediate adhesion between two different types of cells – typically retinal and tectal – might also underly synapse formation. Thus far, most of these experiments have yielded little information concerning the molecular nature of recognition between nerve and muscle. Recently, however, Grumet et al. (1982) observed that a particular macromolecule (which they've termed neural cell adhesion molecule, N-CAM) thought to be presently primarily on neuronal cells was also present on chick muscle cells. Retinal neuron adhesion to muscle cells could be blocked by F(ab') fragments of antibodies directed against N-CAM. The authors suggest that this molecule is involved in at least the initial adhesion between growing nerve processes and muscle cells which presumably occurs prior to actual synapse formation.

3.1.4. Onset of synaptic transmission

On rare occasions, electrical coupling between neurons and muscle cells from normal embryos has been seen in culture (Fischbach, 1972; O'Lague et al., 1974). Under one unusual circumstance – dystrophic muscle cells in tissue culture innervated by spinal cord neurons – the incidence of electrical coupling was quite high, approximately 20% (Peacock and Nelson, 1973). Even these infrequent instances of electrical coupling between nerve and muscle have prompted the suggestion that a stage of electrical communication may occur very early in synapse formation (Fischbach, 1972). For instance, perhaps specific macromolecules are not involved in determining the site of synapse formation. Rather, neurons may receive information about having contacted a muscle cell via gap junctional communication and then initiate steps necessary to form a synapse. This hypothesis is under consideration in spite of the facts that Dennis et al. (1981) tested for electrical coupling in developing rat embryo muscle and found none, and that cells which do not normally contact one another in vivo can still form gap junctions in culture (Lawrence et al., 1978).

Specifying the exact time of chemical synapse formation in culture may also be difficult. There is some evidence that growth cones can release ACh (Cohen, 1980), but it is not known whether all growth cones which release sufficient quantities of ACh to depolarize muscle invariably form stable contacts. This hesitation aside, it is apparent that very soon (a period of a few hours) after initial contacts between neuronal processes and muscle, miniature endplate potentials (MEPPs) and spontaneous end-plate potentials (resulting from spontaneous firing of neurons) can be recorded in contacted muscle fibers (Betz, 1976; Cohen, 1980). The synapses which form resemble adult neuromuscular junctions (Fischbach, 1972) in that (a) they obviously use ACh as a neurotransmitter and (b) ACh is released in quantal packets, although the MEPP frequency and quantal content are both low. Evoked release is dependent upon extracellular calcium, but the release process appears to be less sensitive than that in the adult to perturbations such as increased extracellular potassium, hypertonicity or lanthanum (Kidokoro et al., 1976; but see Peng et al., 1979). In addition, as stated previously, because the neurons spontaneously fire action potentials, muscle fibers are often regularly depolarized by released ACh, commonly to the point of contraction. Thus, many innervated muscle fibers contract at the rate of 2–3 per second without any external stimulation; uninnervated fibers, on the other hand, may contract, but usually irregularly, in a pattern resembling that of fibrillation in adult denervated muscle (Purves and Sakmann, 1974). Other aspects of neuromuscular junction formation, such as AChR clustering and AChE appearance, will be dealt with in later sections.

3.1.5. Ultrastructure of early synapses

Ideally, studies on the ultrastructure of early nerve–muscle contacts would proceed thus: nerve processes would be followed as they course along the culture surface until they contact uninnervated myotubes. Then, as the growth cone proceeds across the myotube surface, intracellular electrodes would be placed in the myotube to monitor for the first signs of ACh release. Once such release were detected, extracellular recording electrodes would be positioned to define (that is, within 1–2 μm) the actual discrete synaptic site, and the culture would be fixed and processed at various times thereafter. The ultrastructure of the physiologically identified site, and not of random nerve–muscle contacts or even of random regions of a long single process known to innervate a muscle fiber, could then be examined. Unfortunately, because of technical difficulties resulting, in large part, from the asynchrony of synapse formation under in vitro conditions, such a study has not been carried out. Below, we summarize available information which must now be considered as suggestive, but not definitive.

Most ultrastructural studies have focused on the time of appearance of certain characteristic features of the neuromuscular junctions: accumulations of synaptic vesicles at their presynaptic release sites (active zones), the junctional basal lamina, postsynaptic membrane thickening and secondary junctional folds. One general conclusion of work carried out thus far is that neuromuscular transmission can seemingly occur at synapses whose structure is quite rudimentary (Nelson et al., 1978; Nakajima et al., 1980). Ultrastructural maturation occurs relatively slowly (perhaps even taking days or weeks) compared to onset of transmission and to the localization of AChRs and AChE (see Sections 3.2 and 3.3 below). As for early structural correlates of electrical synapse formation, Rees (Rees et al., 1976; Rees, 1978), who examined contacts between rat spinal cord neurons and superior cervical ganglion cells, failed to find any evidence for gap junctions. Still, such a stage could be exceedingly transient and difficult to observe in thin section microscopy. A presynaptic accumulation of clear vesicles similar in size to synaptic vesicles is observed relatively early on (within 12–18 h), but these vesicles are not obviously clustered at release sites (Peng et al., 1979; Nakajima et al., 1980; Weldon and Cohen, 1979). Within 1–3 days, such vesicles increase in number and become more closely apposed to specific sites on the presynaptic membrane. Curiously, Cohen and Weldon (1980) observed some accumulation of small vesicles, but no active zones, at sites of contact which were not synapses between *Xenopus* dorsal root ganglion or sympathetic ganglion neuronal processes and muscle cells. Postsynaptic membrane thickening also becomes apparent at some contacts at this later time. Also noticeable then, and even earlier, in spinal cord-superior cervical ganglion cultures is an increase in postsynaptic coated vesicles which are not labeled during incubation with extracellular horseradish peroxidase (Rees et al., 1976). Such coated vesicles may thus represent those in transit *to* the cell surface; these coated vesicles could contain, for instance, AChRs destined for insertion into the plasma membrane. After approximately two weeks in rat nerve–muscle culture, postsynaptic folds appear. In *Xenopus* nerve–muscle cultures, pre- and postsynaptic specializations develop even when cultures are grown in the presence of d-tubocurarine; thus, synaptic activity is not essential for their development (Weldon and Cohen, 1979).

Another area currently of intense interest, primarily because of the work of Marshall et al. (1977), is the initial time of appearance of muscle basal lamina. These workers showed that specific reinnervation of adult muscle might be mediated by the junctional basal lamina. Knowing whether extracellular material is present on muscle cells prior to innervation and might specify the location of synapses is, therefore, important. Unfortunately, this is again a somewhat complex issue. Most workers agree that there is no continuous basal lamina along the surface of myotubes grown in the absence of fibroblasts (Fischbach et al., 1974; Betz, 1976); there are, however, wisps of filamentous material which could be basal lamina precursors (Burrage and Lentz, 1981); see also Section 3.5.1 (1.d). Fairly soon after nerve–muscle contact, such filamentous material is often seen between neuronal endings and muscle and could again represent rudimentary basal lamina (Peng et al., 1979; Nakajima et al., 1980). However, even at certain areas of contact between neuronal processes with no synaptic vesicle accumulation, between neurons and fibroblasts in spinal cord-superior cervical ganglion cultures, (Rees, 1978) and even between cultured fibroblasts (Heaysman and Pegrum, 1973), such filamentous material is seen.

3.2. AChR Clustering at newly-formed synapses

3.2.1. Nerve–muscle synapse formation causes the appearance of new AChR clusters

Since a predominant feature of adult neuromuscular junctions is their high concentration of AChRs, one primary goal of in vitro experimentation has been to determine the origin of synaptic AChRs, and the mechanism determining their appearance. Here we review the clustering of AChRs at sites of transmitter release, and in Section 3.4.2., we discuss the role of synaptic activity and of neurally derived molecular factors in initiation of clustering.

Very early on it became clear that both uninnervated and innervated muscle cells had clusters of AChRs (Fischbach and Cohen, 1973; Sytkowski et al., 1973). Cohen and Fischbach (1977) showed definitively that sites of neurotransmitter release in nerve–muscle co-cultures were associated with relatively large numbers of AChRs. Because, in chick cultures, several different neurites may maintain long regions of contact with myotubes, it was necessary to physiologically identify the actual location of synapses. Cohen and Fischbach recorded intracellularly from innervated myotubes while extracellularly stimulating fine nerve processes in the presence of tetrodotoxin to produce local (non-propagated) depolarization. At discrete sites of stimulation along nerve processes, they were able to record depolarizing end plate potentials in myotubes. Around such release sites they mapped ACh sensitivity using an iontophoretic pipet filled with ACh. In all cases, they observed local peaks of ACh sensitivity across from release sites. Even on innervated myotubes, however, peaks of ACh sensitivity could still exist at sites away from nerve–muscle contacts or at sites at which neurotransmitter release could not be evoked.

As stated in Section 3.1.3., the first question of interest was whether synapses form only at preexisting AChR clusters or whether synapse formation induces the appearance of a new cluster. Frank and Fischbach (1979) carried out careful iontophoretic mapping of ACh sensitivity over extensive areas of uninnervated myotubes. They found that, under normal circumstances, the distribution of AChR clusters remained relatively invariant – no new clusters appearing, for instance – on established myotubes. They then allowed these myotubes to become innervated by processes extending from neurons in spinal cord explants, identified sites of transmitter release physiologically by focal extracellular recording or stimulation and again mapped ACh sensitivity. They found that synapses do not necessarily form at preexisting AChR clusters, but that within hours of the onset of synaptic transmission, new clusters did appear precisely at the sites of transmitter release. The origin of these newly clustered AChRs – whether they were recruited from other regions of the muscle cell or inserted locally into the synaptic region – could not be determined from this study.

Anderson et al. (1977; Anderson and Cohen, 1977) used *Xenopus* nerve–muscle cultures to answer similar questions about AChR appearance at synapses. Instead of using physiological techniques to map AChR distribution, they used a fluorescent derivative of α-bungarotoxin to irreversibly label surface AChRs. Prior to innervation the mononucleated *Xenopus* myocytes had, on occasion, large polygonal AChR clusters. Within 24 h of innervation, muscle cells had a different distribution of AChRs – often long linear tracks along nerve processes – and no such large polygonal clusters. Thus, nerve processes clearly modified the AChR distribution in these cells. Also clear was that nerve processes did not grow preferentially to preclustered regions. Anderson and Cohen also demonstrated that if AChRs were labeled with fluorescent α-bungarotoxin prior to addition of neurons, some of these prelabeled AChRs became localized along the path of nerve–muscle contact. This localization was a specific muscle cell response to motor nerve contact. Cohen and Weldon (1980) subsequently discovered that neurons – those from dorsal root or sympathetic ganglia – which do not innervate muscle cells also did not induce this type of AChR localization. Apparently, then, AChRs present in the muscle cell membrane can be directed to other parts of the cell in response to nerve–muscle synapse formation. These experiments therefore provide information missing from those of Frank and Fischbach: that at least a portion of synaptic AChRs were once extrasynaptic. Anderson's and Cohen's experiments did not provide the detailed physiological information supplied by Frank and Fischbach. So, it remains unclear precisely how the location of AChR clusters is related to the actual sites of neurotransmitter release along the path of nerve–muscle contact. This information is potentially of value in determining how synapse formation triggers AChR localization.

The timing of AChR aggregation with respect to the onset of synaptic transmission is still not exactly clear. For instance, it might occur even prior to initiation of transmission. (As will be discussed later, AChR clustering is not in fact dependent on synaptic activity.) The studies of Frank and Fischbach (1979) in chick cultures and of Anderson and Cohen (1977) in *Xenopus* demonstrated that AChRs localize at synapses within 6–24 h of synapse formation, but could make no more definite statements. Anderson et al. (1979) observed that in *Xenopus* cultures (1) some cells from which they were able to record MEPPs had no detectable AChR aggregation (using fluorescent α-bungarotoxin) and (2) the mean MEPP amplitude was larger in cells which did have noticeable aggregation. They thus suggested that synaptic transmission precedes aggregation and that when AChRs do accumulate the size of synaptic potentials increases. From a few observations on transmitter release made soon after nerve–muscle contact in chick cultures, Cohen (1980) did not eliminate such a sequence. But in spite of these studies, it still might be that AChR aggregation can precede synaptic transmission in some cases.

3.2.2. Is AChR clustering an invariable consequence of nerve–muscle synapse formation?

The vast majority of nerve–muscle synapses in established chick cultures and of nerve–muscle contacts in *Xenopus* cultures are associated with AChR clusters. In fact, even certain sites of contact between neuroblastoma cells and L6 muscle cells, which do not form functional synapses, are associated with AChR accumulation (Harris et al., 1971). Yet, this association has not been universal. Kidokoro et al. (1980) found that spinal cord neurons from older *Xenopus* tadpoles innervated *Xenopus* muscle cells, but did not induce AChR accumulation, at least as evidenced by the lack of an increase in MEPP amplitude with time in culture. Using similar criteria, Kidokoro (1980) concluded that AChR clusters do not occur at synapses formed in cultures of rat spinal cord and muscle cells. Kidokoro and Yeh (1981) found that *Xenopus* muscle cells formed physiologically active synapses with L6 muscle cells, but did not cause AChR clustering. In none of these cases did the investigators first identify the actual sites of stimulation by physiological criteria, such as focal extracellular stimulation or recording, and then look for AChR localization using ^{125}I-α-bungarotoxin autoradiography, a technique more sensitive than fluorescence. In addition, further work will be needed to discover whether changes in culture conditions might permit AChRs to become localized at these synapses. Nonetheless, the possibility remains that cells may interact synaptically in vitro without AChR accumulation – whether because of a failure of the neuron to provide factors necessary for AChR clustering or because of a failure of certain muscle cells to respond to such factors (as might be implied by the results of Kidokoro and Yeh). This possibility was reinforced by the experiments of Busis et al. (1981), who studied interactions between rat skeletal muscle cells and a variety of neuroblastoma-derived cell lines. These workers suggested that synapse formation and induction of AChR aggregation were independent, separable properties of *neurons*.

3.2.3. Redistribution versus local insertion of synaptic AChRs

The results of Anderson and Cohen demonstrated clearly that some AChRs which appear at synapses were present at a prior time elsewhere on the muscle plasma membrane and, hence, were accessible to fluorescent α-bungarotoxin. Whether this means that AChRs diffuse in the plasma membrane from extrasynaptic to synaptic regions as is commonly assumed, or whether they are internalized and redirected from within the cell has not been determined. Similarly, whether this redistribution occurs as a major mechanism in other cell types is not yet known.

Other evidence suggests that, at some point, local insertion of AChRs becomes an important mechanism. Using fluorescence photobleaching techniques, Axelrod et al. (1976) demonstrated that whereas nonclustered AChRs are freely mobile, clustered AChRs are not. An area of diffuse AChRs labeled with fluorescent α-bungarotoxin can be bleached via an intense light flash, but fluorescence will quickly recover because of diffusion of unbleached AChRs from nearby regions. If such bleaching is done within a cluster, fluorescence does not recover; thus, clustered AChRs do not mix. A reasonable conclusion from these results is that synaptic AChRs are initially in part or entirely from extrasynaptic regions, but once the cluster has formed, local insertion of AChRs predominates. In this regard, under conditions in which muscle cells were innervated or induced to have clusters of AChRs (see Section 3.5.1. (1.c)), Bursztajn and Fischbach (1980) found an increase in the number of coated vesicles which could not be labeled during bath perfusion with horseradish peroxidase (HRP), but which could be labeled following membrane permeabilization and incubation with HRP coupled to α-bungarotoxin. These coated vesicles could serve to bring AChRs from the Golgi apparatus to the plasma membrane.

3.2.4. Further modification of synaptic AChRs

There are several established differences between junctional and extrajunctional AChRs, other than their degree of aggregation. As demonstrated by fluctuation analysis, activated junctional AChRs stay open 3–5 times shorter than do activated extrajunctional AChRs in adult muscle (Katz and Miledi, 1972; Dreyer et al., 1976; Neher and Sakmann, 1976). Furthermore, compared to extrajunctional ones, junctional AChRs are quite stable once incorporated into the plasma membrane. Their half-time of turnover is at least 4–5 days, versus about 1 day for extrajunctional AChRs (Berg and Hall, 1975; Chang and Huang, 1975). It was thus of great interest to determine: (1) if whatever stimulus leads to AChR clustering simultaneously causes changes in channel open time and turnover rate or (2) if not, when these modifications occur with respect to clustering. Much related work has been done in vivo. The hope with experiments in culture, however, is that factors causing clustering and other modifications of AChRs can be isolated and studied in more detail (see Section 3.5.1. (1)).

AChRs on uninnervated myotubes in culture clearly have characteristics similar to extrajunctional AChRs in adult muscle. Schuetze et al. (1978) applied low concentrations of ACh from an extracellular recording pipet and determined that both nonclustered and clustered AChRs have long channel open times. Devreotes and Fambrough (1975) studied the AChR turnover rate on uninnervated myotubes in some detail. They labeled cultured myotubes with ^{125}I-α-bungarotoxin and determined that, after a period of incubation, [^{125}I]tyrosine, produced by degradation of AChRs with their bound toxin molecules appears in the culture medium and can be used as a direct and easily accessible measure of AChR turnover. In both chick and rat cultures, the half time of turnover was less than 24 h. They also studied the degradation process itself and determined it to be quite sensitive to metabolic inhibitors such as dinitrophenol and lowered temperature. There is some evidence that lysosomes play a role in the degradation of AChRs, as they do in degradation of other membrane proteins (Libby et al., 1980).

Schuetze et al. (1978) then studied properties of synaptic AChRs in chick nerve–muscle cultures. Again using fluctuation analysis of recordings made at identified synapses via an ACh-filled extracellular pipet, they determined that AChRs at these sites also had long channel open times. They also noted the location of identified synapses, labeled the cultures with ^{125}I-α-bungarotoxin and processed them for autoradiography. Even AChRs at such identified sites were found to have rapid turnover rates. Thus, it seems quite clear that clustering alone, whether on uninnervated cells or at synapses, is not necessarily associated with other changes in AChR properties. These results are quite consistent with those obtained in vivo. They are also consistent with the observations that anti-AChR antibody-induced aggregation of AChRs (Prives et al., 1979) or aggregation induced by binding biotin α-bungarotoxin to AChRs and cross-linking with avidin (Axelrod, 1980) leads not to their stabilization, but to an enhanced rate of degradation.

Thus far, in chick cultures, no maturation of these parameters has been observed. Even junctional AChRs of mature chickens have long channel open times (Schuetze, 1980), so this property would not be expected to change in these cultures. In developing chicken, there is a lag period of about 2 weeks (Burden, 1977b) and in rat a period of days (Steinbach et al., 1979; Reiness and Weinberg,

1981) between AChR clustering and turnover time changes. So, perhaps nerve–muscle cultures have not been maintained for a time sufficient to trigger these developmental modifications.

3.3. AChE appearance at newly-formed nerve–muscle synapses

3.3.1. Chick nerve–muscle cultures
As will become clear, there may be differences in the ways AChE appearance is regulated at different types of nerve–muscle synapses. In this section we discuss the appearance of synaptic AChE in cultures derived from chick embryos. In such cultures, at least, the appearance of discrete patches of enzyme and of a particular high molecular weight form of AChE are dependent on the presence of the neuronal cells. Other systems will be discussed in the next section.

Uninnervated chick muscle cells in culture – even in the myoblast stage – synthesize and secrete AChE. Histochemical studies (Engel, 1961; Rubin et al., 1979) reveal rather uniform distribution of reactive sites in these cultures, so AChE appears to differ from the AChRs, some of which are present in clusters in uninnervated cells. Sucrose density gradient analysis reveals that these cells have two main forms of AChE sedimenting at 7 S and 11 S (Rotundo and Fambrough, 1979; Rubin et al., 1980). The 7 S form has been reported to be a doublet of the globular catalytic subunits, while the 11 S is thought to be a tetramer (reviewed in Massoulié et al., 1980). At least a portion of the AChE appears to be plasma membrane bound and externally disposed since it can hydrolyze ACh added to the culture medium (Rotundo and Fambrough, 1980). The forms of the enzyme secreted by myotubes in culture are also of the lower molecular weight types and were reported by Rotundo and Fambrough (1979) to sediment at 9 S and 15 S. Spinal cord neurons, cultured in the absence of muscle, also synthesize and secrete AChE into the culture medium (Oh et al., 1977).

A high molecular weight form, sedimenting at 19.5 S, is either undetectable (Rotundo and Fambrough, 1979; Rubin et al., 1980) or present in small amounts (Kato et al., 1980) in uninnervated cultures. This form consists of 12 catalytic subunits and a collagen-like tail and is thought to be associated with the synaptic basal lamina (Massoulié et al., 1980). Whether culture or extraction conditions account for differences in the levels of this form measured in various laboratories has not yet been studied in detail.

In an early study, Fischbach et al. (1974) found no evidence for localization of AChE at synapses formed between dissociated spinal cord neurons and myotubes. Their histochemical experiments revealed no patches of reaction product along nerve–muscle contacts. Further, the time course of endplate potentials recorded intracellularly was not affected by application of the cholinesterase inhibitor eserine.

A later study by Rubin et al. (1979) reached different conclusions. Myotubes were innervated by neurites stemming from explants of 14-day spinal cords. Starting about 12–24 h after the initial physiological indication of synapse formation, synapses were identified by extracellular recording. One immediate finding was that there appeared to be three classes of synapses, based on the rate of decay of extracellular synaptic potentials: fast (average rate of decay, τ, at 30 °C, of 1–1.8 ms, a time similar to the average channel open time of AChRs in these cultures – Schuetze et al., 1978), intermediate (1.8 ms $< \tau <$ 2.6 ms) and slow ($\tau >$ 2.6 ms). As mentioned previously, fluctuation analysis demonstrated that AChR channel open times were indistinguishable at all types of synapses, all being of the extrajunctional variety (staying open for a relatively long time once activated). Once synapses were identified and characterized, cultures were fixed and stained histochemically for AChE. Approximately 75% of the fast synapses stained for AChE, whereas none of the slow ones stained. Moreover, application of the cholinesterase inhibitor methanesulfonyl fluoride prolonged the decay at fast synapses, but had no effect at slow synapses. Thus, AChE appears at many of these synapses and functions, as it does at adult neuromuscular junctions, to limit the time course of ACh action in the synaptic cleft. The biochemical correlate of these experiments was the presence of a small amount of 19.5 S AChE in nerve–muscle co-cultures where none was found in neuron-free muscle cultures (Rubin et al., 1980).

Another outcome of these experiments was the development of a physiological assay for AChE at synapses: rapid rates of decay are often associated with the presence of AChE, while slow rates of decay are associated with AChE-deficient synapses. An application of the physiological assay will be described in Section 3.4.3. (1).

What accounts for the different conclusions reached in the studies by Fischbach et al. (1974) and Rubin et al. (1979)? There were technical differences in the two studies – the histochemistry was carried out slightly differently and the earlier study relied on intracellular rather than extracellular recording. Also, Fischbach et al. utilized eserine as an anticholinesterase; Rubin et al. found that this drug produced inconsistent results, apparently because it binds to both AChRs and AChE. Finally, the earlier experiments utilized dissociated spinal cords and the latter used spinal cord explants. Muscle fibers innervated by neurons in explants are normally much more active. This difference in activity might have been, in fact, the crucial difference between the two studies (see Section 3.4.3 (1)).

3.3.2. Other types of cultures
(1) Rat. Koenig and Vigny (1978) found that when myoblasts were removed from 13–14 day embryos, a time prior to that of motor innervation, muscle cells in culture did not develop any high molecular weight (16 S in this case) AChE, possessing only the lower molecular weight 4 S and 10 S forms. Addition of spinal cord neurons led eventually to the appearance of a small quantity of 16 S enzyme.

Whether AChE was actually localized at nerve–muscle synapses in those cultures was not determined, though.

In these aspects, results with rat cultures appear similar to those from chick. A difference appeared if rat myoblasts were taken from 18-day embryos, i.e., past the period of normal motor innervation and AChE localization. These myoblasts and myotubes derived from them were themselves able to synthesize 16 S AChE. This is quite different from chick, in which even myotubes derived from 20-day embryonic myoblasts, are unable, at least in our hands, to synthesize heavy AChE (Rubin, unpublished results). The results of Koenig and Vigny also suggest that there are protein synthetic differences between myoblasts of older and younger embryos and that these differences are maintained in culture, perhaps even after several stages of cell division. Further, these workers advanced the hypothesis that these differences are triggered by the presence of innervating fibers (not even by innervation itself since the myoblasts presumably remain uninnervated). Especially if not restricted to AChE, such a finding would be of profound interest.

(2) Xenopus. Histochemical staining of *Xenopus* myoblasts cultured without neurons showed that most of these cells have clear patches of AChE reaction product (Weldon et al., 1981). Such patches appear to be associated with invaginations of the cell surface, a thickened plasma membrane and extracellular material resembling basal lamina. In addition, these patches are often associated with high concentrations of AChRs detected by fluorescent α-bungarotoxin (Moody-Corbett and Cohen, 1981). When these cells are innervated, preexisting patches of both AChR and AChE are seen to become redirected to sites of nerve muscle contact (Moody-Corbett et al., 1981). It is not yet known which molecular forms of AChE exist in innervated or uninnervated *Xenopus* cultures.

(3) Mouse C2 cells. The mouse muscle cell line C2 was derived initially from adult mouse skeletal muscle (Yaffe and Saxel, 1977). The C2 myoblasts fuse when the concentration of serum in the tissue culture medium is decreased. Myotubes grown in the absence of neurons synthesize relatively large amounts of the 16 S form of AChE (Inestrosa and Hall, 1981; Silberstein et al., 1982). Again the suggestion was advanced that the adult origin of these myoblasts might have been of some importance in explaining the ability of aneural cultures to synthesize this form of AChE and also other molecules which could be identified by immunological techniques and are normally present in adult junctional basal lamina.

3.4. Regulatory events in nerve–muscle synapse formation: dependence on synaptic activity

3.4.1. Onset of synaptic transmission

Quite clearly, synapses between neurons and muscle cells form whether or not synaptic transmission is permitted. If either *Xenopus* or chick nerve–muscle cultures are grown for several days in the continual presence of a high concentration of d-tubocurarine to block ACh binding to its receptors, thereby blocking all synaptic activity, synaptic potentials can be observed in abundance within seconds of drug washout (Cohen, 1972; Rubin et al., 1980). Of course, it is theoretically possible, though highly unlikely, that nerve-muscle contacts stand poised until curare washout and that the brief period of permitted activity triggers the actual formation of synapses. Synapses also form if cultures are grown continually in tetrodotoxin (TTX) levels sufficient to block all propagated activity, but not MEPPs. Absolutely unambiguous in this case is the evidence for synapse formation since MEPPs can be recorded while cultures are still in TTX (Obata, 1977; Rubin et al., 1980); then propagated activity can be recorded within minutes after TTX washout. Thus, neither any activity in the muscle fiber, nor any propagated electrical activity in the neurons, is necessary for synapse formation.

3.4.2 AChR clustering at synapses

Is synaptic transmission involved in the redistribution of AChRs which occurs during synapse formation? A related question – namely, what is the basis of denervation supersensitivity? – has been for some time a central one in neurobiology. It is well known that when the motor nerve of an adult vertebrate muscle is cut, extrajunctional AChRs appear, decreasing the concentration gradient between junctional and extrajunctional regions (Berg et al., 1972; Hartzell and Fambrough, 1972; Brockes and Hall, 1975; Devreotes and Fambrough, 1976). Is it synaptic activity which was responsible for maintaining the concentration gradient or is the motor nerve a source of 'trophic' molecules necessary for maintaining the gradient? In the case of adult denervated muscle, junctional AChR levels actually undergo relatively little change in density (Frank et al., 1975); thus, synaptic activity apparently has little influence on these levels, at least once they've been established. In fact, molecules in the muscle basal lamina may be responsible for maintaining junctional AChR levels (Burden et al., 1979; McMahan et al., 1980). On the other hand, muscle activity clearly does exert a suppressive effect on the number of extrajunctional AChRs (Lomo and Rosenthal, 1972; Lomo and Westgaard, 1975).

In spite of the finding that synaptic activity is not required for maintenance of junctional AChR levels, it conceivably still could be required for inducing the initial localization of synaptic AChRs. This appears not to be the case, though. *Xenopus* cultures grown in the presence of d-tubocurarine sufficient to block synaptic activity still undergo AChR redistribution, as visualized with fluorescent α-bungarotoxin, yielding high AChR levels along sites of nerve-muscle contact (Anderson and Cohen, 1977). Chick cultures grown continually either in curare or in TTX have obvious AChR clusters at physiologically identified synapses (Rubin et al., 1980). Finally, Ziskind-Conhaim and co-workers made use of an elaborate organ culture system consisting of muscle with attached motor nerve and spinal cord segments dissected from 15-day rat embryos. AChR aggregation takes place in this system within 1–2 days of explantation whether or not cultures are grown in TTX (Ziskind-Conhaim and Dennis, 1981; Ziskind-Conhaim and Bennett, 1982). Thus, initial localization of synaptic AChRs and maintenance of these levels in denervated muscle are both independent of synaptic activity.

There may also be an *in vitro* counterpart to the regulation of extrajunctional AChRs by muscle activity. TTX and lidocaine, a local anesthetic, both increase overall AChR levels in uninnervated cultures, possibly by suppressing the small amount of spontaneous contractile activity which occurs in these cultures (Cohen and Fischbach, 1973; Shainberg et al., 1976). Conversely, direct stimulation of these uninnervated cells decreases AChR numbers.

3.4.3. AChE appearance at newly formed synapses
(1) Chick nerve–muscle cultures. Since there is a definite time lag of 12–24 h between the onset of synaptic transmission and the appearance of AChE at synapses (Rubin et al., 1979), a role for synaptic activity in this process appeared quite possible. Rubin et al. (1980) investigated this possibility in chick nerve–muscle cultures. They grew spinal cord explants from 14-day embryos with myotubes for 4 days. Normally, this is sufficient time to allow AChE to appear at approximately 75% of all synapses by histochemical criteria. If, instead, co-cultures were grown in the continual presence of d-tubocurarine, less than 2% of synapses stained for AChE (Rubin et al., 1980). Yet, all of these synapses had, as mentioned previously, obvious AChR clusters. Synapses formed in curare were AChE-deficient by physiological criteria also, the great majority of them having slowly decaying extracellularly recorded synaptic potentials. Approximately 24 h after curare washout, AChE was then detectable at the majority of synapses. Thus, AChE localization at synapses in chick cultures, like the appearance of extrajunctional AChRs in denervated muscle and the accumulation of AChE at ectopic sites in mammalian muscle (Lomo and Slater, 1980), is sensitive to the level of synaptic activity.

To provide a biochemical correlate of these experiments, Rubin et al. (1980) cultured neurons and muscle cells in the presence of curare. They found that total AChE activity was essentially unchanged, but that the appearance of 19.5 S AChE was suppressed. Such an effect of activity on the appearance of heavy AChE is supported by the results of Weinberg and Hall (1979), who studied AChE appearance in rat muscle innervated ectopically.

In a further series of experiments (Rubin et al., 1980), the components of synaptic activity important in inducing synaptic AChE localization were studied. These experiments with curare had simply demonstrated that an event including or subsequent to ACh binding to AChRs was necessary before AChE would appear. If cultures were grown instead in the local anesthetic SKF/525A (proadifen · HCl), which blocks activation of ACh channels without affecting ACh binding to AChRs, AChE did not localize to synapses. Similarly, growing cultures in TTX to block propagated electrical activity, but not MEPPs, was also effective in preventing synaptic AChE appearance; MEPPs alone obviously provided an insufficient level of activity. Further, if cultures were grown in low levels of curare to suppress the size of endplate potentials below that needed to produce muscle action potentials and contraction, AChE appearance was still prevented. These experiments suggested a role for muscle contraction itself.

For a direct examination of this possibility, cultures were grown in high concentrations of curare which was briefly washed out to allow identification and physiological characterization by extracellular recording of a few AChE-deficient synapses on individual myotubes. Curare was replaced in the culture medium to again prevent any synaptic transmission. Myotubes with identified synapses were then stimulated directly for 8–12 h by passing current through an extracellular pipet positioned above the myotube. After the stimulation period, curare was washed out, and the previously identified synapses were again characterized physiologically. These sites now had rapidly decaying extracellular synaptic potentials and were, therefore, by physiological criteria AChE-positive (note that these experiments rely heavily on the previously established nondisruptive physiological assay). Muscle stimulation, in the absence of synaptic transmission, is therefore sufficient to induce the appearance of synaptic AChE. These results are consistent with those of Lomo and Slater (1980) who studied AChE localization at ectopic mammalian synapses. Also important in this regard is the earlier finding of Walker and Wilson (1975) that direct stimulation of uninnervated chick muscle cultures actually decreases total AChE, so such stimulation clearly does not act to induce synaptic AChE by an overall increase in enzyme activity.

(2) Rat muscle cultures. In a related study, Rieger et al. (1980) studied the activity dependence of 16 S AChE appearance in rat embryo muscle cultures. These cells, when derived from older embryos, synthesize this form of the enzyme even when uninnervated. Yet, if these cultures were grown in the presence of tetrodotoxin to block spontaneous contractile activity, synthesis of this form of the enzyme was significantly decreased. The activity-dependence of collagen-tailed AChE in neural rat muscle cultures is similar to that seen in chick nerve–muscle preparations.

(3) Xenopus and C2 cultures. In *Xenopus* muscle cultures, AChE patches are present on uninnervated cells, but become localized along sites of nerve–muscle contact following innervation (Moody-Corbett et al., 1981). This localization occurs even if the cells are grown in curare to block synaptic transmission. Similarly, the existence of the 16 S form of AChE in uninnervated mouse C2 muscle cells is independent of activity, occurring even if cells are grown in TTX to block spontaneous contractile activity (Inestrosa and Hall, 1981). Thus, both the existence of AChE patches in uninnervated cells and the activity-independent nature of AChE localization at nerve–muscle contacts are in contrast with results in chick cultures.

3.5. Molecular mechanisms underlying accumulation of AChE and AChRs at synapses

In this section we review studies designed to identify the molecules involved in determining the activity-dependent and activity-independent aspects of AChR clustering and AChE appearance at synapses. It is important to emphasize at the outset that the changes produced in cells by trophic factors or by activity might, in some cases, be identical or redundant. Thus, apparent differences in regulatory mechanisms underlying, say, the activity dependence of AChE localization at synapses in chick cultures, but its independence of activity in *Xenopus* cultures, might be reconciled.

3.5.1. Activity-independent events
(1) AChR clustering. *(a) AChR clusters in uninnervated cells.* Most studies attempting to determine the mechanism of AChR clustering at synapses involve very basic assumptions about the nature of these mechanisms. Factors – some perhaps very similar, others certainly not – derived from many sources have been shown to alter AChR number or distribution on cultured muscle cells. Only much experimentation will identify those molecules truly involved in AChR localization in vivo. At this point, it is not even obvious that all these factors must induce AChR clustering by identical mechanisms.

Uninnervated cultured muscle fibers from a variety of sources have preexisting AChR clusters and have provided a convenient model system for investigators interested in the mechanisms of AChR clustering. One cautionary note is that extrajunctional clusters have not always been seen on normal uninnervated embryonic muscle fibers (e.g., Burden, 1977a), although there is at least one report of their existence (Jacob and Lentz, 1979). They are also seen on denervated adult fibers (Ko et al., 1977) and on newly innervated embryonic rat muscle fibers maintained in organ culture in the presence of TTX (Ziskind-Conhaim and Bennett, 1982). There is a possibility that the cellular mechanisms involved in organizing or in specifying the location of extrajunctional clusters are quite different from those related to junctional clusters.

One common observation in rat (Axelrod et al., 1976; Bloch and Geiger, 1980) and *Xenopus* (Anderson et al., 1977; Moody-Corbett and Cohen, 1981) muscle cultures is that more AChR clusters form on the bottom surface of cells – in contact with the culture surface – than on the top. This is particularly impressive in *Xenopus* cultures where AChR clusters, AChE, basal lamina and membrane invaginations can be seen at such sites of contact (Moody-Corbett and Cohen, 1981; Weldon and Cohen, 1979). These results have introduced the possibility that the nerve ending simply provides a type of physical contact which can be supplied in vitro by the culture dish.

Bloch and Geiger (1980) studied in some detail AChR clusters in rat muscle cultures. Using interference reflection microscopy together with fluorescent α-bungarotoxin, they determined that AChR clusters on the underside of cells are associated with areas of close contact with the surface. Furthermore, they discovered that vinculin, a protein of molecular weight 130,000 isolated originally from chicken gizzard and present in high concentrations at cellular adhesion points in cultured cells, also often co-distributed with these AChR clusters. When vinculin becomes localized at AChR clusters is not yet known, though, nor is its function in the existence of these clusters.

(b) Maintenance of AChR clusters in uninnervated cells. Several investigators have attempted to study cellular mechanisms underlying maintenance of AChR clusters in uninnervated cells. Prives and collaborators (1980) identified two pools of AChRs with respect to detergent extractability: those rapidly extracted and those relatively slowly extracted. They suggest that the slowly extracted molecules are clustered AChRs associated with the cytoskeleton, also relatively insoluble in detergent. They further found that increasing the number of AChR clusters by treating these cultures with conditioned medium obtained from a neuroblastoma-glioma hybrid also increased the percentage of AChRs in the slowly extractable pool. The hypothesis of these experiments, therefore, is that AChRs are maintained as a cluster via molecular linkage to cytoplasmic filamentous molecules.

Bloch (1979) extensively studied the effects of a variety of agents on the dispersal of clustered AChRs. He found that treating rat myotubes with drugs to inhibit energy production – azide, for example – would effectively break up clusters. Agents which were ineffective included colchicine and cytochalasin B. If azide was washed out, clusters reformed, but not necessarily at their original sites. Cluster reformation was inhibited by colchicine, however, and Bloch also suggested a possible cytoskeletal role in AChR aggregation.

(c) Effect of exogenous neuronally derived factors on AChR aggregation. Since AChR localization at synapses is obviously independent of nerve–muscle activity, one strong possibility is that a molecule secreted from nerve terminals (in a manner independent of electrical activity) serves as a signaling mechanism for the muscle cell. This provides a relatively well-defined example of what has been

traditionally referred to as a trophic interaction.

One somewhat early, but graphic, observation of the effects of neuronally-derived factors was made by Cohen and Fischbach (1977) in their study of synapses formed between neurons of chick spinal cord explants and muscle cells. They noticed that muscle fibers around spinal cord explants had a higher extra-junctional AChR density than did muscle fibers far away from the explants. This suggested that, in addition to causing AChR localization at synapses, cells in the explant were able to exert an overall effect on AChR number in nearby fibers. Several groups have now made similar observations and have attempted to purify active AChR-clustering factors.

Podleski et al. (1978) studied the effects of proximity to rat spinal cord and of spinal cord-derived cell-free factors on AChRs on L6 muscle cells. They found, as did Fischbach and Cohen (1977), that muscle fibers near spinal cord explants had, after approximately 3 days, an increase in AChR number and AChR clusters. They then homogenized spinal cords in physiological saline, spun down membranes, and treated cultures with the soluble material. Such cell extracts at 300–600 μg/day also caused, after 2–3 days, an increase in AChR number and aggregates. Preliminary characterization suggested that the active factor was a trypsin-sensitive protein with molecular weight about 100,000.

Christian et al. (1978) studied the effects of medium conditioned by clonal neuroblastoma–glioma hybrids (NG 108-15), which synthesize and release ACh and can form synapses, on AChRs of cultured mouse muscle cells. After approximately 24 h – although a more recent report places the time for maximal effect at 2 h (Prives et al., 1980), a time more consistent with AChR accumulation during synapse formation – a significant increase in the number of AChR clusters was observed, with little change in the total number of AChRs in cell homogenates. Numbers of cell surface AChRs were not determined. AChR clustering activity was found also in medium conditioned by undifferentiated NG 108-15 cells, as well as by the parent neuroblastoma clone. When AChRs were labeled prior to addition of conditioned medium, an increase in numbers of clusters was still seen, indicating that the medium factors can cause a redistribution of preexisting surface AChRs even in the absence of protein synthesis. Prives et al. (1980) also found that 10–12 h after conditioned medium addition, there was an increase in the proportion of AChRs in the pool which was relatively insoluble in nonionic detergent and suggested that the clustering factor caused increased association between AChRs and the cytoskeleton. A preliminary characterization of the factor indicated that it is a heat-sensitive non-dialyzable molecule.

Jessell et al. (1979) studied the effects of cell-free extracts obtained by homogenizing chick embryo brain and spinal cord on AChRs of cultured chick myotubes. After approximately 4 days of treatment with 300 μg/day, there was a 3- or more fold increase in AChR number, but approximately a 40-fold increase in the number of clusters. Extracts had no effect on the AChR degradation rate.

These extracts also significantly increased the AChE activity of the muscle cultures and stimulated the rate of fibroblast division. Whether these different effects reflect the activities of separate components of the crude extracts is not known. Attempts at characterising the AChR-clustering factors revealed some high molecular weight activity, but also substantial very low molecular weight activity (1,000 or less; G.D. Fischbach, personal communication). This low molecular weight activity was trypsin-sensitive, suggesting that it is a small peptide.

Markelonis et al. (1982) examined the effects of sciatin, an 84,000 dalton protein which they isolated from citrate-soluble extracts of frozen chicken sciatic nerves, on AChRs of cultured chick embryo muscle. They found that sciatin produced a small, transient increase in the level of AChRs after a few days. The number of AChR clusters was also increased.

What conclusions are possible from these studies which are simply representative of the increasingly large numbers of experiments directed at identifying neural factors which localize AChRs? The factor isolated by Podleski et al. (1979) appears to have a high molecular weight and increases both the number of AChRs and the number of clusters. The one studied by Christian et al. (1978) is of undetermined molecular weight (although it is presumably larger than 10–15,000) and acts primarily by increasing cluster frequency. Jessell et al. (1979) found activity with quite low molecular weight which increases both AChR number and cluster frequency, with a larger effect on the latter. Primarily, it is now possible to ask more detailed sorts of questions. Are these factors related in any way? Is the low molecular weight activity derived from the larger molecules? Do the differences in effects reflect differences among the factors or among the postsynaptic cells? Nonetheless, studies of this type should eventually contribute substantially to our understanding of the mode of action of neuronally derived molecules which interact with AChRs.

(d) Extracellular antigens and AChR distribution. The work of McMahan and collaborators (Burden et al., 1979; McMahan et al., 1980) established a role for basal lamina components in the localization of AChRs in regenerating muscle. When the basal lamina appears with respect to synapse formation and AChR accumulation is not known, though. It could be that extracellular material appears prior to AChR accumulation and that the same basal lamina cluster-inducing molecules function in regenerating muscle and at newly formed synapses. On the other hand, it could be that different molecules function in the two situations or that fundamentally the same molecules function in both cases, but do not at first become localized in the basal lamina. Two types of studies have been used to examine extracellular matrix-muscle fiber interaction. The first has been directed at identifying AChR-localizing components of adult basal lamina using in vitro systems. The second has looked for co-distribution of immunologically identified extracellular matrix antigens and AChRs in developing myotubes.

(2) Extracellular material from Torpedo *electric organ.* The experiments which demonstrated the role of muscle basal lamina in AChR aggregation were carried out using regenerating cutaneous pectoris muscles of frog. For a detailed study of the molecular nature of basal lamina components and their mode of action, this preparation would prove terribly cumbersome. The time course of muscle regeneration is slow, occupying 2 or more weeks. Furthermore, detailed biochemical and cell biological experiments would be difficult to carry out. Thus, an in vitro model system offering the potential for interactions between muscle and basal lamina was required.

Rubin and McMahan (1982) carried out a series of experiments demonstrating that AChR distribution on cultured chick myotubes could be altered by a cell fraction enriched in extracellular material (ECM). As a source of ECM, they chose *Torpedo* electric organ which is, compared to vertebrate skeletal muscle, at least 100-fold enriched in synaptic area. A particulate fraction containing ECM was prepared from *Torpedo* electric organ by a modification of the procedures of Meezan et al. (1975), relying primarily on the relative detergent insolubility of ECM. The fraction consisted of amorphous particles ranging from 1–10 μm or so in diameter and was seen, by ultrastructural examination, to be free of plasma membrane, but to contain collagen fibrils. Aliquots of this fraction were added to cultures of chick myotubes whose surface AChRs had been labeled previously with ^{125}I-α-bungarotoxin. Approximately 2 h later, but becoming particularly noticeable after 8 h, an extensive redistribution of AChRs took place, resulting in a large increase in the number of clusters. Such a time period is consistent with that required for AChRs to localize at newly formed synapses (Frank and Fischbach, 1979). The active component of the particulate fraction is presumably a protein, being heat, trypsin and dithiothreitol sensitive. Fractions prepared using similar techniques from either rat or *Torpedo* striated muscle were much less effective in inducing AChR aggregation.

To permit further characterization of the active component, its solubilization was required. This was accomplished by extracting the ECM fraction with 2 M $MgCl_2$, a procedure used also to solubilize collagen-tailed AChE from *Torpedo* electric organ basal lamina. The high salt extraction reduced the clustering activity in ECM, but the high salt extract was then active. When this extract was dialyzed against physiological saline, much of the protein precipitated, but a substantial proportion of AChR clustering activity remained soluble. This soluble fraction also exerted its effects rapidly and without producing significant overall increases in surface AChR levels. The active component in this fraction appears to be a glycoprotein (which binds to wheat germ agglutinin affinity columns; Lappin and Rubin, unpublished results) with apparent molecular weight greater than 50,000. That these soluble molecules are able to induce AChR aggregation implies that they are multivalent and, hence, able to cross-link membrane components – AChRs themselves or perhaps other molecules. Again, the relationship of these molecules, localized originally in detergent insoluble particulate fractions of *Torpedo*, to the soluble molecules of neuronal origin previously discussed is not known.

(3) Extracellular antigens and AChR clusters. Another type of approach has involved the use of primarily immunological procedures to study co-localization of extracellular matrix components and AChR clusters. Many of these studies have utilized uninnervated cultured muscle cells.

Burrage and Lentz (1981) looked ultrastructurally for co-localization of extracellular material and AChR clusters identified using horseradish peroxidase conjugated to α-bungarotoxin in chick myotubes. They found a good correlation between AChR patches and dense plasma membrane and extracellular material. Of course, it is difficult to know the temporal relationship between the two, i.e., whether or not the extracellular material appears prior to AChR clusters and might have specified their location.

Silberstein et al. (1982) studied the development of extracellular matrix antigens in uninnervated cells of the mouse C2 cell line. The antigens they studied were labeled with three different rabbit antisera previously found by Sanes and Hall (1979) to identify molecules localized to the synaptic region of adult muscle. The antisera were raised against: (1) a soluble fraction of anterior lens capsule, (2) collagen from muscle basement membrane and (3) collagen from lens capsule. All appeared on the surface of C2 cells after fusion in culture, and the first two often appeared in patches which co-localized with AChR clusters.

Anderson and Fambrough (1981) prepared monoclonal antibodies against basal lamina from adult *Xenopus laevis*. One antibody recognized determinants present all along the muscle surface, but concentrated at the neuromuscular junction. On cultured *Xenopus* muscle cells, this antibody produced a diffuse labeling pattern with occasional patches which corresponded to AChR cluster. With innervation, both the extracellular matrix antigen and the AChRs localized in high concentrations along the path of the nerve. Similarly, Bayne et al. (1980) used monoclonal antibodies to study an extracellular matrix antigen from chicken muscle which was associated with AChR clusters.

3.5.2. Activity-dependent events
(1) Regulation of synaptic AChE appearance. The assembly and localization of synaptic AChE in chick cultures requires synaptic transmission, specifically muscle contraction (although in the presence of presynaptic nerve terminals). What intracellular events in the muscle cell mediate the effects of contractile activity?

Rubin et al. (1980) grew chick nerve–muscle cultures in curare to block all synaptic activity, thereby producing inactive myotubes. They then tried to induce the appearance of synaptic AChE in innervated, but inactive, cells by treating the cultures with drugs intended to produce specific changes in the cells. Passive depolarization caused by treatment with elevated K^+ or with veratridine, increased Ca^{2+} influx produced by the Ca^{2+} ionophore A23187 or elevated extracellular cyclic AMP levels caused by treating cultures with dibutyryl cyclic AMP; all failed to induce the appearance of the enzyme, by physiological or histochemical criteria. On the other hand, treating the cells with 10–100 μm dibutyryl cyclic GMP was effective in stimulating AChE appearance at synapses, as indicated by both histochemical and physiological assays and also caused assembly of 19.5 S AChE. These results suggest that increases in muscle cyclic GMP levels might be an intermediate essential step between muscle contraction and AChE appearance. In fact, both direct and indirect muscle stimulation have been shown to produce increases in muscle cyclic GMP levels, but not in muscle cyclic AMP (Nestler et al., 1978).

(2) Regulation of 'extrajunctional' AChR levels Enhanced muscle contractile activity is known to decrease, and TTX suppression of contractile activity to increase, overall AChR levels in cultured myotubes (Cohen and Fischbach, 1973; Shainberg and Burstein, 1976). The level of extrajunctional AChRs in denervated adult muscle displays a similar sensitivity to activity; hence, cultured myotubes have been used as a convenient model system in experiments aimed at identifying the molecular basis for the effects of muscle activity.

Betz and Changeux (1979) studied AChR number in myotubes treated with cyclic nucleotides and depolarizing agents. They found that this number was decreased by treatment with veratridine to stimulate ionic flux through Na^+ channels and with dibutyryl cyclic GMP and was increased by dibutyryl cyclic AMP. The effects of increased cyclic GMP appeared to be due to a decreased level of membrane AChR incorporation, rather than to an enhanced rate of degradation. Furthermore, the AChR enhancing effect of TTX was not produced if cultures were simultaneously treated with dibutyryl cyclic GMP. The suggestion was made that the sensitivity of extrajunctional AChRs to muscle concentration might be exerted via a depolarization-induced increase in muscle cyclic GMP. This study and that of Rubin et al. (1980) further imply a common intermediate in the effects of muscle activity on the appearance of synaptic AChE and on suppression of extrajunctional AChRs.

Betz and Changeux also suggested that the effects of muscle contraction might be exerted via increased cytoplasmic Ca^{2+}. Birnbaum et al. (1980) examined the effects on AChR levels of several pharmacological treatments thought to alter muscle Ca^{2+} levels. They suggested that certain agents – the Ca^{2+} ionophore A23187 and sodium dantrolene – which increased sarcoplasmic reticulum Ca^{2+} produced increases in surface AChR levels. Those treatments – muscle stimulation and caffeine – which caused Ca^{2+} release from sacroplasmic reticulum produced a decrease in AChR synthesis. McManaman et al. (1981) found that growing rat muscle culture in medium containing lowered Ca^{2+} caused a decrease in AChR levels, again suggesting that decreased intracellular Ca^{2+} is correlated with a decreased number of AChRs.

Conclusions

In vitro experiments have already proven quite useful in studying muscle development and nerve–muscle interactions and will undoubtedly be even more valuable in the future. Most of these studies appear to complement in vivo ones rather nicely. Experiments in tissue culture demonstrated clearly that myotubes are formed by fusion of the mononucleated myoblasts. How these cells recognize one another, align and fuse is certainly not completely clear, nor is it understood how fusion changes the protein synthetic pattern of the cells. Application of monoclonal antibody and recombinant DNA techniques to cultured cells will soon contribute to our appreciation of these processes.

Nerve–muscle synapse formation has also been studied in detail. A fairly complete description of its sequence of events has been provided. The types of cellular interactions underlying the major steps have been classified, at least on a gross level, in terms of the necessity for electrical activity. Thus, it has become fairly clear that certain stages of synaptogenesis rely strongly on such activity, while other stages occur even in its absence. Many current studies are devoted to an examination of the molecular changes which are involved in each stage. It seems most likely that in the next few years these molecular studies will progress rapidly. Even the once vague notion of trophic interactions between neurons and muscle cells can now be approached in a defined experimental way. We think it quite realistic to expect that many aspects of synaptogenesis will be understood in some detail within a few years.

References

Anderson, M.J. and Cohen, M.W. (1977) J. Physiol. *268*, 757–773.
Anderson, M.J., Cohen, M.W. and Zorychta, E. (1977) J. Physiol. (London) *268*, 731–756.
Anderson, M.J. and Fambrough, D.M. (1981) Abstr. Soc. Neurosci. *7*, 670.
Anderson, M.J., Kidokoro, Y. and Gruener, R. (1979) Brain Res. *166*, 185–190.
Axelrod, D. (1980) Proc. Natl. Acad. Sci. USA *77*, 4823–4827.
Axelrod, D., Ravdin, P., Koppel, D.E., Schlessinger, J., Webb, W.W., Elson, E.L. and Podleski, T.R. (1976) Proc. Natl. Acad. Sci. USA *73*, 4594–4598.
Ball, E.H. and Sanwal, B.D. (1980) J. Cell Physiol. *102*, 27–36.
Bandman, E., Walker, C.R. and Strohman, R.C. (1978) Science *200*, 559–561.
Barbera, A.J., Marchase, R.B. and Roth, S. (1973) Proc. Natl. Acad. Sci. USA *70*, 2482–2486.
Bayne, E.K., Wakshulli, E., Gardner, J. and Fambrough, D.M. (1980) J. Cell Biol. *87*, 260a.
Bekoff, A. and Betz, W.J. (1976) Science *193*, 915–917.

Berg, D.K. and Hall, Z.W. (1975) J. Physiol. (London) 252, 771–789.
Berg, D.K., Kelly, R.B., Sargent, P.B., Williamson, P. and Hall, Z.W. (1972) Proc. Natl. Acad. Sci. USA 69, 147–151.
Betz, H. and Changeux, J-P. (1979) Nature (London) 278, 749–752.
Betz, W. (1976) J. Physiol. (London) 54, 63–73.
Birnbaum, M., Reis, M.A. and Shainberg, A. (1980) Pfluegers Arch. 385, 37–43.
Bischoff, R. and Holtzer, H. (1967) J. Cell Biol. 36, 111–127.
Blau, H.M. and Epstein, C.J. (1979) Cell 17, 95–108.
Blau, H.M. and Webster, C. (1981) Proc. Natl. Acad. Sci. USA 78, 5623–5627.
Bloch, R.J. (1979) J. Cell Biol. 82, 626–643.
Bloch, R.J. and Geiger, B. (1980) J. Cell Biol. 21, 25–35.
Bonner, P.H. (1978) Dev. Biol. 66, 207–219.
Brockes, J.P. and Hall, Z.W. (1975) Proc. Natl. Acad. Sci. USA 72, 1368–1372.
Brunk, C.F. (1979) Differentiation 14, 95–99.
Buckley, P.A. and Konigsberg, I.R. (1974) Dev. Biol. 37, 193–212.
Buckley, P.A. and Konigsberg, I.R. (1977) Proc. Natl. Acad. Sci. USA 74, 2031–2035.
Burden, S. (1977a) Dev. Biol. 57, 317–329.
Burden, S. (1977b) Dev. Biol. 61, 79–85.
Burden. S.J., Sargent, P.B. and McMahan, U.J. (1979) J. Cell Biol. 2, 412–425.
Burrage, T.G. and Lentz, T.L. (1981) Dev. Biol. 85, 267–286.
Burstein, M. and Shainberg, A. (1979) FEBS Lett. 103, 33–37.
Bursztajn, S. and Fischbach, G.D. (1980) Abstr. Soc. Neurosci. 6, 358.
Busis, N.A., Daniels, M.P., Bauer, H.C., Sonderegger, P., Schaffner, A.E. and Nirenberg, M. (1981) Abstr. Soc. Neurosci. 7, 666.
Carmon, Y., Neuman, S. and Yaffe, D. (1978) Cell 14, 393–401.
Chang, C.C. and Huang, M.C. (1975) Nature (London) 253, 643–644.
Chen, L.B. (1977) Cell 10, 393–400.
Chiquet, M., Puri, E.C. and Turner, D.C. (1979) J. Biol. Chem. 254, 5475–5482.
Christian, C.N., Daniels, M.P., Sugiyama, H., Vogel, Z., Jacques, L. and Nelson, P.G. (1978) Proc. Natl. Acad. Sci. USA 75, 4011–4015.
Cohen, M.W. (1972) Brain Res. 413, 457–463.
Cohen, M.W. and Weldon, P.R. (1980) J. Cell Biol. 86, 388–401.
Cohen, S.A. (1980) Proc. Natl. Acad. Sci. USA 77, 644–648.
Cohen, S.A. and Fischbach, G.D. (1973) Science 181, 76–78.
Cohen, S.A. and Fischbach, G.D. (1977) Dev. Biol. 59, 24–38.
Cornell, R.B. and Horwitz, A.F. (1980) J. Cell Biol. 86, 810–819.
Cornell, R.B., Nissley, S.M. and Horwitz, A.F. (1980) J. Cell Biol. 86, 820–824.
Crain, S.M. and Peterson, E.R. (1974) Ann. N.Y. Acad. Sci. 228, 6–34.
Crain, S.M., Alfei, L. and Peterson, E.R. (1970) J. Neurobiol. 1, 471–489.
Croop, J. and Holtzer, H. (1975) J. Cell Biol. 65, 271–285.
Curtis, D.A., Micklem, K.J., Gill, K. and Pasternak, C.A. (1980) Biochem. J. 190, 647–652.
Dahl, G., Schudt, C. and Gratzl, M. (1978) Biochim. Biophys. Acta 514, 105–116.
Delain, D. and Wahrmann, J.P. (1975) Exp. Cell Res. 93, 495–498.
Den, H., Malinzak, D.A., Keating, H.J. and Rosenberg, A. (1975) J. Cell Biol. 67, 826–834.
Dennis, M.J., Ziskind-Conhaim, L. and Harris, A.J. (1981) Dev. Biol. 81, 266–279.
Devlin, R.B. and Emerson, C.P. (1979) Dev. Biol. 69, 202–216.
Devreotes, P.N. and Fambrough, D.M. (1975) J. Cell Biol. 65, 335–358.
Devreotes, P.N. and Fambrough, D.M. (1976) Proc. Natl. Acad. Sci. 73, 161–164.
Doering, J.L. and Fischman, D.A. (1974) Dev. Biol. 36, 225–235.
Dreyer, F., Walther, C. and Peper, K. (1976) Pfleugers Arch. 366, 1–9.

Dribin, L.B. and Barrett, J.N. (1980) Dev. Biol. 74, 184–195.
Dym, H., Turner, D., Eppenberger, H.M. and Yaffe, D. (1978) Exp. Cell Res. 113, 15–21.
Easton, T.G. and Reich, E. (1972) J. Biol. Chem. 247, 6420–6434.
Emerson, C.P. and Beckner, S.K. (1975) J. Mol. Biol. 93, 431–447.
Engel, W.K. (1961) J. Histochem. 9, 66–72.
Ewton, D.Z. and Florini, J.R. (1980) Endocrinology 106, 577–583.
Ewton, D.Z. and Florini, J.R. (1981) Dev. Biol. 86, 31–39.
Fambrough, D.M., Bayne, E.K., Gardner, J.M., Anderson, M.J., Wakshull, E. and Rotundo, R. (1982a) in Neuroimmunology, pp. 259–270, Plenum, New York.
Fambrough, D.M., Rotundo, R., Gardner, J.M., Bayne, E., Wakshull, E. and Anderson, M.J. (1982b) in Disorders in the Motor Unit, pp. 197–212, John Wiley, New York.
Fambrough, D., Hartzell, H.C., Rash, J.E. and Ritchie, A.K. (1974) Ann. N.Y. Acad. Sci. 228, 47–62.
Fear, J. (1977) J. Anat. 124, 437–444.
Fischbach, G.D. (1972) Dev. Biol. 28, 407–429.
Fischbach, G.D. and Cohen, S.A. (1973) Dev. Biol. 31, 147–162.
Fischbach, G.D., Cohen, S.A. and Henkart, M.P. (1974) Ann. N.Y. Acad. Sci. 228, 35–46.
Fischman, D.A. and Masaki, T. (1982) in Perspectives in Differentiation and Hypertrophy, in press, Elsevier, Amsterdam.
Fischman, D.A., Shimada, Y. and Moscona, A. (1967) J. Cell Biol. 35, 445–452.
Fiszman, M.Y. (1978) Cell Differ. 7, 89–101.
Fiszman, M.Y. and Fuchs, P. (1975) Nature (London) 254, 429–431.
Florini, J.R., Nicholson, M.L. and Dulak, N.C. (1977) Endocrinology 101, 32–41.
Frank, E. and Fischbach, G.D. (1979) J. Cell Biol. 83, 143–158.
Frank, E., Gautvik, K. and Sommerschild, H. (1975) Acta Physiol. Scand. 95, 66–76.
Furcht, L.T. and Wendelschafer-Crabb, G. (1978) Differentiation 12, 39–45.
Furcht, L.T., Mosher, D.F. and Wendelschafer-Crabb, G. (1978) Cell 13, 263–271.
Gardner, J.M. and Fambrough, D.M. (1982) Cell (in press).
Garrels, J.I. (1979) Dev. Biol. 73, 134–152.
Gartner, T.K. and Podleski, T.R. (1976) Biochem. Biophys. Res. Commun. 70, 1142–1149.
Giller, E.L., Neale, J.H., Bullock, P.N., Schrier, B.K. and Nelson, P.G. (1977) J. Cell Biol. 74, 16–29.
Giller, E.L., Schrier, B.K., Shainberg, A., Fisk, H.R. and Nelson, P.G. (1973) Science 182, 588–589.
Gospodarowicz, D., Weseman, J., Moran, J.S. and Lindstrom, J. (1976) J, Cell Biol. 70, 395–405.
Gottlieb, D.I., Rock, K. and Glaser, L. (1976) Proc. Natl. Acad. Sci. 73, 410–414.
Grove, B.K., Schwartz, G. and Stockdale, F.E. (1981) J. Supramol. Struct. Cell Biochem. 17, 147–152.
Grumet, M., Rutishauser, U. and Edelman, G.M. (1982) Nature (London) 293, 693–695.
Harris, A.J., Heinemann, S., Schubert, D. and Tarakis, H. (1971) Nature (London) 231, 296–301.
Hartzell, H.C. and Fambrough, D.M. (1972) J. Gen. Physiol. 60, 248–262.
Hauschka, S.D. and Konigsberg, I.R. (1966) Proc. Natl. Acad. Sci. USA 55, 119–126.
Hauschka, S.D., Haney, C., Angello, J.C., Linkhart, T.A., Bonner, P.H. and White, N.K. (1977) Pathogenesis of Human Muscular Dystrophies, pp. 835–855, Excerpta Medica, Amsterdam.
Heaysman, J.L.M. and Pegrum, S.M. (1973) Exp. Cell Res. 78, 71–78.
Henderson, C.E., Huchet, M. and Changeux, J.-P. (1981) Proc. Natl. Acad. Sci. USA 78, 2625–2629.
Holtzer, H., Weintraub, H., Mayne, R. and Mochan, B. (1972) Curr. Top. Dev. Biol. 7, 229–256.
Holtzer, H., Croop, J., Dienstman, S., Ishikawa, H. and Somlyo, A.P. (1975a) Proc. Natl. Acad. Sci. USA 72, 513–517.
Holtzer, H., Rubinstein, N., Fellini, S., Yeoh, G., Chi, J., Birnbaum, J. and Okayama, M. (1975b) Q. Rev. Biophys. 8, 523–556.
Hooisma, J., Slaaf, D.W., Meeter, E. and Stevens, W.F. (1975) Brain Res. 85, 79–85.
Horwitz, A.F., Wight, A., Ludwig, P. and Cornell, R. (1978) J. Cell Biol. 77, 334–357.

Horwitz, A.F., Wight, A. and Knudsen, K. (1979) Biochem. Biophys. Res. Commun. 86, 514–521.
Horwitz, A.F., Neff, N., Sessions, A., Decker, C. (1982) Cold Spring Harbor (in press).
Hynes, R.O. (1976) Biochim. Biophys. Acta 458, 73–107.
Hynes, R.O., Martin, G.S., Shearer, M., Critchley, D.R. and Epstein, C.J. (1976) Dev. Biol. 48, 35–46.
Inestrosa, N.C. and Hall, Z.W. (1981) Abstr. Soc. Neurosci. 7, 203.
Jacob, M. and Lentz, T.L. (1979) J. Cell Biol. 82, 195–211.
Jessell, T.M., Siegel R.E. and Fischbach, G.D. (1979) Proc. Natl. Acad. Sci. USA 76, 5397–5401.
Kalderon, N. and Gilula, N.B. (1979) J. Cell Biol. 81, 411–425.
Kalderon, N., Epstein, M. and Gilula, N.B. (1977) J. Cell Biol. 75, 788–806.
Kantor, H.L. and Prestegard, J.H. (1975) Biochemistry 14, 1790–1795.
Kato, A.C., Vrachliotis, A., Fulpius, B. and Dunant, Y. (1980) Dev. Biol. 76, 222–228.
Katz, B. and Miledi, R. (1972) J. Physiol. 224, 665–699.
Kaufman, S.J. and Lawless, M.L. (1980) Differentiation 16, 41–48.
Kidokoro, Y. (1980) Dev. Biol. 28, 231–241.
Kidokoro, Y. and Yeh, E. (1981) Dev. Biol. 86, 12–18.
Kidokoro, Y., Heinemann, S., Schubert, D., Brandt, B.L. and Klier, F.G. (1976) Cold Spring Harbor Symp. Quart. Biol. 40, 373–388.
Kidokoro, Y., Anderson, M.J. and Gruener, R. (1980) Dev. Biol. 78, 464–483.
Knudsen, K.A. and Horwitz, A.F. (1978) Prog. Clin. Biol. Res. 23, 563–568.
Ko, P.K., Anderson, M.J. and Cohen, M.W. (1977) Science 196, 540–542.
Koenig, J. and Vigny, M. (1978) Nature (London) 271, 75–77.
Konigsberg, I.R. (1963) Science 140, 1273–1284.
Konigsberg, I.R. (1971) Dev. Biol. 26, 133–152.
Konigsberg, I.R. (1977) Regulation of Cell Profileration and Differentiation (Nichols, W.W. and Murphy, D.G., eds.) 105, Plenum, New York.
Konigsberg, I.R., Sollmann, P.A. and Mixter, L.O. (1978) Dev. Biol. 63, 11–26.
Lance-Jones, C. and Landmesser, L. (1980a) J. Physiol. (London) 302, 559–580.
Lance-Jones, C. and Landmesser, L. (1980b) J. Physiol. (London) 302, 581–602.
Landmesser, L. (1971) J. Physiol. (London) 213, 707–725.
Landmesser, L. (1978a) J. Physiol. (London) 284, 371–389.
Landmesser, L. (1978b) J. Physiol. (London) 284, 391–414.
Landmesser, L. and Morris, D.G. (1975) J. Physiol. (London) 249, 301–326.
Lawrence, T.S., Beers, W.H. and Gilula, N.B. (1978) Nature (London) 272, 501–506.
Letinsky, M.S., Fischbeck, K.H. and McMahan, U.J. (1976) J. Neurocytol. 5, 691–718.
Letourneau, P.C. (1975a) Dev. Biol. 44, 77–91.
Letourneau, P.C. (1975b) Dev. Biol. 44, 92–101.
Libby, P. Bursztjan, S. and Goldberg, A.L. (1980) Cell 19, 481–491.
Lim, R.W. and Hauschka, S.D. (1981) J. Cell Biol. 91, 198a.
Linkhart, T.A., Clegg, C.H. and Hauschka, S.D. (1980) J. Supramol. Struct. 14, 483–498.
Linkhart, T.A., Clegg, C.H. and Hauschka, S.D. (1981) Dev. Biol. 86, 19–30.
Lipton, B.H. and Schultz, E. (1979) Science 205, 1292–1294.
Lomo, T. and Rosenthal, J. (1972) J. Physiol. (London) 221, 493–513.
Lomo, T. and Slater, C.R. (1980) J. Physiol. (London) 303, 191–202.
Lomo, T. and Westgaard, R.H. (1975) J. Physiol. (London) 252, 603–626.
Markelonis, G.J., Oh, T.H., Eldefrawi, M.E. and Guth, L. (1982) Dev. Biol. 89, 353–361.
Marshall, L.M., Sanes, J.R. and McMahan, V.J. (1977) Proc. Natl. Acad. Sci. USA 74, 3073–3077.
Masaki, T., Bader, D.M. Reinach, F.C., Shimizu, T., Obinata, T., Safig, S.D. and Fischman, D.A. (1982) Cold Spring Harbor (in press).
Massoulié, J., Bon, S. and Vigny, M. (1980) Neurochem. Int. 2, 161–184.

McMahan, U.J., Sargent, P.B., Rubin, L.L. and Burden, S.J. (1980) in Ontogenesis and Functional Mechanisms of Peripheral Synapses, pp. 345–354, Elsevier, Amsterdam.
McManaman, J.L., Blosser, J.C. and Appel, S.H. (1981) J. Neurosci. 1, 771–776.
Meezan, E., Hjelle, J.T., Brendel, K. and Carlson, E.D. (1975) Life Sci. 7, 1721–1732.
Merlie, J.P., Buckingham, M.E. and Whalen, R.G. (1977) Curr. Top. Dev. Biol. 11, 61–114.
Merlie, J.P., Hofler, J.G. and Sebbane, R. (1981) J. Biol. Chem. 256, 6995–6999.
Merlie, J.P. and Sebbane, R. (1981) J. Biol. Chem. 256, 3065–3068.
Moody-Corbett, F. and Cohen, M.W. (1981) J. Neurosci. 1, 596–605.
Moody-Corbett, F., Weldon, P.R., Davey, D.F. and Cohen, M.W. (1981) Abstr. Soc. Neurosci. 7, 670.
Murray, M.R., (1972) in Structure and Function of Muscle, Vol. 1 (Bourne, G.H., ed.), pp. 237–299, Academic, New York.
Nadal-Ginard, B. (1978) Cell 15, 855–864.
Nakajima, Y., Kidokoro, Y. and Klier, F.G. (1980) Dev. Biol. 77, 52–72.
Neher, E. and Sakmann, B. (1976) J. Physiol. (London) 258, 705–729.
Nelson, P.G. (1975) Physiol. Rev. 55, 1–61.
Nelson, P.G., Christian, C.N., Daniels, M.P., Henkart, M., Bullock, P., Mulinax, D. and Nirenberg, M. (1978) Brain Res. 147, 245–259.
Nestler, E.J., Beam, K.G. and Greengard, P. (1978) Nature (London) 275, 451–453.
Nowak, T.P., Haywood, P.L. and Barondes, S.H. (1976) Biochem. Biophys. Res. Commun. 68, 650–657.
Nurse, C.A. and O'Lague, P.H. (1975) Proc. Natl. Acad. Sci. USA 72, 1955–1959.
O'Lague, P.H., Obata, K., Claude, P., Furshpan, E.J. and Potter, D.D. (1974) Proc. Natl. Acad. Sci. USA 71, 3602–3606.
O'Neill, M.C. and Stockdale, F.E. (1972a) Dev. Biol. 29, 410–418.
O'Neill, M.C. and Stockdale, F.E. (1972b) J. Cell Biol. 52, 52–65.
O'Neill, M.C. and Stockdale, F.E. (1974) Dev. Biol. 37, 117–132.
Obata, K. (1977) Brain Res. 119, 141–153.
Oh, T.H., Chyu, J.Y. and Max, S.R. (1977) J. Neurobiol. 8, 469–476.
Paitridge, T.A., Grounds, M. and Sloper, J.C. (1978) Nature (London) 273, 306–308.
Papahadjopoulos, D., Vail, W.S., Pangborn, M.A. and Poste, G. (1976) Biochim. Biophys. Acta 448, 265–283.
Partridge, T. and Jones, R.E. (1977) Cell Biol. Int. Rep. 1, 271–273.
Partridge, T.A. and Sloper, J.C. (1977) J. Neurol. Sci. 33, 425–435.
Patterson, B. and Strohman, R.C. (1972) Dev. Biol. 29, 113–138.
Peacock, J.H. and Nelson, P.G. (1973) J. Neurol. Neurosurg. Psychiat. 36, 389–398.
Peng, H.B., Bridgman, P.C., Nakajima, S., Greenberg, A. and Nakajima, Y. (1979) Brain Res. 167, 379–384.
Pfenninger, K.H. and Maylié-Pfenninger, M–F. (1981) J. Cell Biol. 89, 536–546.
Pfenninger, K.H. and Rees, R.P. (1976) in Neuronal Recognition, pp. 131–178, Plenum, New York.
Podleski, T.R., Axelrod, D., Ravdin, P., Greenberg, I., Johnson, M.M. and Salpeter, M.M. (1978) Proc. Natl. Acad. Sci. USA 75, 2035–2039.
Podleski, T.R., Greenberg, I., Schlessinger, J. and Yamada, K.M. (1979) Exp. Cell Res. 122, 317–326.
Poole, A.R., Howell, J.T. and Lucy, J.A. (1970) Nature (London) 227, 810–814.
Powell. J.A. and Fambrough, D.M. (1973) J. Cell Physiol. 82, 21–38.
Prives, J. and Shinitzky, M. (1977) Nature (London) 268, 761–763.
Prives, J.M. and Patterson, B.M. (1974) Proc. Natl. Acad. Sci. USA 71, 3208–3211.
Prives, J., Hoffman, L., Tarrab–Hazdai, R., Fuchs, S. and Amsterdam, A. (1979) Life Sci. 24, 1713–1718.

Prives, J., Christian, C., Penman, S. and Olden, K. (1980) in Tissue Culture in Neurobiology, pp. 35–52, Raven, New York.
Puro, D.G., DeMello, F.G. and Nirenberg, M. (1977) Proc. Natl. Acad. Sci. USA 74, 4977–4981.
Purves, D. and Sakmann, B. (1974) J. Physiol. (London) 237, 157–182.
Razin, M. and Ginsburg, H. (1980) Biochim. Biophys. Acta 598, 285–292.
Rees, R.P. (1978) J. Neurocytol. 7, 679–691.
Rees, R.P., Bunge, M.B. and Bunge, R.P. (1976) J. Cell Biol. 68, 240–263.
Reiness, C.G. and Weinberg, C.B. (1981) Dev. Biol. 84, 247–254.
Richler, C. and Yaffe, D. (1970) Dev. Biol 23, 1–22.
Rieger, F., Koenig, J. and Vigny, M. (1980) Dev. Biol. 76, 358–365.
Robbins, J. and Heywood, S.M. (1978) Eur. J. Biochem. 82, 601–608.
Robbins, N. and Yonezawa, T. (1971) Science 172, 395–397.
Rotundo, R.L. and Fambrough, D.M. (1979) J. Biol. Chem. 254, 4790–4799.
Rotundo, R.L. and Fambrough, D.M. (1980) Cell 22, 583–594.
Rubin, L.L., Schuetze, S.M., Weill, C.L. and Fischbach, G.D. (1980) Nature (London) 283, 264–267.
Rubin, L.L., Schuetze, S.M. and Fischbach, G.D. (1979) Dev. Biol. 69, 46–58.
Rubin, L.L. and McMahan, U.J. (1982) in Disorders of the Motor Unit, pp. 187–196, John Wiley, New York.
Rutishauser, U., Thiery, J-P., Brackenbury, R., Sela, B-A. and Edelman, G.M. (1976) Proc. Natl. Acad. Sci. USA 73, 577–581.
Rutter, W.J., Pictet, R.L. and Morns, P.W. (1973) Ann. Rev. Biochem. 42, 601–646.
Sandra, A., Leon, M.A. and Przybylski, R.J. (1977) J. Cell Sci. 28, 251–272.
Sanes, J.R. and Hall, Z.W. (1979) J. Cell Biol. 83, 357–370.
Sanes, J.R., Marshall, L.M. and McMahan, U.J. (1978) J. Cell Biol. 78, 176–198.
Sasse, J., VonderMark, H., Kuhl, U., Dessau, W. and VonderMark, K. (1981) Dev. Biol. 83, 79–89.
Schuetze, S.M. (1980) J. Physiol. (London) 303, 111–124.
Schuetze, S.M., Frank, E.F. and Fischbach, G.D. (1978) Proc. Natl. Acad. Sci. USA 75, 520–523.
Shainberg, A. and Burstein, M. (1976) Nature (London) 264, 368–369.
Shainberg, A., Yagil, G. and Yaffe, D. (1969) Exp. Cell Res. 58, 163–167.
Shainberg, A., Yagil, G. and Yaffe, D. (1971) Dev. Biol. 25, 1–29.
Shainberg, A., Cohen, S.A. and Nelson, P.G. (1976) Pfluegers Arch. 361, 255–261.
Shani, M., Zevin-Sonkin, D., Saxel, O., Carmon, Y., Katcoff, D., Nudel, U. and Yaffe, D. (1981) Dev. Biol. 86, 483–492.
Shimada, Y. (1971) J. Cell Biol. 48, 128–142.
Shimada, Y. and Fischman, D.A. (1973) Dev. Biol. 31, 200–225.
Shimada, Y., Fischman, D.A. and Moscona, A.A. (1969) J. Cell Biol. 43, 382–387.
Silberstein, L., Inestrosa, N.C. and Hall, Z.W. (1982) Nature (London) 295, 143–145.
Steinbach, J.H., Merlie, J., Heinemann, S. and Bloch, R. (1979) Proc. Natl. Acad. Sci. USA 76, 3547–3551.
Stockdale, F.E. (1982) unpublished.
Stockdale, F.E. and Holtzer, H. (1961) Exp. Cell Res. 24, 508–520.
Stockdale, F.E. (1977) Regulation of Cell Proliferation and Differentiation, pp. 165, Plenum, New York.
Stygall, K., Mirsky, R. and Mowbray, J. (1979) J. Cell Sci. 37, 231–241.
Sytkowski, A.J., Vogel, Z. and Nirenberg, M.W. (1973) Proc. Natl. Acad. Sci. USA 70, 270–274.
Teichberg, V.I., Silman, I., Deutsch, D.D. and Resheff, G. (1975) Proc. Natl. Acad. Sci. USA 72, 1383–1387.
Trotter, J.A. and Nameroff, M. (1976) Dev. Biol. 49, 548–555.
VanderBosch, J., Schudt, C. and Pette, D. (1973) Exp. Cell Res. 82, 433–438.
Wahrmann, J.P., Senechal, H., Etienne-Decerf, J. and Winand, R.J. (1980) FEBS Lett. 115, 230–234.

Walker, C.R. and Wilson, B.W. (1975) Nature (London) *256*, 215–216.
Walsh, F.S. and Phillips, E. (1981) Dev. Biol. *81*, 229–237.
Walsh, F.S. and Ritter, M.A. (1981) Nature (London) *289*, 60–64.
Weinberg, C.B. and Hall, Z.W. (1979) Proc. Natl. Acad. Sci. USA *76*, 504–508.
Weldon, P.R. and Cohen, M.W. (1979) J. Neurocytol. 8, 239–259.
Weldon, P.R., Moody-Corbett, F. and Cohen, M.W. (1981) Dev. Biol. *84*, 341–350.
Whalen, R.G., Butler-Browne, G.S. and Gros, F. (1976) Proc. Natl. Acad. Sci. USA *73*, 2018–2022.
Whalen, R.G., Sell, S.M., Butler-Browne, G.S., Schwartz, K., Bouveret, P. and Pinset-Harstrom, I. (1981) Nature (London) *292*, 805–809.
Wright, W.E. and Gros, F. (1981) Dev. Biol. *86*, 236–240.
Yablonka, Z. and Yaffe, D. (1977) Differentiation *8*, 133–143.
Yaffe, D. (1968) Proc. Natl. Acad. Sci. USA *61*, 477–483.
Yaffe, D. (1971) Exp. Cell Res. *66*, 38–48.
Yaffe, D. (1973) Tissue Culture: Methods and Applications, pp. 106–114, Academic, New York.
Yaffe, D. and Saxel, O. (1977) Nature (London) *270*, 725–727.
Yeoh, G.C., Greenstein, D. and Holtzer, H. (1978) J. Cell Biol. *77*, 99–102.
Ziskind-Conhaim, L. and Bennett, J.I. (1982) Dev. Biol. *90*, 185–197.
Ziskind-Conhaim, L. and Dennis, M.J. (1981) Dev. Biol. *85*, 243–251.

CHAPTER 5

Postnatal development of the innervation of mammalian skeletal muscle

R.A.D. O'BRIEN

Department of Anatomy & Embryology, Centre of Neuroscience, University College London, Gower Street, London WC1E 6BT, England

1. Introduction

One of physiology's truisms is that a normal adult mammalian skeletal muscle fibre is innervated at a single end-plate by a single axon terminal. There are some exceptions to this rule, but it is reasonably reliable when considering the major muscles of the trunk and limbs. However, one can hardly fail to notice that the rule is qualified by no less than four adjectives, indicating that different patterns of innervation are found in different circumstances. Abnormal adult mammalian skeletal muscle fibres (ones that are affected by disease, old age or experimental interference) manifest abnormal patterns of innervation, and non-mammalian species have skeletal muscles with a different pattern of innervation; for example, fibres in the slow, tonic muscles of birds and amphibia are innervated at several end-plates distributed along their length. Muscles associated with the autonomic nervous system also have their own type of innervation, and this is described in other chapters of this volume.

Another special case, and one with which this article is principally concerned, is the innervation of skeletal muscles in new born mammals; they obey the law to the extent that the fibres are innervated at a single end-plate, but these end-plates are usually contacted by several nerve terminals (see Plate 1), and one of the most fascinating problems in current research is the study of the transition from this so-called polyneuronal innervation to the single innervation of the adult muscle fibre. Exactly how this transition is achieved remains the subject of heated debate, and there is no lack of hypotheses, but one can nonetheless appreciate the advantages that this transient polyneuronal innervation could confer on the developing neuromuscular system; given a variety of nerve terminals to choose from, the elements of the neuromuscular junction are in a position to select the terminal most suited to function in later life.

This idea of 'selective stabilisation' has been described by Changeux and Danchin (1976), and can be applied to several different neuronal systems. A transient polyneuronal innervation is characteristic of developing autonomic ganglia (Lichtman and Purves, 1980), cerebellar Purkinje cells (Crepel et al., 1976) and the spinal motoneurones themselves (Conradi and Ronnevi, 1975). It is also a feature of reinnervated adult skeletal muscles (McArdle, 1975; Gorio et al., 1980). Whether or not the elimination of the excess terminals is produced by the same mechanism in all of these systems remains to be seen, but it is likely that the rapidly accumulating information from the skeletal neuromuscular junction will provide some clues to events that take place in developing neuro-neuronal systems.

Discussion in this article is restricted primarily to mammalian muscles simply because they are the most widely studied, but work on non-mammalian species will be referred to where appropriate; polyneuronal innervation in chick muscles, for example, is receiving an increasing amount of attention. Various experimental methods currently used to examine polyneuronal innervation are described in an appendix (Section 6), the purpose of which is to introduce these techniques to the prospective researcher and to indicate the advantages and inadequacies which can sometimes affect the interpretation of the results they yield.

The reader's attention is drawn to some reviews which deal with various aspects of the neuromuscular junction; some of them include sections on postnatal development, and this article should be considered as complementary to these. Embryonic neural development is comprehensively covered by Jacobson (1978), and muscle fibre development is reviewed by Fischman (1972); the development of motor innervation in particular is discussed by Curless (1977), Jansen et al. (1978), Landmesser (1980) and Dennis (1981), while various aspects of motor innervation in general are considered in the book edited by Thesleff (1976). Brown et al. (1981a) discuss nerve sprouting, an important phenomenon in development, and Vrbová et al. (1978) focus attention on the mutual influence between nerve and muscle. The most recent review to date is by Grinnell and Herrera (1981), who provide an excellent discussion of many aspects of nerve–muscle plasticity.

Finally, and at the risk of pedantry, a note about terminology. For the purpose of this discussion a distinction is made between 'polyneuronal' and 'multiple' innervation, since the two terms are often confusingly interchanged. 'Polyneuronal' clearly defines the innervation of a muscle fibre as being from two or more motoneurones, while 'multiple' is accepted as describing the innervation of a fibre at several distinct end-plates, as occurs in the tonic muscles of birds and amphibia (Couteaux, 1955). Multiple innervation is often derived from several neurones, so it is tacitly understood as being polyneuronal as well. Normal mammalian skeletal muscle fibres are rarely innervated at several sites; when they are polyneuronally innervated, as at birth, the innervation is at a single end-plate, so the term 'multiple innervation' is not appropriate and can be misleading if used (as, for example, by O'Brien et al., 1977). As will be described later, it appears that each of the multiple end-plates of a tonic avian muscle goes through a phase of polyneuronal innervation during development, just like a mammalian end-plate, although in the chick it sometimes extends into adulthood (Ginsborg, 1959; Atsumi, 1977).

2. Development of muscle fibres and their innervation

The development of muscle fibres and their innervation are, to some extent, mutually dependent processes, so it would be useful to summarise them as background to a discussion of polyneural innervation. These aspects of development are dealt with in far greater detail by some of the reviews listed in Section 1, and the reader is referred to them for further information.

2.1. Muscle fibre development

In vertebrates, skeletal muscle fibres develop by differentiation of mesenchymal cells derived from the lateral plate and somatic mesoderm. Those mesenchymal cells which will eventually differentiate into muscle cells (the presumptive myoblasts) are mitotically active and mononucleated, and they elongate to form myoblasts which contain a few myofilaments and are no longer mitotically active. Groups of myoblasts then align themselves and fuse to produce multinucleated myotubes whose nuclei are located in the central region of the cytoplasm, while a few clusters of myofilaments are placed peripherally. The myotubes eventually develop into true muscle fibres with peripherally located nuclei and a cytoplasm dominated by the presence of organised myofilaments, giving the fibres their characteristic striated appearance.

In the rat the first, or primary, myotubes appear between days 13 and 16 of the 21-day gestation period (Kelly and Zacks, 1969a; Dennis et al., 1981). The primary myotubes serve as the structural foundations along which secondary generations of myotubes are produced, and these clusters become ensheated by the membrane of the basal lamina after day 15. As the muscle fibres mature they split off from the clusters and become enclosed by their own basal lamina.

The generation of muscle fibres is asynchronous, and individual fibres continually appear from about 18 days gestation until 1 or 2 weeks after birth, although the formation of the myotubes ceases by about the time of birth (Kelly and Zacks, 1969a). Light microscopical evidence originally suggested that the number of fibres increases postnatally (e.g., Chiakulas and Pauly, 1965), but the greater resolution of the electron microscope has since shown that the adult complement of muscle fibres is present at birth; most of them are still packed together as clusters and are difficult to discern by light microscopy, and the apparent postnatal proliferation of fibres results from their separation from these clusters (Riley, 1977a; Ontell and Dunn, 1978; Ontell, 1979; Harris, 1981).

Among the cells which do not differentiate before birth are the satellite cells, which stay within the basal lamina of the smaller muscle fibres (Mauro, 1961). Some of these cells fuse with the muscle fibres, adding to the number of their nuclei, while the others remain undifferentiated. The undifferentiated satellite cells are the progenitors of new muscle fibres in the event of damage to the muscle later in life (see chapters by Bischoff, Kelly, Konigsberg and Snow in Mauro, 1979).

The development of muscle cells, in regenerating as well as in developing muscles, is critically dependent on their innervation. Without innervating they can develop as far as the formation of primary myotubes, but no further (Carlson, 1973; Harris, 1981), and if they remain uninnervated the myotubes will eventually atrophy and degenerate (Hunt, 1932). It appears that the mere presence of the nerve is not sufficient – it has to be capable of transmitting activity to the myotube (Drachman, 1968; Giacobini-Robecci et al., 1975; Harris, 1981), so it is clear that even in the early stages of development neuromuscular activity is vital to the survival and differentiation of these cells.

Once the secondary myotubes have been formed the absolute necessity for functional innervation gradually diminishes, and denervation at later stages does not produce such profound destruction of the muscle, although it will retard the production of new muscle fibres and the subsequent growth of the whole muscle (Zelena, 1962; Carlson, 1973). Nevertheless, the postnatal differentiation of muscles into fast or slow contracting types does require neuromuscular activity. Indeed, it is the very pattern of this activity which determines the contractile characteristics of a muscle, and these can be altered by cross-innervation with a nerve from a different type of muscle or by imposing a different pattern of activity on the intact nerve (Buller et al., 1960; Salmons and Vrbová, 1969; Vrbová et al., 1978).

2.2. The first nerve–muscle contact

Motor axons are first seen in the mesenchymal muscle mass just before the myoblasts appear, about day 12 in the trunk muscles of the rat (Filogamo and Gabella, 1967; Dennis et al., 1981). The first physiological signs of innervation (the appearance of nerve-evoked end-plate potentials) appear between days 14 and 15 (Bennett and Pettigrew, 1974a; Dennis et al., 1981), but, curiously enough, morphologically recognisable synapses are not seen until a day later, and have never been seen on myoblasts (Filogamo and Gabella, 1967; Teravainen, 1968; Kelly and Zacks, 1969b). Axons have nevertheless been seen in very close apposition to myoblasts and undifferentiated cells on day 14 (Kelly, 1966), and it would be interesting to know if these primitive contacts are involved in early functional or inductive interactions between nerve and muscle.

Once a synapse has been formed on a myotube, acetylcholine receptors (AChR) and acetylcholinesterase (AChE) begin to accumulate at the same site and the basal lamina starts to develop (Filogamo and Gabella, 1967; Kelly and Zacks, 1969b; Bennett and Pettigrew, 1974a; Bevan and Steinbach, 1977). There is some dispute over whether or not the nerve induces these accumulations; this has generally been assumed to be the case, but Harris (1981) found focal accumulations of AChR and AChE in embryonic rat diaphragm muscles that had been denervated with β-bungarotoxin or paralysed with tetrodotoxin. He suggested that the site of the end-plate may therefore be determined intrinsically by the muscle cell, in which case the processes of innervation, AChR accumulation and AChE localisation could be mutually independent. However, as Harris points out, primitive nerve-muscle contacts (similar, perhaps, to those reported by Kelly (1966)) may have been made prior to denervation in his experiments, and this transient interaction could be enough to mark the site at which these subsequent events take place.

The idea that these accumulations may indeed depend on the presence of the nerve is reinforced by the results of Koenig and Vigny (1978). They found that the end-plate-specific AChE is produced by muscles cultured from 13 to 14-day rat embryos only if they are co-cultured with nervous tissue, whereas muscles cultured from 18-day embryos (in which nerve-muscle contacts will already have been made) could maintain this form of AChE even in the absence of nervous tissue. These observations imply that at least a transient interaction between the nerve and the myotube is required to determine the site at which the AChE is localised. McMahan and his colleagues have shown that, in frog muscle, the end-plate AChE actually accumulates in the part of the basal lamina adjacent to the end-plate, and they proposed that some marker in the basal lamina (not necessarily the AChE itself) directs the accumulation of AChR to the corresponding part of the muscle cell membrane (McMahan et al., 1978; Burden et al., 1979).

Other studies on tissue cultures also indicate that the accumulations of AChE and AChR depend on nerve-muscle contact; it appears that this contact must involve neuromuscular transmission in the case of AChE but not AChR (Obata, 1977; Rubin et al., 1980). For further coverage of this subject the reader is referred to the chapter by Rubin and Barald in this volume.

In mammalian muscles, and in the fast phasic muscles of birds and amphibia, only one end-plate is formed on each myotube under normal conditions, whereas in the slow tonic muscles of these latter vertebrates the myotubes accept multiple end-plates spaced at regular intervals. The factors which determine whether a myotube will accept focal or multiple innervation have been investigated by Vrbová and her colleagues in chicks (see Vrbová et al., 1978). The nerve terminals innervating a tonic muscle fibre have a lower capacity to synthesise and release acetylcholine (ACh) than terminals in a fast muscle (Vyskocil et al., 1971; O'Brien and Vrbová, 1978); consequently, propagating action potentials are not produced in the slow fibers (Kuffler and Vaughan Williams, 1953), but the decremental spread of the subthreshold end-plate potentials (epps) produces a local depolarisation of the muscle fibre membrane in the vicinity of the nerve terminal. It appears that the distance between the multiple end-plates is determined during development by the area of this depolarisation since, if the size of the epp is reduced by curare or hemicholinium, the end-plates are formed closer together. Moreover, multiple innervation can be induced in the fast phasic muscles by the same procedure. These results indicate that, at least in chick muscles, extra end-plates can be established only on parts of the membrane which are not depolarised by the activity of an existing nerve terminal, and that their spacing is determined by the spread of the epp.

2.3. Development of polyneuronal innervation

The formation of polyneuronal connections is restricted to the end-plate region. In a focally innervated muscle cell the first nerve contact is made on a myotube only a few hundred micrometers in length, and the subsequent longitudinal growth of the cell probably determines the ultimate position of the end-plate, approximately half way along the fibre (Bennett and Pettigrew, 1974a). The following axons are not so restricted by the size of the cell, which is elongating at the same time as polyneuronal innervation develops, so there must be other cues to direct them to the end-plate.

A simple but attractive explanation is that these axons follow the path laid down by the original nerve, and are thus led to the end-plate (Harrison, 1910). This cannot entirely explain the observation by Bennett and Pettigrew (1974b) that regenerating axons in new born rats are able to make contact with muscle tissue that is formed subsequent to the time of the nerve injury, where there would be no 'pathfinder axons' for them to follow. Some branches of the regenerating axons also grow to the original end-plates, but never to the extrajunctional regions of the muscle that was present at the time of denervation. This suggests that the new tissue and the old end-plates both contain something that attracts the axons, or that the original extrajunctional region is refractory to innervation (or both, of course).

These notions can be reconciled by supposing that the axons follow lines of contact formed by pathfinder axons or other orienting structures, but that they will only form an end-plate on tissue that has a relatively high density of AChR. Extrajunctional receptor density is already reduced by day 16 in embryonic rat muscles (Bevan and Steinbach, 1977), so the newly arrived axons would preferentially innervate the junctional region. This is also consistent with the formation and subsequent polyneuronal innervation of multiple end-plates in the tonic muscles of chicks (see Section 2.2) because the loss of extrajunctional AChR during development depends on neuromuscular activity (Jones and Vrbová, 1972; Burden, 1977). In the regeneration experiments described above the muscle tissue produced between the times of denervation and reinnervation will presumably have a relatively high receptor density because they have not been activated.

The attraction of growing nerves to areas of high receptor density also seems to be the case in reinnervated adult muscle (see Grinnell and Herrera, 1981), but what component of the receptor is responsible for this attraction is by no means clear, since blocking it with α-bungarotoxin does not prevent end-plate formation (Jansen and Van Essen, 1975).

The formation of polyneuronal innervation requires the growing axons to produce hundreds of branches, but what induces them to do so is still somewhat of a mystery. The phenomenon of nerve sprouting has been extensively examined in adult muscles, in which intact axons can be induced to sprout by partial denervation of a muscle or by suppressing neuromuscular activity, and the stimulus for sprouting has been considered in a review by Brown et al. (1981a). They divide sprouting into two categories – sprouting from nodes of Ranvier and sprouting from the terminal region of the axon – and they originally concluded that nodal sprouts are induced by products of nerve degeneration while terminal sprouts are produced when the muscle undergoes denervation-like changes. The results of subsequent experiments suggested that this distinction might not be true, and that both types of sprouts are induced by denervation-like changes in the muscle (Hopkins and Brown, 1981). It is possible that the same stimulus exists in embryonic muscle since the early myotubes still have a relatively high density of extrajunctional receptors, a feature they share with denervated adult muscle, but another explanation offered by Brown et al. (1981a), and perhaps the simplest when considering the branching of embryonic nerves, is that the axons have an inherent tendency to sprout. Such a tendency could also extend into adult life, and it has been observed in normal muscles of adult cats by Barker and Ip (1966).

Polyneural innervation reaches a maximum by the 18th day of gestation in rats and the 18th day of incubation in chicks (Bennett and Pettigrew, 1974a; Dennis et al., 1981), although the timing for a particular muscle may depend on several factors (discussed in Section 3.1). Estimates of the average number of nerve terminals per end-plate at the time of peak polyneural innervation vary depending on the method used; some end-plates in rat intercostal muyscle are innervated by as many as 6 terminals (Dennis et al., 1981), but an average between 3 and 5 holds for most muscles (e.g., Bennett and Pettigrew, 1974a; Brown et al., 1976). After this time the numbers decrease until, by 2 or 3 weeks after birth, only one nerve terminal is left at each end-plate. There is thus a transition, just before birth, from a proliferation to an elimination of nerve–muscle connenctions. In view of the asynchrony of muscle fibre development it would be surprising if the transition were abrupt, and there may be a period of several hours or days during which both processes occur simultaneously.

3. The elimination of polyneuronal innervation

The histological appearance of nerve terminals and preterminal axons during the elimination period is shown in Plate 2 (see also Riley, 1977b). A prominent feature is the presence at the end-plate of a terminal with well developed arborisations, accompanied by one or more thinner axons with more primitive terminations. As elimination proceeds the thinner axons appear to withdraw from the end-plate while the larger one remains. Ultrastructural observations have confirmed that the supernumerary terminals are eliminated by retraction rather than degeneration, since the presence of degenerating axons is very rare (Korneliussen and Jansen, 1976; Riley, 1981; Bixby, 1981).

A curious feature of the retracting nerve branches is that they terminate in a bulbous enlargement, the 'retraction bulb' (Plate 2). This is probably not the part of the axon that originally contacted the end-plate since it appears to form proximal to the nerve terminal. Riley (1981) has made the elegant suggestion that the retraction bulb is a region of membrane internalisation, by means of which the eliminated nerve branch can withdraw while enabling the parent axon to preserve its membrane material.

Exactly how one nerve terminal is selected for survival at an end-plate when the others are eliminated is an intriguing problem. Various mechanisms have been proposed, and these will be considered in Section 4, but the following sections are concerned with events that occur during elimination and the factors that affect it.

Many of the investigations to be described deal quantitatively with polyneuronal innervation, and it is important to be aware of the errors inherent in the techniques employed. These are outlined in Section 6, but a particularly vexing problem relates to the branching of nerves, and it may be appropriate to reiterate it here: it is obvious that an end-plate innervated by one nerve terminal is ipso facto innervated by one neurone, but it is not necessarily the case that an end-plate with, say, 5 nerve terminals is innervated by 5 neurones, because branches from the same parent axon may innervate the same end-plate. One could argue that the probability of this occurring is slight, but there may in fact be a positive tendency for a nerve sprout to follow the path laid to the end-plate by its siblings (see Section 2.3). In histological investigations care can be taken to see if nerve terminals at an end-plate are derived from the same intramuscular branch point, and this does not occur to any great extent (Riley, 1976), but an axon may branch at any distance proximal to the muscle (Eccles and Sherrington, 1930), and it would be prohibitively difficult to trace the axons throughout the nerve trunk. Electrophysiological measurements usually involve stimulating the distal stump of a cut nerve, and this may again be distal to a branch point, particularly if the stump is short. Clearly, one needs to be wary of this problem, although it is unlikely to have a substantial qualitative influence over the estimation of polyneuronal innervation.

Postnatal development of the innervation of mammalian skeletal muscle

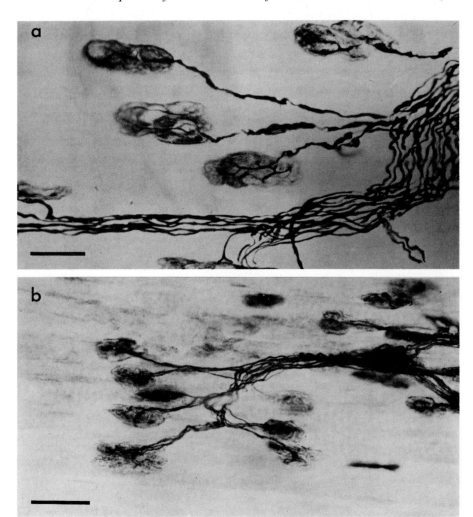

Plate 1. Low power light micrographs of cholinesterase-silver stained end-plates from rat soleus muscles. Calibration bars are 25 μm. (a) adult rat muscle; each end-plate is innervated by a single axon with a well arborised terminal. (b) 10-day-old rat; at least 4 end-plates are innervated by two or more nerve terminals. (From O'Brien et al., 1978a.)

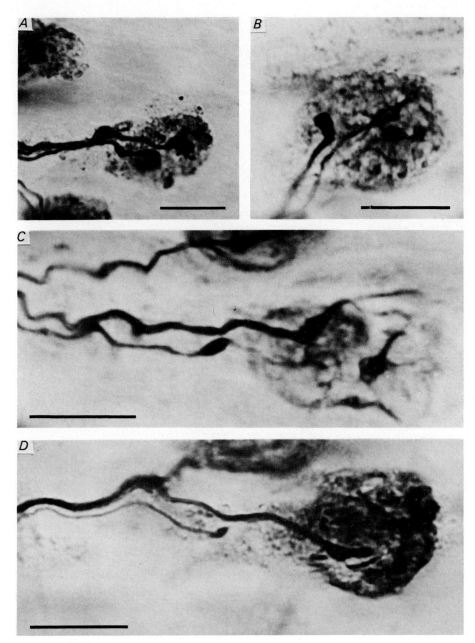

Plate 2. High power light micrographs of cholinesterase-silver stained end-plates from 11 to 13-day-old rat soleus muscles. Calibration bars are 10 μm. (a) An end-plate is contacted by two axons, both terminating in simple bulbous enlargements. (b) One of the terminals innervating this end-plate has developed arborisations, while the other remains bulbous. (c) The thinner axon has developed a bulbous enlargement about 10 μm proximal to its fine terminal branches. (d) The thinner axon terminates in a bulb about 10 μm from the end-plate. (From O'Brien et al., 1978a.)

3.1. Time course of elimination

Temporal changes in the distribution of muscle fibres with different numbers of nerve terminals have been carefully investigated in rat soleus and diaphragm muscles (Bennett and Pettigrew, 1974a; Benoit and Changeux, 1975; Riley, 1977b; Rosenthal and Teraskevich, 1977). Even though polyneuronal innervation reaches a peak just before birth, there are nevertheless some muscle cells which are innervated by only one nerve terminal at that time, while others appear to be innervated by as many as 5 or 6 terminals; there then follows a smooth increase in the proportion of singly innervated fibres until 15 to 20 days after birth, when virtually all the muscle fibres are contacted by one nerve terminal.

The presence of muscle fibres with several inputs even 2 weeks after birth lends support to the idea that a proportion of the fibres do not reach their peak of polyneuronal innervation until some time later than the majority. The singly innervated fibres that are seen before birth could therefore be those relatively immature muscle cells which are just receiving their first nerve terminals; it is unlikely that they are fibres which have already eliminated their supernumerary terminals since that would imply an unusually rapid sequence of proliferation and elimination of nerve endings. Such an argument depends on the assumption that all muscle fibres must undergo a period of polyneuronal innervation at some stage in their development, but this is by no means certain. Willshaw (1981) has pointed out that, if each axon distributes its terminals randomly among the muscle cells during early development, there would inevitably be some fibres that never have more than one input. Virtually no fibres would be left uninnervated if the numbers of terminals per end-plate were distributed normally about a mean of 3 or more (depending on the ratio of axons to muscle fibres), which fits well with experimental observations. Whether or not the embryonic development of motor innervation can validly be described as a random process remains to be answered, but it is likely that true randomness would be disturbed by the asynchrony of development and by mutual influences between nerve and muscle.

The most popular means of examining elimination has been to monitor the appearance of singly innervated muscle fibres with electrophysiological techniques. This is considerably less harrowing than counting the number of inputs at each end-plate, and, although it does not reveal the more subtle changes of innervation, it gives a good indication of the time by which a whole muscle achieves the adult state of one nerve terminal per end-plate. An example of this sort of measurement is shown in Fig. 1.

Elimination is completed earlier in muscles of the rabbit (7 to 11 days postnatal: Bixby and Van Essen, 1979) than in the rat (10 to 21 days postnatal: Redfern, 1970; Bennett and Pettigrew, 1974a; Benoit and Changeux, 1975; Brown et al., 1976; Rosenthal and Teraskevich, 1977; O'Brien et al., 1978a; Betz et al., 1979; Miyata and Yoshioka, 1980; Dennis et al., 1981). In the chick anterior latissimus

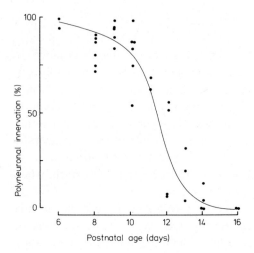

Fig. 1. Time course of the appearance of muscle fibres innervated by single axons in developing rat soleus muscles. Each point represents an animal in which electrophysiological recordings were made from at least 20 fibres in one or both muscles. A fibre was counted as polyneuronally innervated if it produced more than one end-plate potential in response to graded stimulation of the nerve (see Section 6.2 for method), and the results are plotted as a percentage of all the fibres impaled. The curve was fitted by eye. In these rats polyneuronal innervation was eliminated by 16 days of age. (From O'Brien et al., 1978a.)

dorsi muscle, which has multiple end-plates, the adult pattern of innervation is achieved some time between 10 and 30 days after hatching (Bennett and Pettigrew, 1974a), while elimination in the phasic muscles of the hind limb is completed before hatching (Pockett, 1981).

In some of the studies cited above functional polyneuronal innervation (that is, the presence at an end-plate of more than one terminal capable of depolarising the muscle fibre) was seen in a small proportion of mammalian fibres a few days after elimination was complete in the bulk of the muscle. Even these fibres must eventually succumb, for there is no record of polyneuronally innervated fibres in normal adult muscle (e.g., Brown and Matthews, 1960). In tonic muscles of the chick, however, polyneuronal innervation may persist in some fibres well into adult life (Ginsborg, 1959; Atsumi, 1977). Much more common in mammalian muscles is the presence of supernumerary terminals which are non-functional; these can be observed histologically even when physiological measurements show elimination to be complete (e.g., Riley, 1977b; O'Brien et al., 1978a), and they presumably represent nerve branches that have withdrawn from the postsynaptic membrane far enough to preclude neuromuscular transmission while still appearing to be in the region of the end-plate.

There seems to be a correlation between the timing of elimination and the anatomical position of the muscle. Elimination in rats is complete by 10 to 11 days in the intercostal muscles (Dennis et al., 1981), by 14 to 18 days in the diaphragm (Redfern, 1970; Rosenthal and Teraskevich, 1977), by 16 to 20 days in soleus (Benoit and Changeux, 1975; Brown et al., 1976; O'Brien et al., 1978a; Miyata and Yoshioka, 1980) and by 21 days in the lumbrical muscle of the hind foot (Betz et al., 1979). Such timing suggests a rostro-caudal sequence, which is perhaps not surprising since neural development and the onset of muscular activity also proceed rostro-caudally (East, 1932; Angulo Y Gonzalez, 1932; Romanes, 1941). Bixby and Van Essen (1979) systematically tested this in rabbits, and they found that singly innervated fibres appeared earlier in more rostral muscles but that elimination was completed by about the same time in all cases. Thus the rate at which terminals were eliminated (i.e., the number of nerve terminals lost per muscle fibre per day) increased rostro-caudally, from 0.06 terminals/fibre/day in the pronator muscle of the forelimb to 0.14 in soleus. They found no correlation between the rate of elimination and the adult contractile properties (as yet undifferentiated) of the muscles or their proximo-distal relationship within the same limb. This also holds for fast phasic muscles of the chick hind limb (Pockett, 1981), where the rate of elimination is the same in proximal and distal muscles but the onset and completion of elimination are earlier in the proximal muscles. The rate of elimination is faster in rats than in rabbits (0.21 in rat diaphragm: Rosenthal and Teraskevich, 1977), but a rostro-caudal correlation has not been investigated in this species. As will become apparent later, elimination in the rat shows some marked differences from the rabbit, so it would not be surprising if the sequences were different.

3.2. Effect of activity on elimination

Redfern (1970) noticed that the final stages of elimination in rat diaphragm muscle were accompanied by a general increase in the activity of the animal, and a variety of experiments have since been performed to see if changes in neuromuscular activity can influence the time course of elimination. The general conclusion is that elimination does indeed depend on activity.

Benoit and Changeux (1975) and Riley (1978) tenotomised rat soleus muscles soon after birth and showed that this enabled polyneuronal innervation to persist in the operated muscles. Complete elimination eventually took place after some delay, probably because the cut tendons healed. In an adult, tenotomy results in a reduction of muscle activity (Vrbová, 1963), but this may not be the case in neonatal animals because the stretch reflexes are not well developed (Skoglund, 1960; Bursian, 1973). Riley's (1978) explanation is that muscle growth is inhibited by tenotomy, reducing the burden on the motoneurones by obviating the need for their axons to elongate. Cordotomy at 12 days of age (Miyata and Yoshioka, 1980) and cordotomy accompanied by deafferentation at 1 day (Zelena et al.,

1979) also cause polyneuronal innervation to persist, although the effect may be only transient in the case of the combined operation.

More direct methods of reducing activity are easier to interpret. Thompson et al. (1979) treated the nerve with tetrodotoxin, and this again caused a persistence of polyneuronal innervation in rat soleus muscles. Blocking neuromuscular transmission with α-bungatotoxin (Duxson, 1982: Fig. 2) or botulinum toxin (Brown

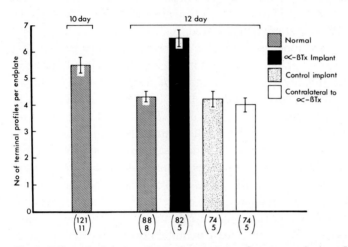

Fig. 2. Effect of α-bugarotoxin on the numbers of nerve terminal profiles at developing rat soleus end-plates. The profiles were counted by quantitative electron microscopy; the numerical relation between profiles and terminals is discussed in Section 6.3.2. In each experimental animal a small silicon rubber strip containing 10 to 15 μg α-bungarotoxin or 0.8 to 1.2 mg NaCl (control) was implanted in one leg at 10 days and both muscles were removed at 12 days of age. Columns are Mean ± 1 S.E.M.; the figures below each column refer to the number of end-plates (top) and number of muscles (bottom) examined. The number of profiles in the 12-day-old toxin-treated muscles was not significantly different from the 10-day normal muscles, while the normal and control muscles at 12 days showed a significant fall in number of profiles. This suggests that the toxin caused the preservation of terminals that would otherwise have been eliminated between 10 and 12 days. (From Duxson, 1982.)

et al., 1981b) also delays elimination in rat muscles, as do curare and α-bugarotoxin in chicks (Srihari and Vrbová, 1978; Sohal and Holt, 1980).

In some of these experiments with reduced activity a transient reversal of elmimination was reported (Thompson et al., 1979; Miyata and Yoshioka, 1980; Brown et al., 1981b); this could be due to sprouting of the axons or to the regrowth of terminals that had been eliminated but had not fully withdrawn from the end-plate. The latter interpretation is consistent with the histologically observed presence of non-functional nerve terminals (Section 4.1), so that the most recently eliminated terminals may be able to return to the end-plate under the appropriate conditions.

It is clear, then, that a reduction of neuromuscular activity retards or weakens the process of elimination; the converse experiment – increasing the activity – yields the same conclusion. Neuromuscular activity has been increased by chronic electrical stimulation of the nerve (O'Brien et al., 1977, 1978a: Fig. 3), by dener-

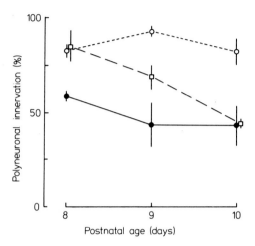

Fig. 3. Effect of increased activity on polyneuronal innervation in developing rat soleus muscles. Stimulating electrodes were implanted around the sciatic nerve in one leg of rats aged 6 to 8 days. Stimulation was at 8 Hz for 4 to 6 h a day for 2 to 4 days, after which polyneuronal innervation was assessed electrophysiologically at the time indicated on the abscissa. Values are Mean ± 1 S.E.M. for between 3 and 6 muscles. Open circles are implanted but unstimulated controls, filled circles are stimulated muscles, and open squares are muscles contralateral to the stimulated muscles. Even the minimum periods of stimulation produced a significant reduction in the proportion of polyneuronally innervated fibres; there was little change after 8 days probably because of the short duration of the stimulation periods. Elimination was eventually accelerated in the contralateral muscles, presumably as a result of reflex activity. (From O'Brien et al., 1978a.)

vating synergistic muscles (inducing compensatory hyperactivity in the muscle: Zelena et al., 1979) and by exaggerating the effects of normal transmission with the cholinesterase inhibitor DFP (M. Duxson, personal communication), and all these procedures accelerate elimination.

3.3. Nerve regeneration in neonatal muscle

The reinnervation and hyperinnervation of developing muscles by regenerating nerves also provides some valuable clues to the mechanism of the elimination process. After crushing the nerve to rat soleus muscles soon after birth, Brown et al. (1976) made the surprising observation that the axons reinnervated the original end-plates rapidly enough to make polyneuronal connections which were eliminated at the same time as in a normal muscle, in spite of the interruption.

However, if the nerve to EDL and tibialis anterior is crushed 4 to 6 days after birth the consequent polyneuronal innervation persists for several months (Domizio et al., 1981). In the latter experiments a large proportion of the muscle fibers atrophied during, but not before, the early stages of reinnervation; many of the surviving fibres became innervated at sevreal end-plates, and this multiple innervation persisted indefinitely. The authors' explanation of these observations (personal communication) is that there may be a critical period during development in which the maturation of the muscle depends on functional innervation; some fibres remained so immature that they could not cope with the activity imposed on them by the reinnervating nerve, and they atrophied, leaving the other fibres effectively hyperinnervated.

Hyperinnervation has also been produced in neonatal rat soleus muscles by reinnervating them with the original nerve and a foreign nerve shortly after birth (Brown et al., 1976). Many of the muscle fibres became functionally innervated by axons from both nerves, but this hyperinnervation persisted only if the foreign nerve formed an end-plate at least 1 mm away from the original, reinnervated end-plate. This spatial restraint also applies to hyperinnervated adult muscles (Jansen et al., 1973; Kuffler et al., 1977; O'Brien et al., 1978b).

These studies indicate that the maturity of the muscle is an important to the success of the elimination process, and that this process can only assert itself over a limited distance from the end-plate.

3.4. Differences between surviving and eliminated nerve terminals

A predominant question relating to the elimination of supernumerary nerve terminals is whether or not there are any characteristics or properties which distinguish the surviving terminals from those that are eliminated. An important observation was made by Brown et al. (1976); they measured an 8-fold variation in the size of motor units in 2 to 3-day-old rat soleus muscles, and this was reduced to a 3-fold variation in the adult. The size of the smallest units was hardly altered, suggesting that the elimination precess primarily affects the terminals belonging to the larger units. Similar observations have been made in lumbrical muscles by Betz et al. (1979).

There is clearly an upper limit to the number of terminals that a motoneurone can support. In a partially denervated adult muscle the intact motor units are unable to expand their territories to more than 4 or 5 times their original size, even if this leaves some muscle fibres uninnervated (Thompson and Jansen, 1977), and regenerating axons cannot expand their territory at all (Gordon and Stein, 1982). These limits may occur because the nerve terminals depend on the parent neurone for their supply, by axonal transport, of materials essential to the maintenance of their structural integrity and functional capacity (see, for example, Heslop, 1975), and there is presumably an upper limit to the neurone's ability to meet the demand. The largest motor units in neonatal rats must be considerably overstretched in this capacity, since they occupy up to 35% of the muscle in soleus, and 80% in the lumbrical muscle (Brown et al., 1976; Betz et al., 1979). One would therefore expect that their terminals receive poorer support than those belonging to smaller units, and could be more susceptible to the elimination process. This idea is complicated by results from rabbit soleus muscles, in which the variation in motor unit size appears to increase postnatally (Van Essen and Gordon, 1981), but this could be due to a greater variation in the abilities of the motoneurones to supply their terminals.

Another interesting finding relates to the relative positions in the spinal cord of motoneurones supplying the same muscle. The rat soleus muscle is innervated by axons from ventral roots L4 and L5, the majority emanating from L5 (Brown et al., 1976). Miyata and Yoshioka (1980) found that the L4 axons innervate virtually all the muscle fibres at birth, declining to about 50% after 2 weeks, but the L5 axons maintain their territory at about 50% throughout this period. These experiments were repeated in our laboratory (O'Brien, 1981; O'Brien et al., 1982), and we found that the axons from L4 together produced 49% of the total muscle contraction in rats aged 1 to 5 days, falling to 15% by 20 days, whereas the contractions produced by L5 axons fell only from 95% to 85% during the same period (Fig. 4). Although these results differ quantitatively from those of Miyata and Yoshioka (probably because the distribution of axons between the

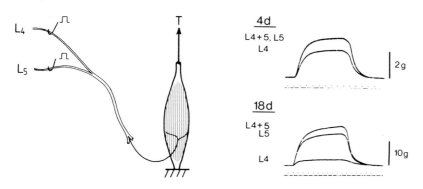

Fig. 4. Overlap of motor units between ventral roots L4 and L5 in neonatal rat soleus muscle. Isometric tensions were recorded in response to tetanic stimulation of the whole roots separately (L4, L5) or together (L4 + L5). At 4 days of age the tension produced by L5 was the same as the tension produced by simultaneous stimulation of the roots, so the whole muscle was innervated by axons from L5 in this example; L4 produced about 60% of the tension, so 60% of the muscle was innervated by axons from both roots. At 18 days (after the elimination of polyneuronal innervation) the sum of the tensions from L4 and L5 corresponded to the tension produced by simultaneous stimulation, so there was no longer any overlap; the L4 axons lost considerably more territory during elimination than the L5 axons.

roots varies between strains of rat), they nonetheless support the conclusion that L4 axons lose more terminals than L5 axons during the elimination period. Preliminary measurements suggest that, at birth, motor units from L4 are larger than those from L5, and that the difference is abolised over the next 2 to 3 weeks (unpublished observations).

We extended these experiments to the measurement of quantal components of transmitter release by axons from either root (O'Brien, 1981; O'Brien et al., 1982) and the results are shown in Fig. 5. In 5-day-old muscles the epps from L5 axons

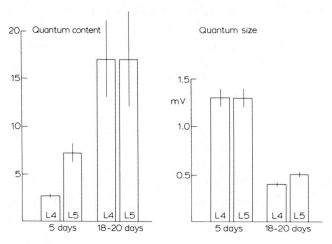

Fig. 5. Quantal components of transmitter release at rat soleus end-plates by nerve terminals of axons from ventral roots L4 and L5. Columns are Mean ± 1 S.E.M. Quantum content (the number of packets of transmitter comprising an epp) was estimated from the variance of epp sizes in trains of 200 epps from curarised muscles. This assumes that transmitter release obeys Poisson statistics, an assumption that is reasonable for the low quantum contents in the 5-day muscles but is unreliable for the older muscles. In the 18-day muscles there is no significant difference in the sizes of the epps obtained from L4 and L5 terminals but the estimated values of quantum content are likely to be inaccurate. Quantum size (the average depolarisation produced by one packet) was estimated from histograms of epp sizes in magnesium-poisoned muscles. The reduction in quantum size between 5 and 18 days after birth is mainly due to the increase in the diameter of the muscle fibres (Diamond and Miledi, 1962).

have more than twice the quantum content of epps from L4 axons, whereas there is no apparent difference at 18 days. At no time is there any difference in the size of the quanta from either root, so there is a large net difference in the capacity of the terminals to relrease ACh at birth. It is presumed that the elimination of some of the 'weaker' L4 axons ensures a better supply of material from the neurone to the remaining terminals, resulting in an equality of transmitter-releasing capacity with the L5 axons when elimination is complete. That transmitter release can so adjust to changes in the size of the motor unit has been shown in frog sartorius muscles by Herrera and Grinnell (1980); they compressed the motor units by forcing the nerve to regenerate into a muscle whose bulk had been reduced by surgery, and found that the output of transmitter increased in proportion to the reduction in motor unit size.

The capacity to release transmitter may also reflect other characteristics that influence a nerve terminal's chance of survival. There is a correlation between transmitter release and nerve terminal size at frog neuromuscular junctions (Kuno et al., 1971); if this also holds for mammalian muscle one would expect the L4 terminals in neonatal rat soleus to be physically smaller than the L5 terminals. This would be consistent with qualitative light microscopical observations that show the thinner axons, with less well developed terminals, to be more prone to elimination (Plate 2). Another interesting idea is that the growth of the terminal may be facilitated by its release of transmitter, since exocytosis involves the fusion of vesicles with the presynaptic membrane, so the higher transmitter output of L5 terminals would further increase their ability to survive elimination. Recycling of the vesicles would tend to diminish this contribution to growth (Heuser and Reese, 1973), but these immature terminals might not yet have developed an efficient recycling system.

Van Essen and Gordon (1981) also reported that terminals from more rostral neurones are more susceptible to elimination in rabbit soleus muscles, although it is less obvious than in the rat. At present one can only guess at the reasons for this gradient. One explanation is that axons from the more rostral neurones might innervate the embryonic muscle mass a little earlier than the more caudal ones because of their earlier development (Romanes, 1941; Harris-Flanagan, 1969); they would then have a greater opportunity to form synapses with the muscle cells and might thus 'overextend' their territories.

4. The mechanism of elimination

The elimination of polyneuronal innervation involves a numerical reduction in the ratio of nerve terminals to muscle fibres, and this could be brought about in a relatively uncomplicated way if the number of muscle fibres were to increase or if terminals were lost by death of the parent neurone.

Motoneurone death is a normal part of development (see chapter by Oppenheim and Chu-Wang in this volume), and it is obvious that the loss of a neurone would result in the elimination of its terminals from the muscle. In avian embryos motoneurone death precedes the period of elimination by several days (Holt and Sohal, 1978), so it cannot account for the loss of terminals in this case. The situation in mammals is more complicated because neurone death may, in some cases, overlap with the early stages of elimination. Nurcombe et al. (1981) reported a loss of about 45% of the neurones in the brachial motor column of the rat between birth and 4 days of age; there was very little loss after this time, whereas elimination continues until at least 10 to 20 days after birth (see Section 3.2). Motoneurone death may therefore account for the loss of some nerve terminals for a few days immediately following the peak of polyneuronal innervation in mammals, but an alternative explanation is that the loss of nerve terminals accounts for the death of the neurones. Brown et al. (1976) found no postnatal reduction in the number of axons innervating the soleus muscle of the rat, so elimination in this muscle may be completely independent of motoneurone death.

If new muscle fibres were produced during the elimination period a transfer of nerve terminals to the new fibres might account for the overall reduction in the terminal/fibre ratio. In most muscles of the rat there appears to be no net production of muscle fibres after birth (see Section 2.1), which would seem to dispense with this idea, but there is an exception: the 4th lumbrical muscle of the hind foot doubles its complement of muscle fibres postnatally, and a transfer of terminals to the new fibres can account for the reduction of polyneuronal innervation for the first 10 days of life (Betz et al., 1979). Nevertheless, these authors found that there is still a net loss of nerve terminals for the following 6 days, and this cannot be accounted for in the same way. The mechanism by which terminals are transferred from one muscle fibre to another still has to be explained, and their removal from an existing end-plate may yet turn out to involve the same processes that are responsible for the net loss of terminals.

Thus motoneurone death or muscle fibre production may, in some instances, accompany the loss of nerve terminals in the early stages of elimination, but perhaps the most critical part of the process is the final stage, when all but one of the terminals are withdrawn from the end-plate. The mechanism that accounts for this and, in most cases, for the loss of terminals throughout the elimination period, is elusive; several models have been proposed, any or none of which may be correct, but their merits can be assessed in the light of the events known to accompany elimination.

The work described in the previous section yields three general conclusions about elimination: (a) it is an activity-dependent process, (b) it exerts its influence over a limited distance from the end-plate, and (c) functionally 'weaker' terminals are preferentially eliminated. One's confidence in these conclusions depends, of course, on one's view of the evidence, but they seem to provide a reasonable platform on which to base a model for the mechanism of the elimination process. An additional, obvious, but most important feature that has to be accounted for is that the end-plate is eventually innervated by one, and only one, nerve terminal.

It is tempting to assume that elimination involves competition between nerve terminals for survival at the end-plate, but the temptation has been resisted by some. The question is this: does the survival of a nerve terminal depend on its absolute 'suitability' or on its performance relative to the other terminals at the same end-plate? Van Essen and Gordon (1981) pointed out that the increase in motor unit variablity during elimination in rabbit soleus muscle is consistent with a random, non-competitive loss of terminals. If so, one would expect to find a population of denervated muscle fibres when elimination is over. Denervated fibres have been sought but not found in rat soleus (Brown et al., 1976), but this muscle manifests a decrease in the variability of motor unit size after birth. Nevertheless, there is evidence that motoneurones have an inherent tendency to withdraw some of their terminals; if a rat muscle is partially denervated soon after birth the remaining motor units still decrease in size despite the reduced competition, although not always to the same extent as in a normal muscle (Brown et al., 1976; Thompson and Jansen, 1977). Such a reduction does not occur in lumbrical muscles, probably because of the postnatal production of muscle fibres (Betz et al., 1980). This tendency to withdraw could still be related to the relative 'strengths' of the nerve terminals if one supposes that they normally compete for survival against some sort of inhibitory agent released by the muscle fibre; the weakened terminals of an overextended motor unit might not even be able to survive alone against such an agent, an increased postnatal production of which would account for the diminishing size of the unit.

It is unlikely that circulating hormones are involved in elimination since the process occurs neither at the same rate nor at the same time in all the muscles of an animal. The possibility that Schwann cells are involved in the selection of the surviving nerve terminal is disputed by Bixby (1981) and Riley (1981); for example, they report that the preterminal axons are not myelinated until some time after elimination is over, so the Schwann cells are not favouring terminals by selective myelination. It has been suggested that 'weaker' terminals could be physically displaced by 'stronger' ones, but this would not explain the inherent tendency of a neurone to retract some of its terminals in a partially denervated muscle. None of these suggestions can account for the limited distance over which the elimination process acts.

The idea that the nerve terminals are competing for a trophic agent is more attractive because it implies that the muscle fibre has some influence in the selection of the surviving terminal. However, one would not expect a motoneurone to retract its terminals from a source of 'survival factor' in the absence of competition, as in experiments on partial denervation, and it is difficult to see how this would prevent synapse formation close to the original end-plate (although it could be argued that the surviving terminals are mopping up the agent within the vicinity of the end-plate).

In this laboratory we favour the idea that the nerve terminals are competing for survival against an inhibitory agent. We have enlarged on this hypothesis by proposing that the inhibitory agent is released into the end-plate region in an activity-dependent fashion, and have suggested that it is a proteolytic enzyme or group of enzymes capable of digesting the nerve terminals. The enzymes could be released from the muscle fibre or from some other cell near the end-plate region.

This suggestion is based on the following evidence:
(1) A wide variety of cell types respond to treatment with ACh by secreting proteolytic enzymes (Ignarro, 1975).
(2) Localised extracellular proteolytic activity has been observed at the end-plates of adult muscles that were stimulated through the nerve or treated with inhibitors of AChE (Poberai et al., 1972).
(3) Treatment of neonatal muscles with ACh in vitro causes a loss of nerve terminal profiles, as observed with the electron microscope; this effect is potentiated by calcium and prevented by protease inhibitors and by curare (O'Brien et al., 1980).
(4) The ultrastructural appearance of muscles so treated is similar to that observed in muscles exposed to AChE inhibitors (Leonard and Salpeter, 1979) or to the calcium ionophore A23187 (Publicover et al., 1978), and is attributed by these authors to the activation of a calcium-dependent neutral protease that exists in the muscle sarcoplasm. The involvement of calcium can also be regarded as consistent with a secretory mechanism.

4.1. A hypothetical model

A model for the sequence of events accompanying the elimination of polyneuronal innervation is represented in Fig. 6. The principle is that proteolytic enzymes are released into the end-plate region in response to the ACh released by the active nerve terminals; all of the terminals will tend to be digested by the enzymes, but those that receive a poorer supply of membrane replacing material from their parent neurones will be digested more quickly than the 'stronger' terminals, and will eventually withdraw from the end-plate. This loss of terminals would result in a better supply of material to the remaining terminals of the motor unit, increasing their chance of survival at other end-plates. Providing that there is always an imbalance, however slight, in the abilities of the terminals at a particular end-plate to survive digestion, competition will eventually result in the survival only of the strongest terminal. On the rare occasions when two terminals might be perfectly matched both would survive, and this could explain the small numbers of polyneuronally innervated fibres that persist in some muscles.

That the last surviving terminal is not removed from the end-plate can be explained if it undergoes a continual process of growth and digestion; if it were to withdraw, the activation of enzyme secretion would cease and the terminal could grow back. A dynamic process of growth and destruction has been mooted by Young (1951) and may be analogous to the turnover of nerve endings seen in adult muscles by Barker and Ip (1966).

Willshaw (1981) has developed a computer model for the elimination process; based on the rules outlined above, the model simulates many of the events known to occur during elimination, and it faithfully produces muscles with singly innervated fibres.

Postnatal development of the innervation of mammalian skeletal muscle | 175

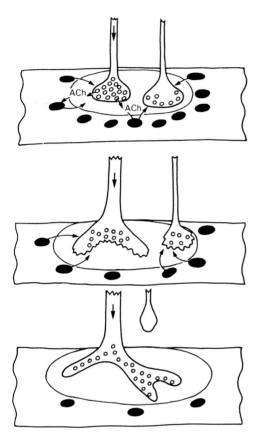

Fig. 6. A model for the mechanism of elimination. The hypothesis is described in Section 4.1; in this case two terminals are competing for survival at an end-plate. Proteolytic enzymes are secreted into the end-plate region in response to the ACh released by nerve activity. The nerve terminals are digested and only the larger terminal survives because it is supplied with enough material to balance the digestion. (From O'Brien et al., 1978a.)

This hypothesis can account for the experimental observations described in previous sections, the more obvious ones being: (a) the rate of enzyme secretion, hence the rate of elimination, would be affected by alterations in neuromuscular activity, (b) being diffusible, the enzymes would act over a limited area within the vicinity of the end-plate, (c) the 'weaker' nerve terminals would be at a competitive disadvantage, (d) as pointed out in the previous section, a terminal could be so weak that it would not even survive alone at an end-plate in a partially denervated muscle, but would be digested by enzymes secreted in response to its own activity.

5. Summary and conclusions

The formation and regeneration of nerve–muscle connections involves a period during which the muscle fibre end-plate is innervated by several axon terminals. In developing rats this polyneuronal innervation reaches a peak just before birth and is reduced to the normal adult ratio of one terminal per end-plate over the following 2 or 3 weeks.

The susceptibility of a nerve terminal to elimination seems ultimately to depend on the size of the motor unit to which it belongs; terminals from large units may in several senses be 'weaker' than those from smaller units, such as in their output of transmitter, and are more likely to be eliminated.

At present, the only certainly about the mechanism of elimination is that it depends on neuromuscular activity. This is little enough to go on when one is trying to propose a model, and, as usual, the number of hypotheses available reflects our ignorance and uncertainty over evidence. Developing motor units may have a tendency to withdraw some of their terminals spontaneously, and some terminals may be lost by neuronal cell death, but it seems likely that the 'fine tuning' depends on competition between terminals for survival at an end-plate and that the muscle fibre itself plays some active part in the selection. Such a mechanism would be the most satisfying because it gives the muscle fibre (whose appropriate function would appear to be one of the goals) some choice in selecting the most suitable, or least unsuitable, nerve terminal as its life-long companion.

Our laboratory is working on the idea that the terminals are competing for survival against an inhibitory agent, probably in the form of proteolytic enzymes, which is released into the end-plate region as a result of neuromuscular activity. This hypothesis can account for many of the events known to accompany the elimination of polyneuronal innervation.

Whatever the mechanism turns out to be, we may be confident that it will explain many of the consequences of experimental or pathological disturbances of nerve–muscle interactions during development and in adult life.

6. Appendix: Techniques for examining polyneuronal innervation

Laboratories that are concerned with the examination of polyneuronal innervation usually address themselves to one of two questions: (a) what proportion of the fibers in a muscle are polyneuronally innervated? and (b) how many axons innervate the muscle fibres?

The first question is the easier to answer because one is simply required to establish whether or not a muscle fibre has more than one input without worrying about how many inputs. If enough fibres are sampled, and the sampling is random, a reasonable estimate can be made of the proportion of polyneuronally innervated fibres in the whole muscle.

The second question is more difficult to answer because none of the techniques to be described is free of criticism when it comes to counting the number of motoneurones that innervate a particular muscle fibre. All are likely to give overestimates or underestimates, and these systematic errors can be minimised only be exercising a great deal of care over the chosen technique.

6.1. Contraction measurements

The motoneurone, its axon, and the muscle fibres it innervates together comprise a motor unit. In an adult mammalian muscle stimulation of the axon, naturally or experimentally, causes the contraction of all the muscle fibres in the unit; stimulation of any other axon will not produce a contraction in these fibres. This is not the case in skeletal muscles from new born animals because the muscle fibres are innervated by more than one axon, and this sharing of fibres between motor units can be exploited to demonstrate polyneuronal innervation.

If a muscle is prepared for contraction measurements and the ventral roots supplying its innervation are carefully dissected to allow the stimulation of individual axons, the tensions produced by single motor units can be recorded. Stimulation of the whole nerve can be employed, but one is then limited to one or two units with the lowest stimulus thresholds, resulting in non-random sampling.

Stimulation of one axon will produce a contraction, say x Newtons. Stimulation of another axon will produce y Newtons and stimulation of both axons together will produce a contraction of z Newtons. In theory, if the two motor units are distinct, as in a normal adult, the individual tensions should exactly summate (i.e., $z = x + y$), but if some of the muscle fibers are innervated by both axons the individual fibre tensions will be contributed once when either axon is stimulated and once (not twice!) during simultaneous stimulation, so $z < (x + y)$. The ratio $(x + y - z)/z$ is termed 'tension excess', 'occlusion' or 'overlap', the latter being the term preferred here because it indicates that the territories of the motor units do indeed overlap. An example of this sort of measurement is given in Fig. 4, although the overlap in this instance is between whole ventral roots rather than individual filaments.

This method of estimating polyneuronal innervation needs to be approached carefully. For example, results obtained with tetanic stimulation often differ from those produced by single shocks; Hunt and Kuffler (1954) found twitch tension overlap in limb muscles of adult cats, and attributed this to polyneuronal innervation, but it was later shown by Brown and Matthews (1960) that the overlap was more likely due to non-linear summation of twitch tensions from individual units. They found very little overlap when tetanic stimulation was used.

The phenomenon of post-tetanic potentiation can also be exploited here. Twitch tension is transiently increased following tetanic stimulation, this being a property of the muscle rather than of the nerve (Brown and von Euler, 1938). Thus tetanic stimulation of one motor unit produces twitch potentiation in another unit if they share some muscle fibres, and the degree of potentiation is proportional to the degree of overlap (Bagust et al., 1973). This should work in undifferentiated muscles (e.g., Uramoto, 1980) but will become complicated in muscles containing a substantial proportion of slowly contracting fibres because these fibres cease to exhibit post-tetanic potentiation (Bagust et al., 1974).

The use of contraction measurements to estimate polyneuronal innervation has two further limitations. Firstly, polyneuronal innervation at single end-plates cannot be distinguished from single innervation at multiple end-plates. Secondly, it is prohibitively difficult to count the number of nerve terminals impingeing on a single muscle fibre by direct means, unless one were prepared to perform a single-fibre dissection and promise that the nerves had not been disturbed. Mathematical estimates can be made, as by Bagust et al. (1973) and Bixby and Van Essen (1979), but these depend on several assumptions about the distribution of innervation, the uniformity of individual muscle fibre tensions, and the ability of each nerve terminal to depolarise its fibre beyond threshold. Once arrived at, the figure is an average for the whole muscle.

6.2. Electrophysiology

Ginsborg (1959) showed that electrophysiological techniques could be used to demonstrate polyneuronal and multiple innervation in chick muscles, and Redfern (1970) used them in the first systematic investigation of polyneuronal innervation in neonatal rat diaphragm.

In the simplest form the method relies on the difference in firing threshold between the axons in a nerve trunk. The muscle is isolated in an organ bath, usually with a centimetre or so of its nerve, and lightly curarised to avoid muscle action potentials (although the curare may sometimes be omitted with embryonic muscles). A muscle fibre is penetrated near the end-plate region by the recording microelectrode and the nerve is stimulated with a suction electrode at a fairly low frequency (about 1/s) and narrow pulse width (about 20 μs). The stimulus current is gradually raised from zero, and at some point an end-plate potential (epp) will be observed, indicating that the stimulus has reached the threshold of the axon producing the epp (see Fig. 7). The current is again increased, and if there is no sudden change in the size of the epp one can conclude that the fibre is innervated by only one axon. If the epp suddenly increases in size it means that the stimulus has reached the threshold of a second axon innervating the fibre, and the epp from that axon has summated with the epp from the first axon to produce a compound epp. Still more axons may be recruited as the current is raised further (Fig. 7).

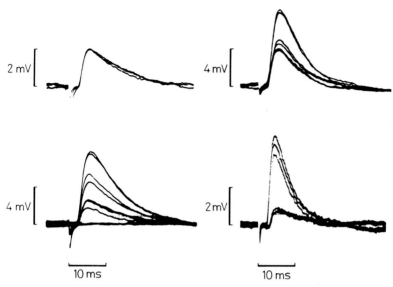

Fig. 7. Electrophysiological demonstration of polyneuronal innervation at neonatal rat soleus endplates. In each case the intensity of nerve stimulation was increased gradually from zero. The nerve stump was about 1 cm in length. No jump in the size of the epp was seen in (a), so it is from a singly innervated muscle fibre. The step-wise increases in the other traces indicate that they are innervated by at least 2 (b and d) or 3 (c) nerve terminals. In these carefully selected traces there is little doubt that fibres (b) to (d) are polyneuronally innervated, but note in (d) that the first epp was small relative to the second; if the recruitment order had been reversed the small epp might easily have been lost in the fluctuation of the large epp, and it might have been concluded that the fibre was singly innervated.

Discrimination of the summated epps is helped by the fact that the first epp in a train is usually larger than the rest, more so if the frequency of the train is increased, so that when an axon is recruited the jump in the size of the compound epp is initially quite large. This is helpful because the amplitude of an epp fluctuates due to the quantal nature of transmitter release; in immature muscles with low quantum contents the fluctuations can be large enough to smother the epps from newly recruited axons.

Such fluctuations can lead to an underestimate of the number of axons innervating the muscle fibre, and the same problem arises if two or more axons to the same fibre have stimulus thresholds that are close enough together to be indistinguishable. Another problem with neonatal muscles is the possibility of impaling a cluster of electrically coupled myotubes; the impalement of one myotube may give a satisfactory resting potential, but epps from the other myotubes will also be recorded, leading to an overestimate of polyneuronal innervation (Ontell, 1979; Harris, 1981). An overestimate might also be obtained if the nerve stump is stimulated distal to a branch point, of which there may be several along a nerve trunk (Eccles and Sherrington, 1930); one could then be discriminating between branches of the same axon going to the same end-plate (see Section 3 for further discussion of this).

These limitations can all be considerably reduced if a long nerve stump is left attached to the muscle, preferably right back to the spinal roots. This would not only ensure that the nerve was stimulated proximal to any branch points, but would also aid the discrimination of epps from their different latencies (e.g., Redfern, 1970), since, in general, higher threshold axons have smaller diameters and slower conduction velocities (Hursch, 1939; Gasser and Grundfest, 1939). Resolution can be improved even further if the ventral roots are dividend so that the contribution from each filament is assesed separately (Rosenthal and Teraskevich, 1977). The use of muscles innervated by separate nerve trunks also helps in this respect (e.g., rat lumbrical muscle: Betz et al., 1979).

The amplitude and rise time of an epp vary with the distance of the recording electrode from the end-plate (Fatt and Katz, 1951), so electrophysiological methods can be used to distinguish between focal and multiple innervation. If the innervation is focal the rise times of the epps should be similar, while multiple innervation will appear as epps of widely varying latencies and rise times. It is thus possible to detect polyneuronal innervation at multiple end-plates (Ginsborg, 1959; Bennett and Pettigrew, 1974a), the only limitation being that epps from end-plates that are distant from the recording electrode may be attenuated to unobservable levels by the time they reach the electrode.

6.3. Microscopy

6.3.1. Light histology

The combined cholinesterase–silver stain of Namba et al. (1967) has been used to great effect in examining polyneuronally innervated muscles, although the standard method needs some modification for use with neonatal material (see O'Brien et al., 1978a). The cholinesterase stain shows up the end-plate region of the muscle fibre, so inspection at low power can reveal the presence of focal or multiple end-plates. The silver impregnates the axons and their terminal arborisations so that the number of preterminal axons innervating a particular end-plate can be counted. One can usually obtain sufficient fibres to estimate the overall degree of polyneuronal innervation in a muscle. Plates 1 and 2 show the sort of pictures obtained with this method.

It is probable that this technique will produce consistent underestimates of the number of axons innervating a muscle fibre because of the difficulty in discriminating between the individual preterminal axons, particularly if they are close together or lie above and below each other in the plane of section. On the other hand, an overestimate might be obtained if two or more of the preterminal axons at a particular end-plate are branches of the same parent axon. With care, one may trace the intramuscular nerve trunk to check for such branches (e.g.,

Riley, 1976), but there may also be branching proximal to the muscle, as alluded to in the previous section.

The zinc iodide–osmium technique (Akert and Sandri, 1968) has also been used for studying these muscles. It produces better pictures of the terminal arborisations because it selectively stains the synaptic vesicles, but it is not generally considered suitable for staining axons, particularly if they are myelinated.

6.3.2. Electron microscopy

For close examination of the pre- and post-synaptic elements of the end-plate electron microscopy is, of course, unparalleled, but caution must be exercised when it is used to quantify polyneuronal innervation. The preparation of immature muscles for electron microscopy and the criteria used to identify nerve terminal profiles are described by Korneliussen and Jansen (1976). They showed that the number of nerve terminal profiles at an end-plate bears a changing relationship to the actual number of nerve terminals during development; this is because the section may pass through several branches of an arborised terminal, and each branch is counted as a nerve terminal profile. When the terminals ramify in early postnatal development the number of profiles seen in the cross section of an end-plate increases, even though other techniques show that the actual number of terminals is decreasing. In rat soleus a peak in the number of profiles per end-plate is reached about the 8th day after birth, after which the number of profiles decreases in line with the elimination of polyneuronal innervation; when elimination is complete the end-plates have an average of two profiles, corresponding to one nerve terminal (Korneliussen and Jansen, 1976).

In spite of this complex relationship, the high counts produced by this technique (up to 12 or more profiles at one end-plate) make it useful for detecting subtle changes of innervation brought about by experimental interference (e.g., O'Brien et al., 1980; Duxson, 1982), providing that measurements are made in muscles that have passed their peak of profiles per end-plate.

Acknowledgements

My warmest thanks go to Marilyn Duxson, Susan Gottlieb, Robert Purves and Gerta Vrbová for suggesting improvements to the manuscript. Marilyn Duxson also unselfishly allowed some of her unpublished material to be used, and Gerta Vrbová has been a constant source of encouragement and imaginative ideas.

References

Akert, K. and Sandri, C. (1968) Brain Res. 7, 286–295.
Angulo Y Gonzalez, A.W. (1932) J. Comp. Neurol. 55, 395–442.
Atsumi, S. (1977) J. Neurocytol. 6, 691–709.
Bagust, J., Lewis, D.M. and Luck, J.C. (1974) J. Physiol. (London) 237, 115–121.
Bagust, J., Lewis, D.M. and Westerman, R.A. (1973) J. Physiol. (London) 229, 241–255.
Barker, D. and Ip, M.C. (1966) Proc. R. Soc. London Ser. B: 163, 538–554.
Bennett, M.R. and Pettigrew, A.G. (1974a) J. Physiol. (London) 241, 515–545.
Bennett, M.R. and Pettigrew, A.G. (1974b) J. Physiol. (London) 241, 547–573.
Benoit, P. and Changeux, J.-P. (1975) Brain Res. 99, 354–358.
Betz, W.J., Caldwell, J.H. and Ribchester, R.R. (1979) J. Physiol. (London) 297, 463–478.
Betz, W.J., Caldwell, J.H. and Ribchester, R.R. (1980) J. Physiol. (London) 303, 265–279.
Bevan, S. and Steinbach, J.H. (1977) J. Physiol. (London) 267, 195–213.
Bixby, J.L. (1981) J. Neurocytol. 10, 81–100.
Bixby, J.L. and Van Essen, D.C. (1979) Brain Res. 169, 275–286.
Brown, G.L. and von Euler, U.S. (1938) J. Physiol. (London) 93, 39–60.
Brown, M.C., Holland, R.L. and Hopkins, W.G. (1981a) Ann. Rev. Neurosci. 4, 17–42.
Brown, M.C., Holland, R.L. and Hopkins, W.G. (1981b) J. Physiol. (London) 318, 355–364.
Brown, M.C., Jansen, J.K.S. and Van Essen, D. (1976) J. Physiol. (London) 261, 387–422.
Brown, M.C. and Matthews, P.B.C. (1960) J. Physiol. (London) 151, 436–457.
Buller, A.J., Eccles, J.C. and Eccles, R.M. (1960) J. Physiol. (London) 150, 417–439.
Burden, S. (1977) Devel. Biol. 61, 79–85.
Burden, S.J., Sargent, P.B. and McMahan, U.J. (1979) J. Cell Biol. 82, 412–425.
Bursian, A.V. (1973) J. Evol. Biochem. Physiol. 9, 525–529.
Carlson, B.M. (1973) Am. J. Anat. 137, 119–149.
Changeux, J.-P. and Danchin, A. (1976) Nature (London) 264, 705–712.
Chiakulas, J.J. and Pauly, J.E. (1965) Anat. Rec. 152, 55–61.
Conradi, S. and Ronnevi, L.-O. (1975) Brain Res. 92, 505–510.
Couteaux, R. (1955) Int. Rev. Cytol. 4, 335–375.
Crepel, F., Mariani, J. and Delhaye-Bouchaud, N. (1976) J. Neurobiol. 7, 567–578.
Curless, R.G. (1977) Prog. Neurobiol. 9, 197–209.
Dennis, M.J. (1981) Ann. Rev. Neurosci. 4, 43–68.
Dennis, M.J., Ziskind-Conhaim, L. and Harris, A.J. (1981) Devel. Biol. 81, 266–279.
Diamond, J. and Miledi, R. (1962) J. Physiol. (London) 162, 393–408.
Domizio, P., Lowrie, M.B. and Vrbová, G. (1981) J. Physiol. (London) 320, 22–23P.
Drachman, D.B. (1968) in Growth of the Nervous System (Wolstenholme, G.E. and O'Connor, M., eds.) pp. 251–273, Churchill, London.
Duxson, M.J. (1982) J. Neurocytol. 11, 395–408.
East, E.W. (1931) Anat. Rec. 50, 201–219.
Eccles, J.C. and Sherrington, C.S. (1930) Proc. R. Soc. London Ser. B: 106, 326–357.
Fatt, P. and Katz, B. (1951) J. Physiol. (London) 115, 320–370.
Filogamo, G. and Gabella, G. (1967) Arch. Biol. (Liege) 78, 9–60.
Fischman, D.A. (972) in The Structure and Function of Muscle (Bourne, G.H., ed.) 2nd Edn., Vol. 1, pp. 75–142, Academic Press, New York.
Gasser, H.S. and Grundfest, H. (1939) Am. J. Physiol. 127, 393–414.
Giacobini-Robecchi, M.G., Giacobini, G., Filogamo, G. and Changeux, J.-P. (1975) Brain Res. 83, 107–121.
Ginsborg, B.L. (1959) J. Physiol. (London) 148, 50–51P.
Gordon, T. and Stein, R.B. (1982) J. Physiol. (London) 323, 307–323.

Gorio, A., Carmignoto, G., Facci, L. and Finesso, M. (1980) Brain Res. *197*, 236–241.
Grinnell, A.D. and Herrera, A.A. (1981) Prog. Neurobiol. *17*, 203–282.
Harris, A.J. (1981) Philos. Trans. R. Soc. London Ser. B: *293*, 254–314.
Harris-Flanagan, A. (1969) J. Morphol. *129*, 281–306.
Harrison, R.G. (1910) J. Exp. Zool. *9*, 789–848.
Herrera, AA. and Grinnell, A.D. (1980) Nature (London) *287*, 649–651.
Heslop, J.P. (1975) Adv. Comp. Physiol. Biochem. *6*, 75–163.
Heuser, J.E. and Reese, T.S. (1973) J. Cell Biol. *57*, 315–344.
Holt, R.K. and Sohal, G.S. (1978) Am. J. Anat. *151*, 313–318.
Hopkins, W.G. and Brown, M.C. (1981) Neuroscience *7*, 37–44.
Hunt, C.C. and Kuffler, S.W. (1954) J. Physiol. (London) *126*, 293–303.
Hunt, E.A. (1932) J. Exp. Zool. *62*, 57–91.
Hursh, J.B. (1939) Am. J. Physiol. *127*, 131–139.
Ignarro, L.J. (1975) in Lysosomes in Biology and Pathology (Dingle, J.T. and Dean, R.T., eds) Vol. 4, pp. 482–523, Elsevier, Amsterdam.
Jacobson, M. (1978) Developmental Neurobiology, Plenum Press, N.Y. 2nd. edition.
Jansen, J., Lomo, T., Nicolaysen, K. and Westgaard, R. (1973) Science 181, 559–561.
Jansen, J.K.S., Thompson, W. and Kuffler, D.P. (1978) Prog. Brain Res. *48*, 3–19.
Jansen, J.K.S. and Van Essen, D.C. (1975) J. Physiol. (London) *250*, 651–667.
Jones, R. and Vrbová, G. (1972) J. Physiol. (London) *222*, 569–581.
Kelly, A.M. (1966) J. Cell Biol. *31*, 58A.
Kelly, A.M. and Rubinstein, N.A. (1980) Nature (London) *288*, 266–269.
Kelly, A.M. and Zacks, S.I. (1969a) J. Cell Biol. *42*, 135–153.
Kelly, A.M. and Zacks, S.I. (1969b) J. Cell Biol. *42*, 154–169.
Koenig, J. and Vigny, M. (1978) Nature (London) *271*, 75–77.
Korneliussen, H. and Jansen, J.K.S. (1976) J. Neurocytol. *5*, 591–604.
Kuffler, D., Thompson, W. and Jansen, J.K.S. (1977) Brain Res. *138*, 353–358.
Kuffler, S.W. and Vaughan Williams, E.M. (1953) J. Physiol. (London) *121*, 289–317.
Kuno, M., Turkanis, S.A. and Weakly, J.N. (1971) J. Physiol. (London) *213*, 545–556.
Landmesser, L.T. (1980) Ann. Rev. Neurosci. *3*, 279–302.
Leonard, J.P. and Salpeter, M.M. (1979) J. Cell Biol. *82*, 811–819.
Lichtman, J.W. and Purves, D. (1980) J. Physiol. (London) *301*, 213–228.
Mauro, A. (1961) J. Biophys. Biochem. Cytol. *9*, 493–504.
Mauro, A. (ed.) (1979) Muscle Regeneration Raven Press, New York.
McArdle, J.J. (1975) Exp. Neurol. *49*, 629–638.
McMahan, U.J., Sanes, J.R. and Marshall, L.M. (1978) Nature (London) *271*, 172–174.
Miyata, Y. and Yoshioka, K. (1980) J. Physiol. (London) *309*, 631–646.
Namba, T., Nakamura, T. and Grob, D. (1967) Am. J. Clin. Pathol. *47*, 74–77.
Nurcombe, V., McGrath, P.A. and Bennett, M.R. (1981) Neurosci. Lett. *27*, 249–254.
Obata, K. (1977) Brain Res. *119*, 141–153.
O'Brien, R.A.D. (1981) J. Physiol. (London) *317*, 89–90P.
O'Brien, R.A.D., Ostberg, A.J.C. and Vrbová, G. (1978a) J. Physiol. (London) *282*, 571–582.
O'Brien, R.A.D., Ostberg, A.J.C. and Vrbová, G. (1978b) J. Physiol. (London) *280*, 38P.
O'Brien, R.A.D., Ostberg, A.J.C. and Vrbová, G. (1980) Neuroscience *5*, 1367–1379.
O'Brien, R.A.D., Ostberg, A.J.C. and Vrbová, G. (1982) in Membranes in Growth and Development (Hoffman, J.F., Giebisch, G.H. and Bolis, L., eds) pp. 247–275, Alan R. Liss, New York.
O'Brien, R.A.D., Purves, R.D. and Vrbová, G. (1977) J. Physiol. (London) *271*, 54–55P.
O'Brien, R.A.D. and Vrbová, G. (1978) Neuroscience *3*, 1227–1230.
Ontell, M. (1979) in Muscle Regeneration (Mauro, A., ed.) pp. 137–146, Raven Press, New York.

Ontell, M. and Dunn, R.F. (1978) Am. J. Anat. *152*, 539–555.
Poberai, M., Savay, G. and Csillik, B. (1972) Neurobiology *2*, 1–7.
Pockett, S. (1981) Devel. Brain Res. *1*, 299–302.
Publicover, S.J., Duncan, C.J. and Smith, J.L. (1978) J. Neuropathol. Exp. Neurol. *37*, 544–557.
Redfern, P.A. (1970) J. Physiol. (London) *209*, 701–709.
Riley, D.A. (1976) Brain Res. *110*, 158–161.
Riley, D.A. (1977a) Exp. Neurol. *56*, 400–409.
Riley, D.A. (1977b) Brain Res. *134*, 279–285.
Riley, D.A. (1978) Brain Res. *143*, 162–167.
Riley, D.A. (1981) J. Neurocytol. *10*, 425–440.
Romanes, G.J. (1941) J. Anat. *76*, 112–130.
Rosenthal, J.L. and Teraskevich, P.S. (1977) J. Physiol. (London) *270*, 299–310.
Rubin, L.L., Schuetze, S.M., Weill, C.L. and Fischbach, G.D. (1980) Nature (London) *283*, 264–267.
Salmons, S. and Vrbová, G. (1969) J. Physiol. (London) *201*, 535–549.
Skoglund, S. (1960) Acta Physiol. Scand. *49*, 299–317.
Sohal, G.S. and Holt, R.K. (1980) Cell Tiss. Res. *210*, 383–394.
Srihari, T. and Vrbová, G. (1978) J. Neurocytol. *7*, 529–540.
Teravainen, H. (1968) Z. Zellforsch. *87*, 249–265.
Thesleff, S. (ed.) Motor Innervation of Muscle, Academic Press, New York.
Thompson, W. and Jansen, J.K.S. (1977) Neuroscience *2*, 523–535.
Thompson, W., Kuffler, D.P. and Jansen, J.K.S. (1979) Neuroscience *4*, 271–281.
Uramoto, I. (1980) Exp. Neurol. *70*, 697–700.
Van Essen, D.C. and Gordon, H. (1981) Soc. Neurosci. Abstr. *7*, 179.
Vrbová, G. (1963) J. Physiol. (London) *166*, 241–250.
Vrbová, G., Gordon, T. and Jones, R. (1978) Nerve-Muscle Interaction, Chapman & Hall, London. pp. 233.
Vyskocil, F., Vyklicky, L. and Huston, R. (1971) Brain Res. *26*, 443–445.
Willshaw, D.J. (1981) Proc. R. Soc. London Ser. B: *212*, 233–252.
Young, J.Z. (1951) Proc. R. Soc. London Ser. B: *139*, 18–37.
Zelena, J. (1962) in The Denervated Muscle (Gutmann, E., ed.) pp. 103–126, Czech. Acad. Sci. Publ. House, Prague.
Zelena, J., Vyskocil, F. and Jirmanova, I. (1979) Prog. Brain Res. *49*, 365–372.

CHAPTER 6

Influence of preganglionic fibres and peripheral target organs on autonomic neuronal development

CARYL E. HILL and IAN A. HENDRY

Department of Pharmacology, John Curtin School of Medical Research, Australian National University, Canberra, ACT 2601, Australia

1. General introduction

The induction, differentiation and maturation of autonomic neurones progresses from the time of the migration of the neural crest cells from the neural tube to their condensation into discrete groups, or ganglia, and the subsequent growth of their processes into prospective target tissues. Over this period of time, interactions occur with a variety of cell types. During the migration of the neural crest cells, interactions are of necessity transitory. After the neural crest cells have arrived in their definitive positions, however, interactions involve the cells with which the neurones are to make long lasting associations; neurones of the central nervous system, glial cells and the cells of prospective target organs. In this chapter, we will deal with the first and last of these three since they are interactions which involve the formation and maintenance of synapses.

Information transfer across synapses may occur in an orthograde and a retrograde direction. In the orthograde direction the neurotransmitter, in addition to its conventional role in impulse transmission, may play a role in the maturation of the postsynaptic cell. In the retrograde direction both cell surface contact phenomena as well as the transfer of soluble material from the postsynaptic cell to the presynaptic cell may be involved in the recognition of appropriate target organs and the consequent formation of synapses. The theoretical group of substances which are synthesized and released by target tissues, taken up and retrogradely transported by the innervating autonomic axons to their cell bodies to permit the survival of the transporting neurones have been called retrophins (Hendry et al., 1981). The protein nerve growth factor (NGF) may be a retrophin for sympathetic neurones. Several recent reports have indicated that other molecules, perhaps at different times during development, may also be implicated in both the neuritic production and ultimate survival of these neurones. In the parasympathetic system the analogous trophic factors are still being identified, purified and characterized.

Interactions involving substances which act specifically on a particular group of neurones, ensure that the pathways established through the synapses so formed are appropriate to the neurone and to the animal. Similar mechanisms may also exist to regulate a series of connections.

2. Development of sympathetic neurones

Studies designed to elucidate the roles played by the pre- and postsynaptic elements in the development of sympathetic neurones have been made using mice, rats and chick embryos (see Black, 1978; Hendry, 1976a; Giacobini, 1978). In mice and rats, synapse formation occurs predominantly during postnatal life and so surgical techniques have frequently been used to approach the problem. As changes in neuronal numbers and in the activity of enzymes involved in the biosynthesis of the neurotransmitters, noradrenaline and acetylcholine, have been used as parameters of postganglionic neuronal and preganglionic synaptic maturation, a brief description of the normal developmental changes is useful.

2.1. Normal biochemical development of sympathetic ganglia

In mice, catecholamine fluorescence, indicating the presence of neurotransmitter, can be detected in the primordia of the sympathetic ganglia at 12 days of gestation (Björklund et al., 1968) and in rats, a day later (de Champlain et al., 1970). In both animals, this is long before the neurones receive synaptic input or themselves innervate target organs. Between 14 and 17 days gestation virtually all neurones in the mouse superior cervical ganglion exhibit fluorescence, while from birth to one week postnatally some neurones are non-fluorescent (Coughlin et al., 1978). After this time, however, fluorescence is again detected in all the neurones (Coughlin et al., 1978).

Tyrosine hydroxylase (TH), the enzyme involved in the rate limiting step in the synthesis of noradrenaline, has been used as a marker for the development of adrenergic neurones as it is restricted to the postsynaptic cell bodies within the ganglion (Black et al., 1971a; Pickel et al., 1975). In mice, the activity of this enzyme increases from 13 days gestation until 3 days postnatally, rapid rises occurring between days 16 and 17 and between birth and day 3 (Black et al., 1971b; Coughlin et al., 1978). Following a plateau in activity to day 7 postnatally, there is a 3-fold increase to adult levels during the second postnatal week (Black et al., 1971b). During the same period, from days 4 to 14 postnatally, there is a 40% increase in the number of neurones in the ganglion (Black et al., 1972a).

The penetration of the ganglion by cholinergic fibres of spinal origin has been examined by assay for the activity of the enzyme, choline acetyltransferase (CAT), and by histological studies. In the mouse superior cervical ganglion, CAT activity remained at a low level from 13 to 16 days gestation, increased 2- to 3-fold between days 16 and 17, remained constant until birth and then increased steadily until 3 weeks postnatally (Black et al., 1971b; Coughlin et al., 1978). The number of synaptic profiles, identified by electron microscopy, increased dramatically between 5 and 11 days postnatally and then more gradually to 60 days (Black et al., 1971b).

2.2. Influence of preganglionic fibres

When these various developmental changes were compared temporally, it was found that the rapid increase in synaptic profiles was immediately followed by the major increase in TH activity (Black et al., 1971b). These observations led to the suggestion that the development of TH activity in the postganglionic neurones might depend upon contact with the preganglionic cholinergic nerves.

2.2.1. Development in the absence of the preganglionic fibres

Section of the preganglionic nerve trunk to the superior cervical ganglion in neonatal mice indeed led to a failure in the normal increase in TH activity in the ganglion (Black et al., 1971b). In the rat, however, such a clear cut relationship between the increases in TH and in CAT activities did not exist and, in fact, the major postnatal increase in TH activity preceded that in CAT activity (Thoenen et al., 1972a). Furthermore, the normal increase in TH activity was not completely abolished by section of the preganglionic nerve trunk (Thoenen et al., 1972b).

The difference between rats and mice in the magnitude of their response to section of the preganglionic nerve trunk may be related to the postnatal changes in neuronal numbers in the two species. In mice, the number of neurones increased by 40% between birth and 2 weeks of age and nerve section prevented this normal increase (Black et al., 1972a). In rats, however, the number of neurones actually decreased by 30% during normal development over the first 3 weeks of postnatal life (Hendry and Campbell, 1976) and it has been stated that nerve section did not further affect this number (Lawrence et al., 1979). In mice, then, the preganglionic fibres may promote the differentiation of immature neurones, in addition to inducing the activity of TH.

2.2.2. Agents responsible for preganglionic influences

Since NGF has been shown to exert both of the above effects on sympathetic neurones (Thoenen et al., 1971; Zaimis, 1972), it was tested for its ability to reverse the effects of preganglionic nerve section (decentralization). In mice and rats, NGF caused an increase in TH activity of both control and decentralized ganglia although it did not abolish the difference in enzyme activity between them (Black et al., 1972a; Hendry, 1973; Thoenen et al., 1972b). On the other hand, administration of the ganglion blocking drugs, chlorisondamine and pempidine, mimicked the effects of decentralization (Black, 1973; Hendry, 1973). Injection of NGF into control and pempidine treated animals increased the enzyme activities of both groups but again failed to eliminate the difference between them (Hendry, 1973). Thus, the neurotransmitter acetylcholine itself, rather than NGF, was more likely to be the factor involved in the trans-synaptic induction phenomenon. The cholinomimetic, carbachol, however, did not cause a premature induction of the activity of TH in neonatal mice (Black et al., 1972a). This may have been due to low circulating levels of glucocorticosteroids in these young animals. Induction of TH activity in rat superior cervical ganglia by reserpine requires the stimulation of both nicotinic and glucocorticoid receptors (Hanbauer et al., 1975), and, following cold stress, is maximal during the periods of highest plasma corticoid concentrations (Otten and Thoenen, 1975). Indeed, induction of TH activity by cholinomimetics in sympathetic ganglia in culture is only possible if the medium is supplied with adequate concentrations of corticosterone (Otten and Thoenen, 1975).

2.2.3. Trophic effects of sympathetic neurones on preganglionic fibres
The sympathetic neurones themselves influence the development of the presynaptic, cholinergic nerve terminals. During early postnatal development there is a 30% loss of postganglionic neurones (Hendry and Campbell, 1976) and a 50% loss of axons in the preganglionic nerve trunk (Aguayo et al., 1973). Selective sympathetic postganglionic target organ removal leads to a 28% decrease in neuronal number within the ganglion and a 35% decrease in CAT activity in the presynaptic fibres (Dibner et al., 1977). Destruction of the majority of the postganglionic adrenergic neurones by axotomy, 6-hydroxydopamine or antisera to NGF, results in a failure in the normal increase in acetylcholinesterase (Klingman and Klingman, 1969) and CAT activity within the ganglion (Black et al., 1972b; Hendry, 1975a) and a 70% decrease in the number of preganglionic fibres (Aguayo et al., 1976). These changes in the preganglionic nerve trunk may represent the death of preganglionic neurones following a reduction in the number of their target cells. Degeneration of preganglionic neurones occurs in the cervical, but not the thoracic region of the spinal cord during development (Levi-Montalcini, 1950), while neuroblasts resembling those of the preganglionic columns appear in pieces of cervical spinal cord when transplanted to thoracic levels (Shieh, 1951).

An increase in the size of the target cells of the preganglionic fibres following the injection of NGF into neonatal animals (Levi-Montalcini and Booker, 1960a) results in an increase in the number of preganglionic axons, the number of preganglionic synapses per neurone (Schaefer et al., 1980) and the activity of CAT within the ganglion (Thoenen et al., 1972c). Thus the number of preganglionic fibres and the maturation of their neurotransmitter synthetic mechanisms are regulated by the number and size of the postganglionic sympathetic neurones.

2.2.4. Preganglionic fibres and sympathetic cholinergic neurones
Experiments in tissue culture have shown that sympathetic neurones from neonatal rats can be induced to become cholinergic rather than proceed to mature as adrenergic (O'Lague et al., 1974; Johnson et al., 1976; Furshpan et al., 1976; Hill and Hendry, 1977). This change in transmitter type is due to a substance released by non-neuronal cells such as those within the ganglion itself (Patterson and Chun, 1974) or the cells comprising target organs such as the heart (Patterson and Chun, 1977). Sympathetic neurones can be prevented from responding to these agents by treatment with depolarizing agents or direct electrical stimulation in vitro (Walicke et al., 1977). These results suggest that in the absence of the preganglionic fibres, the sympathetic neurones may become cholinergic in response to factors from non-neuronal cells. Transection of the preganglionic nerve trunk in neonatal rats, however, did not result in an increase in the intrinsic CAT activity of the ganglion (Hill and Hendry, 1979).

2.3. Influence of target tissues on sympathetic neuronal development

2.3.1. Neurite outgrowth
In order to have a close interaction with potential target tissues, the developing neurones must extend processes. What, then, permits the growth of these processes and is this growth random or is it oriented in the direction of the target organ? Experiments designed to investigate the part played by target tissues in the growth of nerve fibres in vitro have shown that a variety of tissues can stimulate the growth of sympathetic nerve fibres, with tissues which are normally densely innervated in vivo being the most potent (Chamley et al., 1973b; Chamley and Dowel, 1975). Furthermore, the stimulation of nerve fibre outgrowth appeared to be directed towards the target tissue explants in that the fibres on the side of the ganglion facing the target were longer than those on the side away (Levi-Montalcini et al., 1954; Chamley et al., 1973b; Cook and Peterson, 1974; Ebendal and Jacobson, 1977a; Ebendal, 1981). An apparently similar unidirectional stimulation of fibre outgrowth occurred when sympathetic ganglia were transplanted 0.5 mm to 3 mm distant from iris explants under the kidney capsule in cats (Carruba et al., 1974).

Nerve growth factor was named for its effect in stimulating nerve fibre growth of sensory and sympathetic neurones (Cohen et al., 1954). The demonstration that nerve fibre growth could be oriented towards capillary tubes containing NGF (Charlwood et al., 1972) led to the suggestion that NGF may be released by target tissues to direct the ingrowth of sympathetic nerve fibres. Indeed, sympathetic nerve fibres are capable of responding to local concentrations of NGF (Campenot, 1977). When neonatal rat sympathetic neurones are grown in a three chamber culture system, nerve fibres will not penetrate chambers in which the medium lacks NGF (Campenot, 1977). Furthermore, the withdrawal of NGF from chambers containing nerve fibres leads to the cessation of growth and eventual degeneration of the fibres (Campenot, 1977). While NGF does not affect the site of axon initiation nor fibre growth rate in a dose dependent fashion (Letourneau, 1978), it appears that it can orientate the tips of nerve fibres and hence direct axon growth up a concentration gradient (Letourneau, 1978; Gunderson and Barrett, 1979, 1980). Evidence for a chemotaxis by sympathetic axons in vivo in response to NGF is seen with the spurious growth of peripheral sympathetic axons into the central nervous system following intracerebral injections of NGF in newborn rats (Menesini-Chen et al., 1978).

A major criticism of the involvement of NGF in neurite outgrowth is that it is impossible to separate these effects of NGF from those simply permitting cell survival. Since NGF has a generalized anabolic effect on sympathetic neurones (Bradshaw, 1978), a secondary effect may be to increase the number of processes produced by that neurone or to hasten recovery from a previous injury. The real issue, then, is whether sympathetic neurones will grow axons as long as they survive or whether a specific factor is required for neurite production.

In one case, NGF has been shown to be associated with neurite outgrowth but not neuronal survival. Treatment with antiserum to NGF does not destroy the neurones of the hypogastric and some other prevertebral ganglia (Vogt, 1964; Zaimis et al., 1965; Levi-Montalcini and Angeletti, 1968; Gorin and Johnson, 1980a) but it does prevent the regeneration of axons destroyed by 6-hydroxydopamine (Bjerre and Rosengren, 1974). However, assessment of the presence of fibres in peripheral tissues was made using only fluorescence histochemistry and assay for noradrenaline. The results, then, could be interpreted as a reduction in neurotransmitter content of the fibres due to NGF depletion rather than an actual absence of fibres. Indeed, exogenous NGF causes little, if any, changes in fibre density in the vas deferens but increases in other peripheral tissues (Olson, 1967; Bjerre and Rosengren, 1974).

2.3.2. Factors other than NGF influencing neurite outgrowth
Chick embryo heart contains a factor which stimulates nerve fibre outgrowth from cultured sympathetic neurones and whose action is not abolished by antiserum to mouse NGF (Ebendal and Jacobson, 1977b; Helfand et al., 1978; Ebendal, 1979; Ebendal et al., 1979; Lindsay and Tarbit, 1979; Obata and Tanaka, 1980; Varon et al., 1981). While these results are open to the criticism that there may be only limited cross reactivity between chick NGF and antisera to mouse NGF if, for example, the molecules only share the biologically active site in common (see also Harper and Thoenen, 1980), the failure of anti-mouse NGF to prevent

the effects due to target tissues from mice (Coughlin et al., 1978,1981) strongly suggests the existence of growth factors, other than NGF, affecting sympathetic neurones. The identification of the precise role of these factors during development in vivo must await their purification (Ebendal et al., 1979; Coughlin et al., 1981). It is interesting that the sources of these factors also support neuronal survival (Coughlin et al., 1981; Varon et al., 1981).

2.3.3. Development of the innervation pattern in target organs
The penetration of target tissues by sympathetic nerve fibres occurs predominantly during the early postnatal period in rats and mice. Following preparation of the vas deferens, iris and heart for the demonstration of catecholamines, only a few faintly fluorescent fibres are found at birth (de Champlain et al., 1970; Owman et al., 1971). The plexus formed by these fibres becomes more extensive during the second week postnatally with an increase in the number of fibres and the catecholamine content of their non-terminal regions (de Champlain et al., 1970). By the end of the third week, the plexus is similar to that of the adult following an increase in the catecholamine content of the terminals and a concomitant decrease in the non-terminal regions (de Champlain et al., 1970; Owman et al., 1971). Biochemical studies have shown that the endogenous noradrenaline content of the heart increases from 5% to 80% of adult levels over the first 3 weeks of life (Iversen et al., 1967). Similarly, in the iris, the TH activity increases from birth to 60 days postnatally (Black and Mytilineou, 1976a). The ability to take up exogenous noradrenaline also increases postnatally in both the heart and iris (Sachs et al., 1970) although at birth noradrenaline uptake is already between 10% and 20% of the adult capacity (Iversen et al., 1967; Black and Mytilineou, 1976a). Since noradrenaline uptake is primarily intraneuronal (Sachs et al., 1970), then the catecholamine uptake pump in the nerve membrane appears to be functional before the ability to synthesize noradrenaline.

2.3.4. Factors influencing the innervation pattern in target organs
The factors influencing the development of the pattern of innervation within a particular target organ have been investigated by transplanting tissues to the anterior chamber of the eyes of adult rats (Olson and Malmfors, 1970). Target organs are placed adjoining the host iris, where they become revascularised and reinnervated by the sympathetic fibres present within the host iris. Alternatively, sympathetic ganglia are positioned on the host iris and study is made of the penetration of fibres from the transplants into the already innervated host iris, or into the denervated host iris following extirpation of the host superior cervical ganglion (Olson and Malmfors, 1970). Several important conclusions arise from these experiments. The innervation pattern of various transplanted target organs is typical of the organ transplanted rather than of the host iris. Sympathetic ganglia from different positions in the paravertebral chain innervate the host iris

in a similar manner. The presence of a normal innervation pattern within the host iris prevents hyperinnervation due to penetration of the iris by fibres from transplanted ganglia. Thus, the specificity of the innervation pattern appears to be imposed on the ingrowing sympathetic fibres by the target organ.

The conclusions from these transplantation experiments pertain to the reinnervation of adult tissues rather than the innervation of developing tissues and it is possible that the previous exposure of the target tissue to nerve fibres may have imprinted on it a certain selectivity for axons of a similar kind. Differences do indeed exist between reinnervation and de novo innervation. For instance, after decentralization of the adult superior cervical ganglion, regenerating preganglionic nerves reoccupy the old postsynaptic sites (Raisman et al., 1974) while in the neonate the preganglionic nerves form new synapses (Smolen, 1981). Furthermore, developing postganglionic parasympathetic neurones are innervated by preganglionic fibres in spite of the ablation of their peripheral targets (Landmesser and Pilar, 1974a), while reinnervation of adult postganglionic sympathetic neurones is dependent on an intact connection with their peripheral targets (Purves, 1975). Similar results to those of the anterior eye chamber transplants have, however, been obtained in vitro using tissues from newborn rats. The dilator of the iris was penetrated by adrenergic fibres and the sphincter innervated by cholinergic fibres from both cervical and thoracic sympathetic ganglia (Hill et al., 1976). Furthermore, cardiac myocytes were innervated by sympathetic neurones via muscarinic synapses (Furshpan et al., 1976), while skeletal myotubes were innervated by the same neurones via nicotinic synapses (Nurse and O'Lague, 1975).

NGF has been implicated in the extent to which sympathetic nerve fibres ramify within explants of iris in culture. Penetration of fibres was increased by exogenous NGF and decreased by antisera to NGF or preincubation of the irides in culture medium, a procedure resulting in a decrease in their endogenous NGF content (Silberstein et al., 1971; Johnson et al., 1972). These changes, however, may have simply reflected changes in the total number of surviving neurones within the ganglionic explants. The failure to observe hyperinnervation of irides in anterior eye chamber transplants of sympathetic ganglia may be due to a decrease in free NGF within the iris following uptake and retrograde axonal transport of NGF to the cell bodies (Hendry et al., 1974b). NGF release from irides does indeed increase following denervation in vivo (Ebendal et al., 1980).

2.3.5. Influence of target organs on neuronal survival. A role for NGF
The importance of the interaction between sympathetic neurones and their target organs on the long term survival of the neurones has been studied either by the removal of a target organ or the severence of the major postganglionic nerve trunks.

Unilateral ablation of the salivary glands in both adult and neonatal mice results in a reduction in the TH activity in the superior cervical ganglion on the operated side (Hendry and Iversen, 1973). When the removed gland is replaced by a depot preparation of NGF, however, the TH activity of the ganglion on the operated side is not decreased. This is not simply due to diffusion of NGF, since an NGF preparation containing 30% bound to cellulose and 70% in soluble form causes an increase in TH activity of ganglia on both operated and unoperated sides. The proposal was therefore forwarded that sympathetic axons take up NGF from their target organs and this NGF permits their future survival and consequent differentiation (Hendry and Iversen, 1973).

Similar studies on target organ removal in neonatal rats have also revealed a reduction in the normal developmental increase in TH activity in the ganglion on the operated side (Dibner and Black, 1976). Histological studies show that neuronal numbers are reduced to 72% of control following sialectomy and iridectomy (Dibner et al., 1977). This compares favourably with the estimate of 17% for the proportion of neurones that can be shown to innervate the eye and salivary glands (Iversen et al., 1975). Furthermore, the effects in the ganglion are specific to the neurones which had previously innervated the ablated organ since sialectomy alone results in an inhibition of ganglionic TH by 30% but has no effect on TH activity in the iris (Dibner et al., 1977). Thus, the neurones innervating the iris are not affected by the removal of the salivary glands.

Axotomy of the neurones in the superior cervical ganglion of 6 day old rats prevents the normal increase in TH activity in the ganglion (Hendry, 1975b). The severity of the effects decrease with increasing age at which the operation is performed. Thus, there is no reversal of the ptosis in the eye following axotomy at 6 days of age but full recovery after axotomy at 6 weeks of age (Hendry, 1975b). Anatomical studies show that, during the normal development of the ganglion, there is a 30% loss of neurones between birth and 3 weeks of age and the loss amounts to 90% following axotomy (Hendry and Campbell, 1976). Failure to contact appropriate target organs within a specific period of time thus leads to cell death. Because of the effects of depot preparations of NGF following sialectomy (Hendry and Iversen, 1973), NGF was injected into rats and mice which had one of their superior cervical ganglia axotomized (Hendry, 1975c; Hendry and Campbell, 1976; Banks and Walter, 1977). In rats, the longer the NGF treatment from day 6, when unilateral axotomy was performed, the more similar the TH and protein levels were to control (Hendry, 1975c). Cell counts verified that NGF prevented the loss of neurones due to the operation, while on the control side NGF also prevented the 30% loss of cells which normally occurs during the first 3 weeks of life (Hendry and Campbell, 1976). Thus, NGF was able to overcome the effects of axotomy and, in control ganglia, was also able to maintain the 30% of neurones which would normally have died during the period of target organ synaptogenesis. Similar experiments in mice (Banks and Walter, 1977) have unfortunately been performed at a time when the TH activity of the superior cervical ganglion has already reached adult levels and the postnatal increase in neuronal number has also occurred (Black et al., 1972a). The conclusion that prolonged treatment with NGF was not able to prevent the effects of axotomy must therefore be tempered by the knowledge that the experiments were done outside the period of rapid biochemical differentiation of the adrenergic neurones. Considering the complications due to the nature of the operation and the probability of causing some cell death due to the severence of small blood vessels in the nerve trunks themselves, these studies are strongly suggestive of the participation of NGF in the interactions between sympathetic axons and effector cells. Indeed, injection of antibodies to NGF into developing rats, prevented the normal postnatal increase in TH in the neurones of the superior cervical ganglion (Goedert et al., 1978). Furthermore, the effects of the antiserum became less pronounced with increasing age at which the rats were first treated (Goedert et al., 1978) and this time course was almost identical to that seen following axotomy at different ages (Hendry, 1975b).

2.3.6. Transport of NGF to the nerve cell soma
Since sympathetic neurones could respond to preparations of NGF placed at their terminals, it became necessary to demonstrate the mechanism by which NGF had its effect on the neuronal cell bodies. Injection of ^{125}I-labelled NGF into the anterior eye chamber of mice led to a significantly greater accumulation of radioactivity in the superior cervical ganglion on the injected, compared to the non-injected, side and this difference was abolished by axotomy or colchicine (Hendry et al., 1974b). Autoradiography showed that the label was found in the iris one hour after injection and in the cell bodies of the neurones of the ganglion eleven hours later (Iversen et al., 1975). Thus, NGF was transported in a retrograde fashion by the axons of the sympathetic neurones. The specificity and characteristics of the transport system have since been examined in detail (Hendry et al., 1974a; Stöckel et al., 1974). Most importantly, the NGF transported by the neurones has been shown to be biologically active. Intraocular injection of insolubilized NGF in young rats, leads to an increase in the TH activity in the superior cervical ganglion on the injected side (Stöckel and Thoenen, 1975; Hendry, 1977). Oxidation of NGF so as to render it biologically inactive is accompanied by a reduction in its ability to bind to the NGF receptor (Banerjee et al., 1973; Cohen et al., 1980) and to be retrogradely transported (Stöckel et al., 1974). Furthermore, when neurones, whose terminals have been exposed to insolubilized NGF for 7 days, are labelled by injection of iodinated NGF into the same eye, the diameter of the labelled neurones is significantly greater than that of the surrounding unlabelled cells (Hendry, 1977). Therefore, NGF appears to remain within the neurones which transport it and its biological action is also restricted to these cells. The rescue by NGF of the 30% of neurones which normally die during the postnatal development of the rat superior cervical ganglion further supports the

theory that those neurones not making adequate connections or perhaps even those making inappropriate connections and hence not receiving adequate NGF do not survive.

2.3.7. Is NGF in target organs?
The hypothesis that NGF is a target tissue derived substance required for sympathetic neuronal survival depends upon the demonstration that target organs of the sympathetic nervous system synthesize and release NGF. At the present time there is conflicting evidence on this point.

Early investigations of NGF distribution employed radioimmunoassays and indicated the presence of NGF in a variety of tissues including such sympathetic target tissues as vas deferens and heart (Johnson et al., 1971; Hendry, 1972). Furthermore, the levels observed in the hearts of mice decreased from postnatal week 1 to postnatal week 4 (Johnson et al., 1971). While these results were suggestive of a developmental regulation, denervation did not cause any change in the NGF content per organ (Johnson et al., 1971). The results discussed above were obtained using either one-site (Johnson et al., 1971) or two-site (Hendry, 1972) radioimmunoassays. The one-site assay measures the decrease in radioactivity resulting from the displacement, by the sample NGF, of labelled NGF from antibody. On the other hand, the two-site assay measures the increase in radioactivity due to the binding of labelled antibody to the sample NGF which has been previously bound to a solid phase antibody, thus requiring two antigenic sites for the detection of activity.

Results of one-site assays have provided high values for NGF content and these values have been questioned on the basis of the displacement of the labelled NGF by components other than NGF in serum (Suda et al., 1978). Such binding then produces spuriously high estimates of NGF levels in the sample to be analyzed. On the other hand, the presence of these binding components in serum gives a falsely low estimate of NGF in the two-site assay due to the lower amount of NGF available to bind in the first stage of the assay. The estimations of NGF in peripheral tissues by both one- and two-site radioimmunoassays have been further queried since both studies employed commercial antisera contaminated with anti-mouse gammaglobulin (Carstairs et al., 1977). While this was a damaging comment on the results of the one-site assay, the antibodies used in the second phase of the two-site assay were labelled when bound to β-NGF (Hendry, 1972) thereby reducing the probability of spurious results. Purified antibodies have been used in both stages of the two-site assay to minimise the problems of impurities in the antisera (Suda et al., 1978). However, due to the binding of NGF by components in serum, the estimate of NGF in serum in this study was below the limit of detection of the assay. Thus, at present there is no satisfactory radioimmunoassay demonstrating the levels of NGF in tissues and in serum; one-site assays have provided falsely high estimates of NGF and two-site assays misleadingly low values.

More recent attempts to demonstrate NGF in sympathetic target tissues have relied on the ability of pieces of tissue, or of tissue homogenates, to stimulate nerve fibre outgrowth from embryonic chick sensory ganglia in tissue culture, as originally demonstrated by Levi-Montalcini et al. (1954). Using this type of bioassay, NGF has been shown to be synthesized and released by a variety of tissues when grown in culture but no such activity has been detected in homogenates of the same tissues (Harper et al., 1976; Harper et al., 1980a; Harper et al., 1980b; Ebendal et al., 1980). While these results are suggestive of a production of NGF in vitro by tissues which do not make it in vivo, perhaps a good deal of the blame can be laid on the assay itself. The criticism of serum contaminants altering the level of NGF detected by radioimmunoassay can be equally well levelled at the bioassay since the sensory ganglia are grown in serum-containing media. Indeed, estimates of a known concentration of NGF are lower in the presence of serum than in its absence (Suda et al., 1978). Furthermore, the sensitivity of the bioassay varies from laboratory to laboratory. Using intact embryonic chick sensory ganglia, Harper et al. (1980a) were unable to detect neurite promoting activity in any embryonic chick tissue including the heart. On the other hand, Ebendal (1979) observed a significant stimulation of neurites from similarly cultured sensory ganglia evoked by explants of heart obtained from embryos of age from 4 days of incubation to post hatching. Irrespective of whether or not these responses were sensitive to antisera to NGF, one would still reach opposite conclusions from these two studies in regard to the ability of explants of embryonic chick heart to stimulate neurite outgrowth from sensory ganglia. In the bioassay of Harper et al. (1980a,b) the limit of sensitivity for the detection of NGF was, according to the authors, 6 ng/ml. Greene (1974) has described a bioassay for NGF based on the production of neurites by neurones dissociated from 11 day old chick embryo sympathetic ganglia. This assay can detect NGF at concentrations as low as 0.01 to 0.03 ng/ml of the β or 2.5 S NGF preparations with a maximum response occurring with a concentration of 0.3 to 0.5 ng/ml (Greene, 1974; Greene, 1977a). Similarly, neurones dissociated from 8-day-old embryonic chick sensory ganglia show a maximum response at a concentration of 0.5 ng/ml 2.5 S NGF (Greene, 1977b). Perhaps these more sensitive bioassays should be used in those cases where the classical bioassay is too insensitive to detect activity.

2.3.8. Effects of innervation on NGF release from target organs
The fact that tissues in situ do not contain NGF in concentrations detectable in the classical bioassay does not eliminate NGF from contention as a mediator in nerve–muscle interactions. The presence of significantly lower levels of NGF in innervated tissues in situ than in denervated tissues in vitro is consistent with the

results of Ebendal et al. (1980). These authors found that freshly excised rat irides killed by freezing and thawing did not contain NGF detectable by bioassay. Twenty-four hours after transplantation to the anterior eye chamber or to a culture dish, however, NGF could be detected (Ebendal et al., 1980). Furthermore, freshly excised rat irides, which had been denervated ten days previously, now contained detectable NGF activity. Thus the level of NGF in the iris was increased by denervation resulting from either explantation in culture, transplantation to the anterior eye chamber or severence of the normal innervation to the iris. These results support the hypothesis that NGF acts as the messenger between sympathetic neurones and their target organs during reinnervation. Analysis of NGF content in target tissues during development using more sensitive assays or histochemistry is necessary to demonstrate that NGF plays such a role during de novo innervation. It is possible, however, that the appropriate amount of NGF is released by the target tissues only in response to a stimulus from the ingrowing axons. Indeed, the cellular content and release of NGF by C6 glioma cells in culture can be increased following β-adrenergic receptor stimulation (Schwartz and Costa, 1977; Schwartz et al., 1977) and the growth cones of cultured sympathetic neurones do contain adrenergic vesicles (Landis, 1978). One would then predict that alterations in neuronal activity would produce alterations in NGF synthesized and released by target tissues. Neither increases in activity due to cold stress nor decreases due to decentralization or ganglion blocking drugs have any effect on the retrograde axonal transport of labelled NGF by neurones of the superior cervical ganglion in rats and guinea pigs (Freeman et al., 1978; Stöckel et al., 1978; Johnson et al., 1979; Lees et al., 1981). In these studies, however, the amount of labelled NGF (50 ng to 4 μg) injected into the eye would have been much higher that that existing within the iris itself (estimates of 5 to 10 ng NGF/g wet weight of iris rising to 80 to 160 ng/g wet weight after denervation have been made by Ebendal et al. (1980)) and so displacement of labelled NGF from the transport system due to small changes in the endogenous content of NGF would not have been detected. Experiments are required where truly tracer amounts of NGF are used (see Johnson et al., 1978).

2.3.9. Factors other than NGF influencing neuronal survival
While the preceding sections have indicated that NGF may be a retrophin for sympathetic neurones, recent studies in vitro have suggested that other molecules can be released by target tissues in culture to permit the survival and biochemical differentiation of sympathetic neurones (Coughlin et al. 1978, 1981; Edgar et al., 1981; Varon et al., 1981). The responses are not blocked by antisera to mouse NGF (Coughlin et al., 1978, 1981; Edgar et al., 1981; Varon et al., 1981) and parasympathetic neurones are also affected (Coughlin et al., 1981) while in 2 studies, conspecific sensory neurones are not (Coughlin et al., 1981; Varon et al., 1981).

Reliable classification of target neurones for a particular factor must, however, await purification of that factor as partially pure preparations may still contain more than one, non-NGF, factor.

2.3.10. Developmental changes in requirements for survival factors
Studies of the effects of antisera to NGF on developing sympathetic neurones led to the conclusion that sympathetic neurones were more susceptible to NGF and its antibody at certain stages of development (Klingman, 1966; Klingman and Klingman, 1967). Thus, injection of pregnant mice with antisera to NGF during early stages of gestation appeared to be less effective in decreasing noradrenaline levels in target tissues than treatment during later stages (Klingman, 1966; Klingman and Klingman, 1967). These studies were refined and extended by Kessler and Black (1980) who injected foetuses directly with antisera to NGF, or NGF itself, over a range of gestational ages from 12 days onwards and assessed the effect of the treatment by TH assay in the superior cervical ganglion. They found that, while the neurones responded to NGF at all ages with an increase in TH activity, the greatest effects of the antisera were obtained with embryos from 17 days onwards. Similar results were obtained with superior cervical ganglia grown in tissue culture. Ganglia from 13- to 15-day mouse embryos grew for 2 to 3 days without exogenous NGF and then degenerated (Coughlin et al., 1977). During the first 2 days in vitro, nerve fibres regenerated and the TH activity increased three fold but anti-NGF had no effect on these changes. On the other hand, ganglia from 18-day-old foetuses showed virtually no axonal outgrowth and a 50% reduction in TH activity in the absence of NGF. Ganglia from both ages, however, responded to exogenous NGF (Coughlin et al., 1977).

In the early postnatal period, the sympathetic neurones were still susceptible to destruction with anti-NGF, 95% to 98% of the neurones disappearing (Levi-Montalcini and Booker, 1960b). With increasing age up to 3 weeks, there was a progressively decreasing effect of antisera on sympathetic neuronal survival (Goedert et al., 1978). Injection of antisera to NGF into adult animals caused reversible neuronal damage (Angeletti et al., 1971). Long term immunization of adult rats, however, has been reported to result in either no cell death (Otten et al., 1979) or a 35% to 40% decrease in the number of sympathetic neurones (Gorin and Johnson, 1980a). In the latter study, only those animals with persistently high serum titres were used. Thus, sympathetic neurones in adult animals appear to be relatively insensitive to antibodies to NGF when compared to those from neonates.

The relative importance to the sympathetic neurones of prenatal and postnatal exposure to NGF has also been studied by raising litters from rats immunized with NGF (Gorin and Johnson, 1979). By cross fostering control and immunized litters with control and immunized mothers, the effects of exposure in utero can be compared with postnatal exposure to antibodies from milk. The effects on TH activity in the superior cervical ganglion of exposure to anti-NGF in utero were more variable and sometimes less extensive than those arising from postnatal exposure in milk (Gorin and Johnson, 1979, 1980b), although serum titres of foetuses as determined at birth were considerably lower than those of the young animals in the milk group (Gorin and Johnson, 1979). The effects seen were due to cell death since rats evaluated at maturity still showed reduced enzyme levels (Gorin and Johnson, 1980a).

These various experiments demonstrate that sympathetic neurones in rodents go through a period of development during which they require NGF for survival. For the majority of neurones this period begins about 17 days of gestation and ceases by 2 to 3 weeks postnatal.

In culture, neurones from 14- to 15-day-old foetuses responded to target tissues, such as heart or salivary glands, with an increase in survival, neurite outgrowth and TH activity and these effects were not prevented by antisera to NGF (Coughlin et al., 1978, 1981). Thus, at early times, factors other than NGF are likely to be important for survival and differentiation. Sympathetic neurones cultured from chick embryos of different ages have also been shown to pass through periods when they require NGF for survival (Edgar et al., 1981) or neurite outgrowth (Partlow and Larrabee, 1971) and periods when they require other factors from media conditioned by heart or glial cells (Edgar et al., 1981).

2.4. Regulation of synaptic connections

A series of neurones which form a pathway may be regulated as a unit. Events occurring at one synapse may cause trans-synaptic modifications to neurones of the entire series.

2.4.1. Effect of the loss of synapses with target organs on preganglionic innervation

Axotomy of postganglionic sympathetic axons causes a depression of synaptic transmission through the ganglion (Brown and Pascoe, 1954). While changes in postsynaptic sensitivity to acetylcholine (Brown and Pascoe, 1954; Brenner and Martin, 1976) account for some decrease in transmission, there is also a decrease in the number of ganglionic synapses amounting to about 70% in sympathetic ganglia (Matthews and Nelson, 1975; Purves, 1975). In adult rats and guinea pigs, postganglionic nerve crush results in a 50% loss of neurones (Purves, 1975; Hendry, 1976b). Only 30% of the neurones survive a chronic ligation and these remaining cells fail to show synaptic recovery (Purves, 1975). Interruption of axonal transport with colchicine also results in synaptic depression, chromatolytic changes within the neurones (Pilar and Landmesser, 1972) and a loss of synapses (Purves, 1976). These results led to the suggestion that the integrity of the ganglionic synapses depends upon contact of the postganglionic axons with peripheral target tissues. An alternative explanation, however, is that the withdrawal of synapses is due to a loss of specializations for attachment on the postganglionic neuronal surface (see also Matthews and Nelson, 1975). Regenerating preganglionic nerves penetrating the adult rat superior cervical ganglion reoccupy the old postsynaptic thickenings which persist for several months following decentralization (Raisman et al., 1974). These postsynaptic structures disappear after axotomy (Matthews and Nelson, 1975).

Since NGF is essential for sympathetic neuronal survival and consequent maturation (Levi-Montalcini and Angeletti, 1968) and is also retrogradely transported from the nerve terminals to the cell bodies (Hendry et al., 1974b; Stöckel et al., 1974), NGF was tested for its ability to reverse the effects of axotomy, and anti-NGF to mimic axotomy. An NGF impregnated silicone pellet placed in contact with the ventral surface of the ganglion did reduce the depression of synaptic transmission and the loss of preganglionic synapses (Njå and Purves, 1978a). On the other hand, injection of antiserum to NGF caused a 50% decrease in the number of synapses per unit area, depression of synaptic responses and a loss of post-synaptic specialization (Njå and Purves, 1978a). From these results it was concluded that target cells provide specific trophic factors, in this case NGF, to the neurones which innervate them and, following retrograde axonal transport of these factors, synapses on these neurones are also maintained (Njå and Purves, 1978a). The maintenance of the preganglionic synapses may be due to the production and secretion of an agent analogous to NGF, since injection of labelled NGF into the anterior eye chamber results in the accumulation of label in the ganglion cell bodies but not in the preganglionic nerve terminals (Schwab, 1977; Schwab and Thoenen, 1977). The agent released by the ganglionic cell bodies may then be retrogradely transported by preganglionic axons and be important for the survival of these cells (see also Purves and Lichtman, 1978). In this way a series of synapses may be regulated as a unit by the production of appropriate trophic agents which are taken up and transported in a retrograde direction.

Regulation of sequential synapses might account for the specificity of connections through the superior cervical ganglion. Stimulation of the preganglionic input to the ganglion at different spinal levels elicits different peripheral responses in the head and neck (Langley, 1892). This specificity of connections is regained on the regeneration of the preganglionic fibres after ligation (Langley, 1895). Intracellular neuronal recordings show that each neurone is innervated by a group of preganglionic nerve fibres and these fibres arise from contiguous segments, one of which is dominant, the influence of the other fibres decreasing with increasing distance from the dominant segment (Njå and Purves, 1977). Moreover, the antero-posterior origin of the preganglionic innervation of a ganglion cell is matched to the rostrocaudal position of its peripheral target (Lichtman et al., 1979). Thus neurones innervating target organs at the same peripheral position are contacted by preganglionic fibres arising from the same spinal segments (Lichtman et al., 1979). Even during the early stages of synapse formation following ligation of the preganglionic nerve trunk, neurones can still be shown to be innervated by fibres from contiguous segments (Njå and Purves, 1978b). Thus, the peripheral target tissue may specify the neurone so that it receives an appropriate synaptic input.

2.4.2. Effect of the loss of preganglionic synapses on peripheral innervation
When the preganglionic trunk to the superior cervical ganglion is sectioned in 2- to 3-day-old rats, the nerve fibre plexus within the iris and pineal gland fails to develop normally (Black and Mytilineou, 1976a; Mytilineou and Black, 1978; Lawrence et al., 1979). The greatest differences between operated and unoperated sides are seen for denervation at 2 to 4 days of age, the magnitude of the effects decreasing with increasing age at which the operation is performed (Mytilineou and Black, 1978). This phenomenon is not due to neuronal death within the ganglion (Lawrence et al., 1979) nor to a deficit of NGF (Black and Mytilineou, 1976b). Injection of NGF into rats whose ganglia were denervated at day 2, had no effect on the reduced nerve density or TH activity of the irides (Black and Mytilineou, 1976b). The conclusion from these studies was that the development of the innervation pattern of the iris requires an intact cholinergic innervation within the ganglion (Black and Mytilineou, 1976b). This is in contrast to cell culture studies in which adrenergic sympathetic neurones form contacts on heart cells in the absence of a cholinergic input (Furshpan et al., 1976).

Ganglia decentralized in 3-day-old rats and placed in culture one day later in the presence of sufficiently high concentrations of NGF to maintain neuronal survival, fail to exhibit the increase in TH activity seen in ganglia grown in vitro for the entire period (Hendry and Hill, 1980a). The development of sympathetic neurones which contact the periphery but fail to receive preganglionic innervation is inhibited.

2.4.3. Effect of spinal transection on the development of sympathetic neurones
In young rats, section of the axons which innervate the cell bodies of the preganglionic fibres to the sixth lumbar sympathetic ganglion results in an impairment of the normal developmental increase of both CAT and TH activities within the ganglion (Black et al., 1976; Hamill et al., 1977). This surgery did not, however, destroy preganglionic cells of the intermediolateral column (Hamill et al., 1977). The integrity of these central synapses therefore appears to be important for the maturation of the neurones of the intermediolateral column and, via their synapses, the maturation of the postganglionic sympathetic neurones.

3. Development of parasympathetic neurones

3.1. The avian ciliary ganglion

The avian ciliary ganglion comprises two morphologically and functionally distinct types of neurone in approximately equal numbers, the ciliary and the choroid (Marwitt et al., 1971). The larger ciliary neurones are myelinated and receive calyciform terminals in young birds and bouton terminals in adults (Marwitt et al., 1971). Synaptic transmission is both chemical and electrical (Martin and Pilar, 1963). The choroid neurones are unmyelinated and receive bouton type terminals from slower conducting fibres (Marwitt et al., 1971). The ciliary neurones innervate the iris and ciliary muscles and the choroid neurones the vasculature in the choroid coat (Marwitt et al., 1971). Both pre- and postsynaptic neurones employ the neurotransmitter, acetylcholine.

3.2. Influence of preganglionic fibres

During development, synaptic transmission through the ganglion begins at stage $26\frac{1}{2}$ ($4\frac{1}{2}$ to 5 days of incubation; Hamburger and Hamilton, 1951) and is 100% by stage 33 ($7\frac{1}{2}$ to 8 days; Landmesser and Pilar, 1972). Electrical coupling first appears in the ciliary cells at stage 41 (15 days) and increases to 80% by 1 to 2 days post hatching (Landmesser and Pilar, 1972). From the earliest time recorded, the preganglionic fibres innervating the ciliary cells are of lower threshold and faster conduction velocity than those innervating the choroid cells (Landmesser and Pilar, 1972). Following section of the preganglionic nerve in the adult pigeon ciliary ganglion, the neurones become reinnervated in a selective fashion, i.e., the ciliary neurones are innervated at first chemically then later chemically and electrically, and the choroid neurones only chemically, by slower conducting fibres (Landmesser and Pilar, 1970). This selectivity is not imposed by prior contact with peripheral target tissues as discussed in the previous section, since the two cell types can be distinguished and are innervated normally in ciliary ganglia which have had their peripheral targets removed early in development (stages 16 and 20, 2 to 3 days; Landmesser and Pilar, 1974a).

Section of the preganglionic nerves to the ciliary ganglia in adult birds results in a decrease in acetylcholine in the postsynaptic neurones (Pilar et al., 1973). This decrease in acetylcholine content appears to be mediated by a decrease in the CAT in the postsynaptic cells (Giacobini et al., 1979). During development, there is an increase in both CAT and acetylcholinesterase in the ganglion (Pilar et al., 1974; Sorimachi and Kataoka, 1974; Chiappinelli et al., 1976). By determining the relative contributions of the pre- and postsynaptic elements to these enzyme activities and by comparing the timing of the enzymic increases with the development of synaptic transmission through the ganglion, the authors concluded that the formation of synapses increased the activities of acetylcholinesterase in the postsynaptic cells and CAT in the presynaptic cells (Pilar et al., 1974; Chiappinelli et al., 1976). The appearance of nicotinic cholinergic receptors as demonstrated by alpha bungarotoxin binding increases in the ganglion between stages 31 and 37 (days 7 and 11) and therefore also appears to increase on the postsynaptic cell in response to synapse formation (Chiappinelli and Giacobini, 1978). Blockade of these nicotinic receptors with chlorisondamine results in a decrease in ganglionic volume and a decrease in CAT activity (Chiappinelli et al., 1978) due to a 24% reduction in both the ciliary and choroid populations (Wright, 1981).

3.2.1. Trophic effects of parasympathetic neurones on preganglionic neurones
Destruction of more than 90% of the neurones of the ciliary ganglion due to the early removal of the optic vesicle is followed by the degeneration of the cell bodies of the preganglionic fibres within the accessory oculomotor nucleus (Cowan and Wenger, 1968). Elimination of only 24% of the ganglion neurones following ganglionic blockade, however, had no detectable effect on the survival of the preganglionic cells (Wright, 1981). These results suggest that the preganglionic cells are still able to receive sufficient trophic support in spite of the blockade of the nicotinic receptors and the loss of one quarter of their target cells. On the other hand, an increase of 8% to 27% in the number of ciliary neurones was accompanied by an increase of 9% to 33% in the number of cells in the accessory oculomotor nucleus (Narayanan and Narayanan, 1978).

3.3. Influence of target organs

3.3.1. Neurite outgrowth
Small unmyelinated axons are seen in the iris at stage 25 ($4\frac{1}{2}$ days; Landmesser and Pilar, 1972) although neuromuscular junctions are not functional until stage 34 to 38 (8 to 12 days; Landmesser and Pilar, 1974b). The initiation of axons and their growth towards the periphery do not appear to be dependent on target tissues since ciliary ganglion cells develop normally and grow axons even when their prospective targets are ablated very early in development (Landmesser and Pilar, 1974a). Secondary collateral sprouting, however, which normally occurs between stages 34 and 36 (8 to 10 days) is not seen in the nerves of peripherally deprived neurones (Landmesser and Pilar, 1976).

3.3.2. Target tissue derived factors influencing neurite outgrowth
Target tissues release substances which promote neurite outgrowth from cultured parasympathetic neurones (Coughlin, 1975b; Helfand et al., 1976; Ebendal and Jacobson, 1977b; Collins, 1978; McLennan and Hendry, 1978; Ebendal, 1979; Adler and Varon, 1980). Factors that cause neurite outgrowth can be separated under appropriate circumstances from factors that cause neuronal survival (Col-

lins, 1978; Adler and Varon, 1980). The neurite promoting factors bind to culture substrata and are then apparently more active (Collins, 1978). It is not known if these factors are only important for the attachment of neurites to various substrata or whether they have a physiological role.

3.3.3. Development of the innervation pattern in target organs

The branching of neurites within the submandibular glands parallels the morphogenetic development of the submandibular gland epithelium itself (Coughlin, 1975a). Tissue culture studies have shown that neurite outgrowth from the submandibular ganglion occurs predominantly towards the epithelium. Stimulation of fibre outgrowth is still seen when a 0.1 μm pore filter separates the salivary epithelium from the ganglion (Coughlin, 1975b). Subsequent studies, however, demonstrated that glandular homogenates and conditioned media did not stimulate axon growth while live and formalin fixed glands did (Coughlin and Rathbone, 1977). These experiments did not, then, distinguish between an epithelial product which was highly labile and diffusible or one which was only active when bound to the substrate.

3.3.4. Influence of target organs on neuronal survival

A decrease in the number of neurones in a nerve centre coincident with the onset of functional synapse formation by those neurones is a phenomenon which has been described for a number of systems (Hughes, 1968; Prestige, 1970; Cowan, 1973). Ablation of the target cells for those neurones leads to an even greater degeneration (Hamburger and Levi-Montalcini, 1949), while transplantation of additional peripheral tissue causes a reduction in the normally occurring cell loss (Hollyday and Hamburger, 1976) and promotes the differentiation of the innervating neurones (Hamburger and Levi-Montalcini, 1949).

In the avian ciliary ganglion, the number of neurones is reduced by 50% between stages 35 and 39 (8 to 13 days; Landmesser and Pilar, 1974b). This cell death coincides with the establishment of functional neuromuscular junctions in the iris muscle (Landmesser and Pilar, 1974b). Removal of the optic cup early in development augments this cell death so that only 8% of neurones remain on the operated side by stage 39 (Amprino, 1943; Cowan and Wenger, 1968; Landmesser and Pilar, 1974a). Axotomy of the major postganglionic nerves at stage 32 ($7\frac{1}{2}$ days) also results in the rapid death of the majority of the neurones (Pilar et al., 1980). Thus, the parasympathetic neurones can survive without a periphery up until the normal time of neuromuscular junction formation. The current debate now centres on whether the neurones which die in the presence of the target tissue are those which were unable to make any connections or had made incorrect connections, or whether they had simply not made enough connections for survival. In a series of studies, Landmesser and Pilar have shown that the number of axons in the postganglionic nerves exceeds the number of ganglion cells and that between stages 36 and 40 ultrastructural changes indicative of increased synthetic activity occur in all the ganglion cells but are not seen in neurones lacking a periphery (Landmesser and Pilar, 1976; Pilar and Landmesser, 1976). These results are interpreted as indicating that interaction with the periphery triggers the neurone into a secretory state and cell death may be due to a build up of materials following the formation of an insufficient number of synapses (Pilar and Landmesser, 1976). Measurements of conduction velocities before and after neuromuscular junction formation show that few ciliary axons grow down the choroid nerves and vice versa (Landmesser and Pilar, 1972; Landmesser and Pilar, 1976). Therefore, naturally occurring cell death is not designed to remove cells making incorrect peripheral connections. Narayanan and Narayanan (1981), however, maintain that, since cell death occurs at the same time as the increase in the number of synaptic terminals in the iris, those neurones dying are the ones which do not make contact with the periphery. These two hypotheses, however, need not be mutually exclusive.

An increase in the size of the periphery does lead to a reduction in the naturally occurring cell death suggesting that under normal conditions some aspect of the periphery is limiting. Transplantation of an additional optic primordium early in development results in an increase from 8% to 27% in the number of ciliary neurones (Narayanan and Narayanan, 1978). Axotomy of two of the three major branches of the ciliary nerves before the period of cell death results in an increase in the peripheral field for the ciliary axons in the third branch (Pilar et al., 1980). Counts of the number of horseradish peroxidase labelled cells projecting through this branch on the control and axotomised sides after the period of cell death showed that there is a 100% increase on the side of the operation (Pilar et al., 1980). Furthermore, the axons of these surviving cells were larger and more matured than control suggesting that the cells develop faster when competition is reduced (Pilar et al., 1980; see also Hamburger and Levi-Montalcini, 1949).

3.3.5. Factors influencing neuronal survival

Studies of parasympathetic neurones grown in vitro have shown that factors permitting cell survival occur in extracts of target tissues and media conditioned by such cells in culture (Helfand et al., 1976; Nishi and Berg, 1977, 1979; Landa et al., 1980; Adler et al., 1979; Bennett and Nurcombe, 1979; Tuttle et al., 1980; Bonyhady et al., 1980; Coughlin et al., 1981; Hill et al., 1981). Optic tissues grown in culture synthesize and release proteins which are retrogradely transported by the neurones of the avian ciliary ganglion in vivo (Hendry and Hill, 1980b). Partially purified factor from bovine heart is also retrogradely transported by parasympathetic neurones (Hendry et al., 1981) and the isoelectric points of the transported material and of the biological activity of the partially purified factor coincide at pH 5 (Hendry, Bonyhady and Hill, unpublished results). While these results suggest that the factor permitting parasympathetic neuronal survival in vitro is present in target tissues in vivo and can be synthesized and released in vitro, the role of such a factor in vivo is yet to be determined.

3.3.6. Effects of synapse formation on nerve and muscle maturation
Correlation of the developmental increases in the activity of acetylcholinesterase and neuromuscular transmission in the iris had led to the suggestion that the formation of synapses induces the development of the enzyme in the muscle cells (Pilar et al., 1974; Chiappinelli et al., 1976; Narayanan and Narayanan, 1981). Alpha bungarotoxin binding indicative of the appearance of acetylcholine receptors in the muscle also increases in the iris soon after functional innervation is first observed (Chiappinelli and Giacobini, 1978). Following partial axotomy, no differentiated muscle cells are seen in areas where nerves have not penetrated (Pilar et al., 1980). Neuromuscular junctions therefore appear to be essential for the normal maturation of the muscle. CAT activity, which occurs within the nerve terminals, is also increased after synapse formation perhaps due to an increase in the enzyme in the ganglion cell bodies and its transport to the terminals (Pilar et al., 1974).

4. General discussion

Results of experiments using embryonic chick sensory neurones in culture suggest that axon initiation, elongation and the number of axons produced by a neurone can be promoted with the use of particular substrates (Letourneau, 1975). These effects were discussed in terms of an increased adhesion of the neurone and its processes to the substrate (Letourneau, 1975). Similar changes in cell substrate adhesivity have been observed for sensory neurones cultured in the absence of serum (Ludueña, 1973). More recently, there has been demonstration of the deposition of substances onto certain substrates thereby permitting neurite adhesion and extension (Helfand et al., 1976; Collins, 1978; Hawrot, 1980; Adler and Varon, 1981). If growth of axons in vivo is also favoured along certain pathways due to the deposition of substances on cell surfaces or on the extracellular matrix, it is most likely that these molecules will be produced by cells of ganglionic origin or those occurring along the pathway, rather than by the cells of the target tissue, since neurones lacking a periphery still send axons out towards the usual position of the target tissue (Landmesser and Pilar, 1974a).

During the growth towards the periphery, a certain selectivity has been observed for the paths of ciliary versus choroid neurones of the parasympathetic ciliary ganglion (Landmesser and Pilar, 1976). This apparent specificity of nerve fibres to the correct path could be explained by a preferential adhesion amongst ciliary axons and amongst choroid axons. Such a cell adhesion molecule has been described, purified and characterized from neural retina (Brackenbury et al.,

1977; Thiery et al., 1977). In cultures of sensory neurones, this molecule has also been demonstrated to be important in neurite–neurite adhesion, resulting in fasciculation (Rutishauser et al., 1978). Indeed, a role for such a neurite adhesion molecule in the development of fibre tracts towards target tissues has been suggested following the reduction in neurite asymmetry from sensory ganglia in response to an NGF point source when antibodies to the cell adhesion molecule are present in the medium; with higher NGF concentrations, more neurites are produced (perhaps due to improved neuronal vitality) and these bind together into bundles due to the cell adhesion molecule (Rutishauser and Edelman, 1980).

The final approach of the neurites to the target tissue and their growth within it may be under the influence of diffusible factors produced and released by the target tissues, such as NGF, which has been shown to orient growth cones (Gunderson and Barrett, 1979, 1980), or substrate bound factors, which could improve the rate of growth of fibres towards the target. Neurite promoting factors which bind to certain substrates have been described in media conditioned by target tissues (Helfand et al., 1976; Collins, 1978; Adler and Varon, 1980). Ramification of fibres within the target tissue may also be promoted by substrate bound factors. Increased adhesion and fibre branching has been observed in tissue culture when sympathetic nerve fibres grow over smooth and cardiac muscle cells but not fibroblasts (Chamley et al., 1973a; Mark et al., 1973).

The interaction between the nerve terminal and the cells of the target tissue leads to functional synapse formation. This interaction may involve release of neurotransmitter from the nerve and concomitant release of trophic factor from the target cells. Coincident with, or slightly earlier, the preganglionic fibres have made functional connections with the cell bodies of these nerve fibres. In both the ganglion and the periphery, the effect of synapse formation is to stimulate the differentiation and maturation of the postsynaptic cells and permit the survival of the presynaptic cell. The latter effect is probably mediated by trophic molecules released by the target tissue, taken up into the nerve terminals and retrogradely transported to the neuronal cell bodies. Experimental isolation of autonomic neurones from their targets over this important period of synapse formation leads to degeneration of more than 90% of the neurones in the ganglion. Chronic trophic deprivation of sympathetic neurones in adult animals, following either axotomy or antibody treatment, leads to a 30% to 50% cell death. Thus, the requirement for target-tissue derived trophic molecules may persist to adulthood, at least in some neurones.

Acknowledgements

The authors are pleased to acknowledge the support of the National Heart Foundation of Australia in the form of a Research Fellowship and a Grant-in-Aid to C.E.H.

References

Adler, R. and Varon, S. (1980) Brain Res. *188*, 437–448.
Adler, R. and Varon, S. (1981) Dev. Biol. *81*, 1–11.
Adler, R., Landa, K.B., Manthorpe, M. and Varon, S. (1979) Science *204*, 1434–1436.
Aguayo, A.J., Terry, L.C. and Bray, G.M. (1973) Brain Res. *54*, 360–364.
Aguayo, A.J., Peyronnard, J.M., Terry, L.C., Romine, J.S. and Bray, G.M. (1976) J. Neurocytol. *5*, 137–155.
Amprino, R. (1943) Arch. Ital. Anat. Embriol. *49*, 261–300.
Angeletti, P.U., Levi-Montalcini, R. and Caramia, F. (1971) Brain Res. *27*, 343–355.
Banerjee, S.P., Snyder, S.H., Cuatrecasas, P. and Greene, L.A. (1973) Proc. Natl. Acad. Sci. USA *70*, 2519–2523.
Banks, B.E.C. and Walter, S.J. (1977) J. Neurocytol. *6*, 287–297.
Bennett, M.R. and Nurcombe, V. (1979) Brain Res. *173*, 543–548.
Bjerre, B. and Rosengren, E. (1974) Cell Tissue Res. *150*, 299–322.
Björklund, A., Enemar, A. and Falck, B. (1968) Z. Zellforsch. *89*, 590–607.
Black, I.B. (1973) J. Neurochem. *20*, 1265–1267.
Black, I.B. (1978) in Annual Reviews of Neuroscience (Cowan, W.M., ed.) Vol. 1, pp. 183–214, Annual Rev. Inc., California.
Black, I.B. and Mytilineou, C. (1976a) Brain Res. *101*, 503–521.
Black, I.B. and Mytilineou, C. (1976b) Brain Res. *108*, 199–204.
Black, I.B., Hendry, I. and Iversen, L.L. (1971a) Nature New Biol. *231*, 27–29.
Black, I.B., Hendry, I.A. and Iversen, L.L. (1971b) Brain Res. *34*, 229–240.
Black, I.B., Hendry, I.A. and Iversen, L.L. (1972a) J. Neurochem. *19*, 1367–1377.
Black, I.B., Hendry, I.A. and Iversen, L.L. (1972b) J. Physiol. (London) *221*, 149–159.
Black, I.B., Bloom, E.M. and Hamill, R.W. (1976) Proc. Natl. Acad. Sci. USA *73*, 3575–3578.
Bonyhady, R.E., Hendry, I.A., Hill, C.E. and McLennan, I.S. (1980) Neurosci. Lett. *18*, 197–201.
Brackenbury, R., Thiery, J.-P., Rutishauser, U. and Edelman, G.M. (1977) J. Biol. Chem. *252*, 6835–6840.
Bradshaw, R.A. (1978) Ann. Rev. Biochem. *47*, 191–216.
Brenner, H.R. and Martin, A.R. (1976) J. Physiol. (London) *260*, 159–175.
Brown, G.L. and Pascoe, J.E. (1954) J. Physiol. (London) *123*, 565–573.
Campenot, R.B. (1977) Proc. Natl. Acad. Sci. USA *74*, 4516–4519.
Carruba, M., Ceccarelli, B., Clementi, F. and Mantegazza, P. (1974) Brain Res. *77*, 39–53.
Carstairs, J.R., Edwards, D.C., Pearce, F.L., Vernon, C.A. and Walter, S.J. (1977) Eur. J. Biochem. *77*, 311–317.
Chamley, J.H. and Dowel, J.J. (1975) Exp. Cell Res. *90*, 1–7.
Chamley, J.H., Campbell, G.R. and Burnstock, G. (1973a) Dev. Biol. *33*, 344–361.
Chamley, J.H., Goller, I. and Burnstock, G. (1973b) Dev. Biol. *31*, 362–379.
Charlwood, K.A., Lamont, D.M. and Banks, B.E. (1972) in Nerve Growth Factor and its Antiserum (Zaimis, E. and Knight, J., eds.) pp. 102–107, Athlone Press, University of London.
Chiappinelli, V.A. and Giacobini, E. (1978) Neurochem. Res. *3*, 465–478.

Chiappinelli, V., Giacobini, E., Pilar, G. and Uchimura, H. (1976) J. Physiol. (London) *257*, 749–766.
Chiappinelli, V.A., Fairman, K. and Giacobini, E. (1978) Dev. Neurosci. *1*, 191–202.
Cohen, P., Sutter, A., Landreth, G., Zimmermann, A. and Shooter, E.M. (1980) J. Biol. Chem. *255*, 2949–2955.
Cohen, S., Levi–Montalcini, R. and Hamburger, V. (1954) Proc. Natl. Acad. Sci. USA *40*, 1014–1018.
Collins, F. (1978) Proc. Natl. Acad. Sci. USA *75*, 5210–5213.
Cook, R.D. and Peterson, E.R. (1974) J. Neurol. Sci. *22*, 25–38.
Coughlin, M.D. (1975a) Dev. Biol. *43*, 123–139.
Coughlin, M.D. (1975b) Dev. Biol. *43*, 140–158.
Coughlin, M.D. and Rathbone, M.P. (1977) Dev. Biol. *61*, 131–139.
Coughlin, M.D., Boyer, D.M. and Black, I.B. (1977) Proc. Natl. Acad. Sci. USA *74*, 3438–3442.
Coughlin, M.D., Dibner, M.D., Boyer, D.M. and Black, I.B. (1978) Dev. Biol. *66*, 513–528.
Coughlin, M.D., Bloom, E.M. and Black, I.B. (1981) Dev. Biol. *82*, 56–68.
Cowan, M.W. (1973) in Development and Aging in the Nervous System (Rockstein, M., ed.) pp. 19–41, Academic Press, New York.
Cowan, W.M. and Wenger, E. (1968) J. Exp. Zool. *168*, 105–124.
De Champlain, J., Malmfors, T., Olson, L. and Sachs, Ch. (1970) Acta Physiol. Scand. *80*, 276–288.
Dibner, M.D. and Black, I.B. (1976) Brain Res. *103*, 93–102.
Dibner, M.D., Mytilineou, C. and Black, I.B. (1977) Brain Res. *123*, 301–310.
Ebendal, T. (1979) Dev. Biol. *72*, 276–290.
Ebendal, T. (1981) J. Embryol. Exp. Morph. *61*, 289–301.
Ebendal, T. and Jacobson, C.-O. (1977a) Exp. Cell Res. *105*, 379–387.
Ebendal, T. and Jacobson, C.-O. (1977b) Brain Res. *131*, 373–378.
Ebendal, T., Belew, M., Jacobson, C.-O. and Porath, J. (1979) Neurosci. Lett. *14*, 91–95.
Ebendal, T., Olson, L., Seiger, A. and Hedlund, K-O. (1980) Nature (London) *286*, 25–29.
Edgar, D., Barde, Y.-A. and Thoenen, H. (1981) Nature (London) *289*, 294–295.
Freeman, C.G., Chubb, I.W. and Rush, R.A. (1978) Proc. Aust. Physiol. Pharmacol. Soc. *9*, 132P.
Furshpan, E.J., MacLeish, P.R., O'Lague, P.H. and Potter, D.D. (1976) Proc. Natl. Acad. Sci. USA *73*, 4225–4229.
Giacobini, E. (1978) in Maturation of Neurotransmission. Biochemical Aspects (Vernadakis, A., Giacobini, E. and Filogamo, G., eds.) pp. 41–64, Karger, Basel.
Giacobini, E., Pilar, G., Suszkiw, J. and Uchimura, H. (1979) J. Physiol. (London) *286*, 233–253.
Goedert, M., Otten, U. and Thoenen, H. (1978) Brain Res. *148*, 264–268.
Gorin, P.D. and Johnson, E.M. (1979) Proc. Natl. Acad. Sci. USA *76*, 5382–5386.
Gorin, P.D. and Johnson, E.M. (1980a) Brain Res. *198*, 27–42.
Gorin, P.D. and Johnson, E.M. Jr. (1980b) Dev. Biol. *80*, 313–323.
Greene, L.A. (1974) Neurobiology *4*, 286–292.
Greene, L.A. (1977a) Dev. Biol. *58*, 96–105.
Greene, L.A. (1977b) Dev. Biol. *58*, 106–113.
Gundersen, R.W. and Barrett, J.N. (1979) Science *206*, 1079–1080.
Gunderson, R.W. and Barret, J.N. (1980) J. Cell Biol. *87*, 546–555.
Hamburger, V. and Hamilton, H.L. (1951) J. Morphol. *88*, 49–67.
Hamburger, V. and Levi-Montalcini, R. (1949) J. Exp. Zool. *111*, 457–501.
Hamill, R.W., Bloom, E.M. and Black, I.B. (1977) Brain Res. *134*, 269–278.
Hanbauer, I., Lovenberg, W., Guidotti, A. and Costa, E. (1975) Brain Res. *96*, 197–200.
Harper, G.P. and Thoenen, H. (1980) J. Neurochem. *34*, 5–16.
Harper, G.P., Pearce, F.L. and Vernon, C.A. (1976) Nature (London) *261*, 251–253.
Harper, G.P., Al-Saffar, A.M., Pearce, F.L. and Vernon, C.A. (1980a) Dev. Biol. *77*, 379–390.
Harper, G.P., Pearce, F.L. and Vernon, C.A. (1980b) Dev. Biol. *77*, 391–402.

Hawrot, E. (1980) Dev. Biol. 74, 136–151.
Helfand, S.L., Smith, G.A. and Wessells, N.K. (1976) Dev. Biol. 50, 541–547.
Helfand, S.L., Riopelle, R.J. and Wessells, N.K. (1978) Exp. Cell Res. 113, 39–45.
Hendry, I.A. (1972) Biochem. J. 128, 1265–1272.
Hendry, I.A. (1973) Brain Res. 56, 313–320.
Hendry, I.A. (1975a) Brain Res. 86, 483–487.
Hendry, I.A. (1975b) Brain Res. 90, 235–244.
Hendry, I.A. (1975c) Brain Res. 94, 87–97.
Hendry, I.A. (1976a) Rev. Neurosci. 2, 149–194.
Hendry, I.A. (1976b) Brain Res. 107, 105–116.
Hendry, I.A. (1977) Brain Res. 134, 213–223.
Hendry, I.A. and Campbell, J. (1976) J. Neurocytol. 5, 351–360.
Hendry, I.A. and Hill, C.E. (1980a) Brain Res. 200, 201–205.
Hendry, I.A. and Hill, C.E. (1980b) Nature (London) 287, 647–649.
Hendry, I.A. and Iversen, L.L. (1973) Nature (London) 243, 500–504.
Hendry, I.A., Stach, R. and Herrup, K. (1974a) Brain Res. 82, 117–128.
Hendry, I.A., Stöckel, K., Thoenen, H. and Iversen, L.L. (1974b) Brain Res. 68, 103–121.
Hendry, I.A., Hill, C.E. and Bonyhady, R.E. (1981) in Development of the Autonomic Nervous System (Ciba Foundation Symposium 83) pp. 194–206, Pitman Medical, London.
Hill, C.E. and Hendry, I.A. (1977) Neuroscience 2, 741–749.
Hill, C.E. and Hendry, I.A. (1979) Neurosci. Lett. 13, 133–139.
Hill, C.E., Purves, R.D., Watanabe, H. and Burnstock, G. (1976) Pflügers Arch. 361, 127–134.
Hill, C.E., Hendry, I.A. and Bonyhady, R.E. (1981) Dev. Biol. 85, 258–261.
Hollyday, M. and Hamburger, V. (1976) J. Comp. Neurol. 170, 311–320.
Hughes, A. (1968) Aspects of Neural Ontogeny, Academic Press, New York.
Iversen, L.L., de Champlain, J., Glowinski, J. and Axelrod, J. (1967) J. Pharmac. Exp. Ther. 157, 509–516.
Iversen, L.L., Stöckel, K. and Thoenen, H. (1975) Brain Res. 88, 37–43.
Johnson, D.G., Gorden, P. and Kopin, I.J. (1971) J. Neurochem. 18, 2355–2362.
Johnson, D.G., Silberstein, S.D., Hanbauer, I. and Kopin, I.J. (1972) J. Neurochem. 19, 2025–2029.
Johnson, E.M., Andres, R.Y. and Bradshaw, R.A. (1978) Brain Res. 150, 319–331.
Johnson, E.M., Blumberg, H.M., Costrini, N.V. and Bradshaw, R.A. (1979) Brain Res. 178, 389–401.
Johnson, M., Ross, D., Meyers, M., Rees, R., Bunge, R., Wakshull, E. and Burton, H. (1976) Nature (London) 262, 308–310.
Kessler, J.A. and Black, I.B. (1980) Brain Res. 189, 157–168.
Klingman, G.I. (1966) Int. J. Neuropharmacol. 5, 163–170.
Klingman, G.I. and Klingman, J.D. (1967) Int. J. Neuropharmacol. 6, 501–508.
Klingman, G.I. and Klingman, J.D. (1969) J. Neurochem. 16, 261–268.
Landa, K.B., Adler, R., Manthorpe, M. and Varon, S. (1980) Dev. Biol. 74, 401–408.
Landis, S.C. (1978) J. Cell Biol. 78, R8–R14.
Landmesser, L. and Pilar, G. (1970) J. Physiol. (London) 211, 203–216.
Landmesser, L. and Pilar, G. (1972) J. Physiol. (London) 222, 691–713.
Landmesser, L. and Pilar, G. (1974a) J. Physiol. (London) 241, 715–736.
Landmesser, L. and Pilar, G. (1974b) J. Physiol. (London) 241, 737–749.
Landmesser, L. and Pilar, G. (1976) J. Cell Biol. 68, 357–374.
Langley, J.N. (1892) Phil. Trans. R. Soc. B, 183, 85–124.
Langley, J.N. (1895) J. Physiol. (London) 18, 280–284.
Lawrence, J.M., Black, I.B., Mytilineou, C., Field, P.M. and Raisman, G. (1979) Brain Res. 168, 13–19.
Lees, G., Chubb, I., Freeman, C., Geffen, L. and Rush, R. (1981) Brain Res. 214, 186–189.

Letourneau, P.C. (1975) Dev. Biol. *44*, 77–91.
Letourneau, P.C. (1978) Dev. Biol. *66*, 183–196.
Levi-Montalcini, R. (1950) J. Morphol. *86*, 253–284.
Levi-Montalcini, R. and Angeletti, P.U. (1968) Physiol. Rev. *48*, 534–569.
Levi-Montalcini, R. and Booker, B. (1960a) Proc. Natl. Acad. Sci. USA *46*, 373–383.
Levi-Montalcini, R. and Booker, B. (1960b) Proc. Natl. Acad. Sci. USA *46*, 384–391.
Levi-Montalcini, R., Meyer, H. and Hamburger, V. (1954) Cancer Res. *14*, 49–57.
Lichtman, J.W., Purves, D. and Yip, J.W. (1979) J. Physiol. (London) *292*, 69–84.
Lindsay, R. and Tarbit, J. (1979) Neurosci. Lett. *12*, 195–200.
Ludueña, M.A. (1973) Dev. Biol. *33*, 470–476.
McLennan, I.S. and Hendry, I.A. (1978) Neurosci. Lett. *10*, 269–273.
Mark, G.E., Chamley, J.H. and Burnstock, G. (1973) Dev. Biol. *32*, 194–200.
Martin, A.R. and Pilar, G. (1963) J. Physiol. (London) *168*, 443–463.
Marwitt, R., Pilar, G. and Weakly, J.N. (1971) Brain Res. *25*, 317–334.
Matthews, M.R. and Nelson, V.H. (1975) J. Physiol. (London) *245*, 91–135.
Menesini-Chen, M.G., Chen, J.S. and Levi-Montalcini, R. (1978) Arch. Ital. Biol. *116*, 53–84.
Mytilineou, C. and Black, I.B. (1978) Brain Res. *158*, 259–268.
Narayanan, C.H. and Narayanan, Y. (1978) J. Embryol. Exp. Morphol. *44*, 53–70.
Narayanan, Y. and Narayanan, C.H. (1981) J. Embryol. Exp. Morphol. *62*, 117–127.
Nishi, R. and Berg, D.K. (1977) Proc. Natl. Acad. Sci. USA *74*, 5171–5175.
Nishi, R. and Berg, D.K. (1979) Nature (London) *277*, 232–234.
Njå, A. and Purves, D. (1977) J. Physiol. (London) *264*, 565–583.
Njå, A. and Purves, D. (1978a) J. Physiol. (London) *277*, 53–75.
Njå, A. and Purves, D. (1978b) J. Physiol. (London) *281*, 45–62.
Nurse, C.A. and O'Lague, P.H. (1975) Proc. Natl. Acad. Sci. USA *72*, 1955–1959.
Obata, K. and Tanaka, H. (1980) Neurosci. Lett. *16*, 27–34.
O'Lague, P.H., Obata, K., Claude, P., Furshpan, E.J. and Potter, D.D. (1974) Proc. Natl. Acad. Sci. USA *71*, 3602–3606.
Olson, L. (1967) Z. Zellforsch. *81*, 155–173.
Olson, L. and Malmfors, T. (1970) Acta Physiol. Scand. Suppl. *348*, 1–112.
Otten, U. and Thoenen, H. (1975) Proc. Natl. Acad. Sci. USA *72*, 1415–1419.
Otten, U., Goedert, M., Schwab, M. and Thibault, J. (1979) Brain Res. *176*, 79–90.
Owman, C., Sjöberg, N.-O. and Swedin, G. (1971) Z. Zellforsch. *116*, 319–341.
Partlow, L.M. and Larrabee, M.G. (1971) J. Neurochem. *18*, 2101–2118.
Patterson, P.H. and Chun, L.L.Y. (1974) Proc. Natl. Acad. Sci. USA *71*, 3607–3610.
Patterson, P.H. and Chun, L.L.Y. (1977) Dev. Biol. *56*, 263–280.
Pickel, V.M., Joh, T.H., Field, P.M., Becken, C.G. and Reis, D.J. (1975) J. Histochem. Cytochem. *23*, 1–12.
Pilar, G. and Landmesser, L. (1972) Science *177*, 1116–1118.
Pilar, G. and Landmesser, L. (1976) J. Cell Biol. *68*, 339–356.
Pilar, G., Jenden, D.J. and Campbell, B. (1973) Brain Res. *49*, 245–256.
Pilar, G., Chiappinelli, V., Uchimura, H. and Giacobini, E. (1974) Physiologist *17*, 307.
Pilar, G., Landmesser, L. and Burstein, L. (1980) J. Neurophysiol. *43*, 233–254.
Prestige, M.C. (1970) in The Neurosciences (Schmitt, F.O., ed.) pp. 73–82, Rockefeller University Press, New York.
Purves, D. (1975) J. Physiol. (London) *252*, 429–463.
Purves, D. (1976) J. Physiol. (London) *259*, 159–175.
Purves, D. and Lichtman, J.W. (1978) Physiol. Rev. *58*, 821–862.
Raisman, G., Field, P.M., Ostberg, A.J.C. Iversen, L.L. and Zigmond, R.E. (1974) Brain Res. *71*, 1–16.

Rutishauser, U. and Edelman, G.M. (1980) J. Cell Biol. *87*, 370–378.
Rutishauser, U., Gall, W.E. and Edelman, G.M. (1978) J. Cell Biol. *79*, 382–393.
Sachs, Ch., De Champlain, J., Malmfors, T. and Olson, L. (1970) Eur. J. Pharmacol. *9*, 67–79.
Schaefer, T., Schwab, M.E. and Thoenen, H. (1980) Neurosci. Lett. Suppl. *5*, 122.
Schwab, M.E. (1977) Brain Res. *130*, 190–196.
Schwab, M.E. and Thoenen, H. (1977) Brain Res. *122*, 459–474.
Schwartz, J.P. and Costa, E. (1977) Naunyn Schmiedeberg's Arch. Pharmacol. *300*, 123–129.
Schwartz, J.P., Chuang, D-M. and Costa, E. (1977) Brain Res. *137*, 369–375.
Shieh, P. (1951) J. Exp. Zool. *117*, 359–395.
Silberstein, S.D., Johnson, D.G., Jacobowitz, D.M. and Kopin, I.J. (1971) Proc. Natl. Acad. Sci. USA *68*, 1121–1124.
Smolen, A.J. (1981) Dev. Brain Res. *1*, 49–58.
Sorimachi, M. and Kataoka, K. (1974) Brain Res. *70*, 123–130.
Stoeckel, K. and Thoenen, H. (1975) Brain Res. *85*, 337–341.
Stöckel, K., Paravicini, U. and Thoenen, H. (1974) Brain Res. *76*, 413–421.
Stöckel, K., Dumas, M. and Thoenen, H. (1978) Neurosci. Lett. *10*, 61–64.
Suda, K., Barde, Y.-A. and Thoenen, H. (1978) Proc. Natl. Acad. Sci. USA *75*, 4042–4046.
Thiery, J.-P., Brackenbury, R., Rutishauser, U. and Edelman, M. (1977) J. Biol. Chem. *252*, 6841–6845.
Thoenen, H., Angeletti, P.U., Levi-Montalcini, R. and Kettler, R. (1971) Proc. Natl. Acad. Sci. USA *68*, 1598–1602.
Thoenen, H., Kettler, R. and Saner, A. (1972a) Brain Res. *40*, 459–468.
Thoenen, H., Saner, A., Kettler, R. and Angeletti, P.U. (1972b) Brain Res. *44*, 593–602.
Thoenen, H., Saner, A., Angeletti, P.U. and Levi-Montalcini, R. (1972c) Nature New Biol. *236*, 26–27.
Tuttle, J.B., Suszkiw, J.B. and Ard, M. (1980) Brain Res. *183*, 161–180.
Varon, S., Skaper, S.D. and Manthorpe, M. (1981) Dev. Brain Res. *1*, 73–88.
Vogt, M. (1964) Nature (London) *204*, 1315–1316.
Walicke, P.A., Campenot, R.B. and Patterson, P.H. (1977) Proc. Natl. Acad. Sci. USA *74*, 5767–5771.
Wright, L. (1981) Dev. Brain Res. *1*, 283–286.
Zaimis, E. (1972) in Nerve Growth Factor and its Antiserum (Zaimis, E. and Knight, J., eds.) pp. 59–70, The Athlone Press, University of London.
Zaimis, E., Berk, L. and Callingham, B.A. (1965) Nature (London) *206*, 1220–1222.

CHAPTER 7

Nerve Growth Factor and the neuronotrophic functions

SILVIO VARON and STEPHEN D. SKAPER

Department of Biology and School of Medicine, University of California, San Diego, La Jolla, CA 92093 USA

1. Introduction

In recent years, the concept of extrinsic factors regulating neuronal behaviors has expanded considerably and currently forms the basis of one of the most active fields in neurobiological research. The inspiration for this recent expansion has derived from the Nerve Growth Factor (NGF) phenomenon, but also from several other sources. Their combined contributions have been the subject of several recent reviews (cf., Varon and Adler, 1980, 1981; Varon et al., 1982b).

Inherent to a concept of neuronal plasticity is a distinction between the intrinsic capabilities of neural cells and their susceptibility to extrinsic regulation – that is, the extent to which such capabilities may be expressed under the complex influences of their humoral and cellular environments. Extrinsic influences may be divided into three categories: *trophic* influences are defined as directed to a quantitative regulation of cellular anabolism, *specifying* (or instructive) influences determine what specific cell programs are to benefit from a trophic drive, and *permissive* influences presumably are needed for the actual execution of a given program so that ancillary cell activities may be carried out (Varon, 1977; Varon and Bunge, 1978; Varon and Adler, 1980). In vitro techniques provide ways to investigate neural cell performances in contexts that are simpler than their in vivo counterparts, and to manipulate humoral environment, physical surfaces and the cellular societies with which the neural cells under study are to operate (Murray, 1965; Varon, 1970, 1975b; Sato, 1973; Varon and Saier, 1975; Fedoroff and Hertz, 1977; Giacobini et al., 1980; Pfeiffer, 1982).

1.1. Sources of neuronotrophic factors

1.1.1. Target-derived factors

One of the most dramatic processes in neural morphogenesis is the natural occurrence in situ of neuronal cell death (Prestige, 1970; Cowan, 1973; Hollyday and Hamburger, 1976; Landmesser and Pilar, 1978; Hamburger et al., 1981). Within a restricted period of embryonic or perinatal life, 50% to 80% of all the postmitotic neurons in a particular neuronal subset will die. Generally, death of the neurons coincides with the time of arrival of their axons at their target territory, and their innervation of it. In several instances, neuronal death has been shown to be increased by prior removal of the innervation territory, and to be decreased by pre-implantation of additional target tissue. The lethal effects of target deprivation are less dramatic when the latter is imposed after neural connections are established, and are progressively delayed as target removal is further postponed. It appears, then, that developmental death occurs in those neurons which, on reaching their innervation territory, fail to make or consolidate synaptic connections there (cf., Varon and Adler, 1980).

The following hypothesis, based on the above observations, has been proposed to explain the dependence of neuronal survival on appropriate connections with the postsynaptic target: (1) target tissues produce and release appropriate neuronotrophic factors, (2) the nerve terminals compete in some manner for the available factor, (3) the 'successful' neurons take up the factor at their terminals and transport it retrogradely through their axons and (4) on arrival at the neuronal soma, the factor elicits life-sustaining activities. In the adult, disconnection from a peripheral target (e.g., by axotomy) also leads to neuronal alterations, which may result in neuronal death or axonal degeneration. Axotomy is mimicked by localized application of colchicine (a blocker of axonal transport), suggesting once again a retrograde transport of trophic factors (e.g., Purves, 1976).

The hypothesis of target-derived neuronotrophic factors has been validated recently by the actual isolation of such a soluble macromolecule for ciliary ganglionic (CG) neurons (Varon et al., 1979; Adler et al., 1979; Landa et al., 1980; Manthorpe et al., 1980). CG neurons innervate the choroid, ciliary body and iris tissues of the eye and, in the chick embryo, display all the expected features of the developmental death phenomenon (Landmesser and Pilar, 1978). In vitro, dissociated CG neurons will not survive even 24 h unless supplied with special medium supplements (Helfand et al., 1976; Tuttle, 1977). In our laboratory, we were able to show that: (i) soluble Ciliary Neuronotrophic Factors (CNTFs) occur at much higher specific activities in extracts from the embryonic eye than from the whole embryo, (ii) within the eye, CNTF is restricted to the very tissues that constitute the innervation territory for CG neurons, (iii) ocular CNTF activity reaches a maximum over the very developmental times at which the fate of those neurons is decided in vivo, and (iv) the CNTF activity is associated with an acidic protein, now partially purified. Additional studies have shown that CG neurons from chick embryos younger than 7 days do not depend on ocular CNTF for their survival and neurite production in vitro, just as they appear to be independent of their ocular target in vivo (Manthorpe et al., 1981a).

1.1.2. Other peripheral sources and glial sources
Several peripheral tissues are under investigation in various laboratories (reviewed in Varon and Adler, 1981) for neuronotrophic activities directed mainly to ganglionic neurons or spinal cord neurons. Tissue extracts and media conditioned over corresponding cell cultures are both used as trophic source materials. Some of these sources are the appropriate innervation territory of the test neurons, e.g., skeletal muscle for spinal cord cells. In other cases, one might speculate that the sources examined could serve as 'surrogate' innervation territories: such might be skeletal and heart muscle for the CG neurons (which are both cholinergic and motor). In yet other studies, however, the case for a target tissue relationship between sources and test neurons could hardly be made, bringing into question the source-specificity of neuronotrophic factors in general (see Section 1.4.1). Very recent studies on peripheral dependence of dorsal root ganglion neuronal survival (Hamburger et al., 1981) suggest that the early source of neuronotrophic factors may be the membrane traversed by growing axons on their way to their final destination, rather than or in addition to the final territory.

Both in the central and the peripheral nervous systems, the most immediate environment of a neuron is represented by the glial cells and the extracellular fluids interposed between glia and neurons. Neurons and glial cells operate in close partnership, each affecting the other via surface interactions and/or exchanges of macromolecular as well as micromolecular signals. The general field of neuron-glia interactions has been recently reviewed (Varon and Somjen, 1979; Varon and Manthorpe, 1982). Glial cells are likely to play guidance roles for growing or re-growing axons in vivo, and conditioned media from peripheral glial cultures contain substratum-binding neurite promoting factors for peripheral and spinal cord neurons (Adler et al., 1981; Longo et al., 1982). Peripheral glial cells may represent additional sources of neuronotrophic factors in vivo, perhaps in partial replacement of target territories in the older animal (cf., Varon and Bunge, 1978), and are among the in vitro sources of conditioned media that contain neuronotrophic factors for ganglionic (Varon et al., 1981) and spinal cord (Longo et al., 1982) neurons. Banker (1980) has recently reported that conditioned medium from primary astroglial cells promotes survival and growth in vitro of hippocampal neurons from the rat.

1.2. Nerve Growth Factor as a model factor

Nerve Growth Factor traditionally designates a particular protein (or a group of related proteins) which (i) is required for the *survival* of neurons from sympathetic ganglia (SG) and dorsal root ganglia (DRG), (ii) elicits an extensive *neurite production* from its target neurons, and (iii) controls their production of *transmitter-synthesizing enzymes* and other expressions of differentiated functions. All such effects of NGF have been observed in vivo as well as in vitro. Throughout its history, NGF has been viewed as a prototype for other hypothetical macromolecular factors directed to the control of neuronal maintenance, growth, functional activities and, possibly, regeneration.

The more than a quarter century of NGF investigations have been repeatedly and extensively reviewed (Levi-Montalcini, 1966, 1975, 1976, 1979; Levi-Montalcini and Angeletti, 1968; Varon, 1975a; Hendry, 1976; Bradshaw and Young, 1976; Mobley et al., 1977; Varon and Bunge, 1978; Ikeno and Guroff, 1979; Harper and Thoenen, 1980; Greene and Shooter, 1980; Varon and Adler, 1980, 1981; Varon and Skaper, 1980a,b, 1981a). In the remaining part of this section, therefore, we shall summarize only some background information on NGF sources and target cells, in order to proceed and discuss in subsequent sections three main questions that are still unsettled, namely: (1) what does NGF do? (2) where in the cell does it do it? and (3) how might it do it?

1.2.1. NGF sources
Much of what we know about the NGF protein comes from extensive investigations of NGF derived from the *adult male mouse submaxillary gland* (Greene and Shooter, 1980). In the gland extract, NGF occurs in the form of a 7 S complex containing the active subunit, Beta NGF, two other proteins (Alpha and Gamma) and zinc ions. The abundance of NGF in the mouse submaxillary gland has made this tissue the most practical starting material for purified NGF. Other NGF-rich sources have been found to be certain snake venoms (cf., Angeletti and Bradshaw, 1979), the prostate and its secretions in certain animal species (cf., Harper et al., 1979; Harper and Thoenen, 1980), and the human placenta (e.g., Goldstein et al., 1978). The physiological sources of NGF, particularly during early development, however, still remain to be identified. Much of the problem may lie with the detection methodology applied (reviewed by Harper and Thoenen, 1980).

An important question is whether NGF sources relate to the target territories innervated by the NGF-dependent neurons (see Section 1.3.1). Favorable circumstantial evidence is offered by the ability of NGF to be taken up by nerve terminals and be retrogradely transported to the neuronal bodies and the observed consequences on the neuron of an interruption of its retrograde transport capabilities. A widespread distribution of NGF among peripheral tissues would be in keeping with the extensive territories of sensory and sympathetic neurons: in addition, source specificity of neuronotrophic factors (NGF as well as others) need not be sought in terms of unique locations of the factor but rather in terms of unique *opportunities* for a neuron to encounter its trophic factor(s) in vivo in the right physiological place and at the right physiological time. Such a widespread NGF distribution was in fact reported by some investigators (e.g., Angeletti and Vigneti, 1971; Johnson et al., 1971; Hendry and Iversen, 1973), but not confirmed by others (Harper et al., 1980b). The same also applies to a presence of NGF in serum, about which reports have been conflicting over several years (cf., Hendry and Iversen, 1973; Harper and Thoenen, 1980; Skaper and Varon, 1982). A variety of tissue and cell cultures including muscle (Murphy et al., 1979) have also been found to release and even synthesize NGF in vitro (e.g., Oger et al., 1974; Bradshaw and Young, 1976; Young et al., 1976; Harper et al., 1980a). However, it may be that the expression of a 'latent' or 'masked' competence to produce NGF is a regulated performance, only brought forth by particular treatments or circumstances (Harper et al., 1980a,b). A recent report has indeed shown that rat iris (an authentic innervation target for sensory and sympathetic neurons) displays NGF content only in culture or after denervation in situ (Ebendal et al., 1980).

During ganglionic development, the enlargement of neuronal cell bodies and axons leads to a numerical increase in satellite and Schwann cells for a complete glial ensheathment of the neuronal surfaces (cf., Pannese, 1980; Varon and Manthorpe, 1982). These glial cells might then decrease the neuronal dependence on the innervation territories by providing the neurons in part with the required neuronotrophic factors (e.g., Varon, 1975a; Lasek and Hoffman, 1976; Varon and Bunge, 1978). Schwann cells can indeed produce neuronotrophic factors different from NGF, for several ganglionic neuronal populations (Varon et al., 1981; Longo et al., 1982). Glial cells may also be capable of producing NGF, at least in vitro. When mouse DRG neurons are deprived of most of their nonneuronal partners, their survival in culture requires either exogenous NGF (Varon et al., 1973) or the restitution of ganglionic nonneural cells (Varon et al., 1974a,b). Treatment with antibody against mouse NGF renders the ganglionic nonneurons incompetent to provide trophic support to the neurons, and their competence is resumed after antibody withdrawal (Varon et al., 1974c,d). This NGF-like competence was also observed between nonneurons and neurons from chick and rat DRG, and was best displayed between ganglionic cells of the same species or age (Burnham et al., 1972; Varon et al., 1974b). Glial production of genuine NGF has been subsequently reported for several central glial populations, both normal (Ebendel and Jacobson, 1977) and neoplastic (e.g., Longo and Penhoet, 1974; Schwartz et al., 1977; Perez-Polo et al., 1977; Longo, 1978). Nerve Growth Factor has also been detected in normal and hormone-treated brains (e.g., Walker et al., 1979).

1.2.2. NGF-responsive cells

Success or failure to define a cell population as responsive to NGF is strongly influenced by several considerations. First, one must distinguish between dependence on NGF (expression of a given behavior only if NGF is present) and responsiveness to NGF (NGF alters one or more cell behaviors, but may not be essential for them). Secondly, test conditions must be both permissive (to allow the response to take place, particularly in vitro) and restrictive (so that a behavior is not already expressed fully in the absence of exogenously supplied NGF). Thirdly, developmental age is most critical: as we shall repeatedly see, cells may display a particular response to NGF only within resticted time spans, or may be supported by factors other than NGF at different ages. Species of origin appears not to be limiting for responses to NGF (except with regard to different age requirements) but may have considerable bearing when antibody treatments are involved (cf., Harper and Thoenen, 1980). With such considerations in mind, three NGF-responsive cell populations have been firmly identified: primary neurons from DRG and sympathetic ganglia (SG), and adrenal chromaffin cells of normal and tumoral origin. Also reported to respond to NGF have been certain other ganglionic neurons (e.f., Haas et al., 1972; Ebendal, 1979) and, more controversially, some central nerve cells (cf., Freed, 1976; Turner et al., 1980).

Two subsets of *DRG neurons*, ventrolateral (VL) and dorsomedial (DM), have been distinguished in the chick embryo by their relative positions during early development, as well as by their functional associations (cf., Hamburger and Levi-Montalcini, 1949, Levi-Montalcini and Hamburger, 1951; Levi-Montalcini, 1958, 1963, 1966). In vivo studies had initially suggested that only DM neurons respond to NGF with increased neuronal survival (cf., Hendry, 1976), enhanced cell size (hypertrophy), neurite production, neurite orientation (toward the NGF sources), and accelerated differentiation (cf., Levi–Montalcini and Hamburger, 1951; Levi-Montalcini, 1966; Levi-Montalcini and Angeletti, 1968). More recently, however, Hamburger et al. (1981) have shown that daily injections of NGF into the yolk sac caused the rescue from developmental death of *both* VL and DM neurons, provided that the NGF treatment started early enough (embryonic day 3.5, or E3.5). In fact, a VL response was already detectable within a day of treatment onset (E4.5), the earliest neuronal response yet reported in vivo to NGF. Given such an early response, these investigators suggested that the peripheral source of NGF for the VL neurons in vivo may be the mesechyme interposed between it and the neuronal somata, rather than the not yet reached peripheral innervation territory. In any case, for both VL and DM neurons, developmental age was clearly critical for a demonstration of NGF action: VL neurons no longer showed such responses beyond E7.5, while DM neurons no longer differed in NGF-treated and control ganglia after about E11. NGF may not be altogether irrelevant beyond such ages: in postnatal rats injected with NGF, DRG neurons will retrogradely transport it (Stoeckel and Thoenen, 1975), and display a rise in their substance P content (Kessler and Black, 1980).

The age dependence of DRG responses to NGF is also conspicuous in vitro. The traditional neurite halo displayed by E8 chick DRG in explant cultures (Levi-Montalcini et al., 1954; Varon et al., 1972) is decreasingly elicited by NGF as the DRG age increases (e.g., Winick and Greenberg, 1965; Herrup and Shooter, 1975; Ebendal, 1979). Even more explicit is the behavior of DRG neurons in dissociated cell cultures where (i) surviving neurons can be directly quantitated, and (ii) ganglionic nonneurons can be largely removed by a differential cell attachment step (Varon et al., 1973; Greene, 1977b; Barde et al., 1980), thereby offering a more quantitative and objective approach to the bioassay of NGF (e.g., Varon et al., 1974a, 1981; Greene, 1977b; Manthorpe et al., 1981b) or of other trophic factors directed to these neurons (e.g., Varon et al., 1981b – see Section 2.1.2).

Sympathetic ganglion neurons as well as DRG ones, respond in vivo to NGF with numerical increases, hypertrophy, neurite outgrowth and orientation, and accelerated differentiation. Nerve Growth Factor sensitivity persists well into postnatal life, though to progressively declining extents (as shown mainly using rodent superior cervical ganglia, or SCG). Cell cultures of dissociated SG, like those from DRG, can yield purified neuronal populations (Varon and Raiborn, 1972a; Mains and Patterson, 1973; McCarthy and Partlow, 1976; Greene, 1977a; Edgar et al., 1981), and a similar age-related dependence was shown using purified E8–18 chick SG neurons (Edgar et al., 1981).

Sympathetic ganglion neurons are typically noradrenergic, and transmitter-promoting effects of NGF (or other agents) on their transmitter function can be readily evaluated by assessing transmitter enzymes (e.g., tyrosine hydroxylase, TOH) and uptake or storage of catecholamines. Some such studies in vivo have stressed the point that appearance and persistence of transmitter function may, in fact, involve different extrinsic factors (Cochard et al., 1978, 1979). In vitro, explant cultures of mouse SCG were used by Coughlin et al. (1977, 1978) to demonstrate that early (E14) ganglia do not require NGF (nor are they sensitive to antiserum against NGF) for neurite outgrowth or TOH expression, although they respond to NGF with an increase in both. Dissociated neurons from the same source display near complete dependence on exogenous NGF for both survival and neurite production (cf., Black et al., 1979).

A third neural crest derivative, the *chromaffin cell* of the adrenal medulla, has recently emerged as a potential target of NGF. Aloe and Levi-Montalcini (1979) have shown that injection of NGF into prenatal rats causes chromaffin cells to assume neuronal-like morphologies, while injection of antiserum against NGF causes cellular degeneration at early enough developmental stages. Rat (and bovine) chromaffin cells in vitro also extend neurites when NGF is supplied (e.g., Unsicker et al., 1978, 1980). Greene and coworkers have developed from a rat pheochromocytoma (the neoplastic derivative of the adrenal medulla chromaffin cells) a *clonal line, PC12*, which responds to NGF with the acquisition of several neuronal behaviors (for reviews, see Greene 1978; Greene et al., 1980; Greene and Shooter, 1980). If PC12 cultures are treated with NGF over several days, NGF causes progressive mitotic arrest, cell hypertrophy, extension of neurites, acquisition of electrical and chemical excitability (cf., Dichter et al., 1977), increased storage and depolarization-inducible release of both catecholamines and acetylcholine (Greene and Rein, 1977a,b), and even the ability to form synapses on muscles cells (Schubert et al., 1977). The NGF effects are reversible, starting within 1 to 2 days from NGF withdrawal and proceeding with a temporal pattern that mirrors their appearance under NGF treatment (Greene, 1978).

2. NGF effects and roles

The main consequences on neurons for which NGF is traditionally known are: (i) survival and hypertrophy, (ii) neurite growth, and (iii) enhanced transmitter production. We will be concerned here only with the first two. The effects of NGF on transmitter production have been described in numerous other studies (Chun and Patterson, 1976; Johnson et al., 1981; Patterson, 1978; Potter et al., 1981; Rohrer et al., 1978; Thoenen et al., 1971, 1979; Weber, 1980). The PC12 cell system has added to this list a new NGF consequence, namely the induction of neuronal behaviors *(priming)*.

2.1. Neuronal survival and general growth

2.1.1. NGF effects

In vivo, administration of NGF to chick embryos rescues both VL and DM neurons if NGF is applied early enough, and only DM neurons if applied at later times (Section 1.4.2). Conversely, continued presence of NGF-antibodies during fetal life – by use of pre-immunized mothers – causes neuronal death in both DRG and SG of rat, rabbit or guinea pig (Gorin and Johnson, 1979; Johnson et al., 1980a), while antibody administration postnatally only affected SG neurons (Johnson et al., 1980a). In rodents, postnatal injection of antibody causes a massive 'immunosympathectomy' – the crucial evidence for a physiological, not just pharmacological role of NGF: this immunosympathectomy effect declines with increasing age of the treated animal (cf., Steiner and Schönbaum, 1972). Target territory removal in neonatal rodents causes death of the corresponding sympathetic neurons, which can be prevented by NGF administration (Hendry and Iverson, 1973; Dibner and Black, 1976; Dibner et al., 1977). Counteracting effects of NGF have been similarly observed in adult animals subjected to mechanical or functional axotomy (Hendry and Campbell, 1976; Purves 1976; Purves and Nja, 1976; Banks and Walker, 1977; Nja and Purves, 1978), or to destruction of their terminals with 6-hydroxydopamine (Hendry, 1975; Levi-Montalcini et al., 1975).

In vitro, chick DRG neurons require NGF for survival over the E6–15 age period, but not beyond (Levi-Montalcini and Angeletti, 1963; Cohen et al., 1964; Greene, 1977b; Barde et al., 1980). Greene (1977b) reported that neuronal survival was totally dependent on NGF over the E8–11 age period while over the E13–18 age span, survival in the absence of NGF increased to match eventually that observed in the presence of NGF (thereby demonstrating an increasing irrelevance of NGF). In contrast, Barde et al. (1980) reported that purified DRG neurons from different ages failed to survive in cultures lacking exogenous trophic supplement. NGF promoted survival optimally between E10–12, but decreasingly so beyond that age and nearly not at all by E16, thereby demonstrating an increasing incompetence of NGF. An age-related dependence on NGF was similarly shown with purified neurons from chick SG over the E8–17 age span (Edgar et al., 1981), with mouse SCG starting from E14 onward (cf., Coughlin et al., 1978), and with postnatal rat SCG neurons (Lazarus et al., 1976; Chun and Patterson, 1977).

The PC12 cell system may offer additional insights into the survival role of NGF. PC12 cells survive and grow in the absence of NGF provided serum is supplied (Greene, 1978). Without serum PC12 cells cease proliferation and die over the next two weeks: under these conditions, NGF fully replaces the serum requirement for the PC12 cell survival (Greene, 1978).

A trophic factor should promote not only survival but also *general growth* (hypertrophy) if supplied at adequate levels. In vivo, and in vitro, NGF-dependent neurons increase their size markedly in response to the factor (Levi-Montalcini and Angeletti, 1978; O'Lague et al., 1978). PC12 cells also respond to NGF with an increase in size (cf., Greene and Shooter, 1980).

2.1.2. NGF supplements and NGF surrogates
In vitro studies have also shown that, even when required, NGF is *not sufficient* by itself for neuronal survival (Bottenstein et al., 1980; Skaper et al., 1979a; Varon and Skaper, 1981b). Cultured neurons also require serum or the serum-replacing, chemically defined *N1 supplement* (a mixture of insulin, transferrin, putrescine, progesterone and selenite). Neither serum nor the N1 supplement supports survival without the concurrent presence of NGF. Different N1 constituents are needed by different neurons: E8 chick DRG cells tolerate only the omission of selenite, while selenite and insulin are the only critical constituents for E11 chick SG (Varon and Skaper, 1981b). No information is yet available on (i) the role that each relevant N1 constituent plays in neuronal survival, or (ii) the relationship between them and NGF in this respect.

Recent in vitro studies have further revealed the possibility of *replacing NGF* in its survival role. The level of external potassium has been reported to improve culture survival of various neuronal populations (Scott and Fisher, 1970, 1971; Scott 1971, 1977; Lasher and Zagon, 1972), and to prevent a decline and actually increase the number of recognizable neurons in E10 chick DRG cultures, even in the absence of NGF (Chalazonitis and Fischbach, 1980). Thus, high K^+ appears to both substitute for and enhance the survival effect of NGF (see also Section 4.3.2). Other studies have shown that NGF can be replaced with macromolecular factors (ganglionic neuronotrophic factors, or GNTFs) from a variety of sources including muscle cells and glial cells, and that these factors cannot be blocked by NGF antibody, hence differ from the traditional NGF (cf., Varon and Adler, 1981). Some of these studies indicate that the same subset of DRG or SG neurons is supported equally well by GNTFs as by NGF (Varon et al., 1981; Manthorpe et al., 1981b). Other studies reveal that the relative dependence on NGF and GNTFs varies with the age of the neurons (e.g., Barde et al., 1980; Edgar et al., 1981). Ganglionic neurons, therefore, may comprise different subsets that are (i) sensitive only to one or the other factor, (ii) responsive to either factor or (iii) receptive to both factors but requiring concurrent treatment for their survival (cf., Varon et al., 1982a). PC12 cells also appear to respond to GNTFs present in various culture-conditioned media with increased choline acetyltransferase (CAT) activity (Edgar et al., 1979a,b) and neurite formation (DeVellis, 1982).

2.2. Neurite growth

Neurite extension, at the very least, must require (i) an intrinsic program to confer neurite competence to the cell, (ii) a local trigger to initiate and sustain growth cone activity, and (iii) an anabolic support to provide the needed building materials. Thus, extrinsic agents that would promote any one of these three aspects could stimulate neurite formation provided the other two aspects are independently met. Since NGF promotes neurite formation in both primary neurons and in PC12 cells, one may ask the question: does NGF address only one of the above aspects, or does it promote all three?

2.2.1. NGF effects on PC12 and primary neurons
The *PC12 cell system* offers a useful illustration of the problem (Greene and Tischler, 1976; Luckenbill-Edds et al., 1979; Greene et al., 1980). These cells display neurites only after being treated with NGF and therefore must be 'taught' to do so by NGF, a task called 'priming' (Burstein and Greene, 1978). PC12 cells treated with NGF over 1 to 2 weeks extend neurites on a collagen substratum. Treatment of PC12 cells with NGF under conditions where neurite extension will not occur (e.g., on a non-adhesive substratum), followed by transfer to neurite-permissive conditions will result in much prompter (24 h) neurite extension and only if NGF remains available. These observations indicate that NGF (i) has

'taught' neurite production to the cells by inducing a neurite program (priming), and (ii) also provides a continuous trigger for the execution of the program. The slow development of the priming effect depends on RNA and protein synthesis, while the execution of neurite growth depends on neither.

Primary ganglionic neurons are usually examined in culture at a time when they have already grown neurites in vivo. Thus, the NGF effects on their neurite production in vitro concern regeneration rather than de novo induction: the neurite 'program' is presumably already in place, and the NGF action must be either at the level of local trigger or as a special expression of its general growth promotion, or in both such capacities. Campenot (1977) has indeed demonstrated a local action of NGF. Sympathetic ganglion neurons survived with NGF available to only the neuronal soma or the nerve ending, whereas the neurites required local availability of NGF for their very maintenance as well as further extension. Other studies have shown that primary neurons, like primed PC12 cells, do not require ongoing RNA synthesis for neurite extension (e.g., Partlow and Larrabee, 1971). Protein synthesis is not required for initial neurite production (Bloom and Black, 1979), but remains necessary for its support under NGF stimulation (Partlow and Larrabee, 1971), suggesting the additional involvement of NGF for anabolic support.

2.2.2. Substratum-bound neurite promoting factors
Growth cones operate via adhesive interactions with their substratum (e.g., Letourneau 1975, 1979; Johnston and Wessells, 1980). Thus, the local trigger for neurite extension could be supplied by either changing the adhesiveness of the neuronal membrane or improving the adhesiveness of the culture substratum (See Varon and Adler, 1981). NGF is reported to increase adhesiveness of DRG neurons (Varon and Raiborn, 1972b; Varon et al., 1973) and PC12 cells (e.g., Schubert and Whitlock, 1977; Greene et al., 1979; Chandler and Herschman, 1980). There is now considerable evidence for a class of *polyornithine-binding neurite promoting factors* (PNPFs), which are devoid of survival or general growth promoting competence but confer to a polycationic substratum competence for neurite stimulation. Such PNPFs were first reported to be present in heart-conditioned media and to elicit neurite extension from E8 chick ciliary ganglionic neurons (Collins, 1978a,b): those neurons would otherwise not grow neurites on polyornithine, even when supplied with their own ciliary neuronotrophic factor (Varon et al., 1979; Adler and Varon, 1980). Subsequently, PNPFs were found in a variety of conditioned media (including those from glial cells), and shown to promote neurite extension even from the NGF-responsive ganglionic neurons which already do so to some extent on untreated polyornithine (Adler et al., 1980). Ciliary ganglia explanted on polyornithine generate their own PNPF to coat the surrounding substratum, and their neurite outgrowth remains restricted to the coated region (Adler and Varon, 1981a). On collagen (to which PNPFs do not

bind), they will grow out neurites only after nonneuronal cells have migrated out, and only on the nonneuronal cell surface itself (Adler and Varon, 1981b). These findings have encouraged the speculation that PNPFs are normal constituents of cell surfaces, involved in vivo in a neurite-promoting role with regard to neurons that send out their processes outside the CNS (Varon and Adler, 1980, 1981; Collins, 1980), including the NGF-responsive DRG and SG ones (cf., also Hawrot, 1980).

2.2.3. Neurite directional guidance
Besides promoting neurite extension, NGF may be also involved in neuritic guidance. Directional guidance of a growing neurite could occur via 'restrictive' (or tactic) influences limiting the growth to neurite-permissive pathways such as those provided by PNPF, or via 'attractive' (or tropic) influences emanating from the intended destination site (cf., Varon and Adler, 1981). A chemotropic action of NGF could be exerted if NGF (i) were to be produced in the innervation territory (or along the path leading to it), (ii) distributed itself along a gradient increasing with increasing proximity of its source, and (iii) were recognized by the advancing growth cone as an advantage (neurite-stimulating or trophic roles of NGF). Chemotropic effects of NGF have been detected in vivo (cf., Levi-Montalcini, 1966; Levi-Montalcini and Angeletti, 1968; Levi-Montalcini, 1976; Menesini-Chen et al., 1978), as well as in vitro (Ebendal and Jacobson, 1977; Gundersen and Barrett, 1979; Letourneau, 1978).

3. Modes of action of NGF

As our perception increases of the consequences of NGF action, one is struck by the multiplicity of the roles that NGF appears to play toward its target cells. The question therefore arises whether such diverse roles of NGF involve multiple sites of action within the target cells, each concerned with a different molecular mechanism, or whether a single site of action and a single mechanism may serve as the common trigger to a variety of different consequences. At present, investigations of these questions center around three distinctions: (i) NGF binding to cell surface receptors versus NGF internalization for intracellular action, (ii) RNA-dependence or independence of NGF consequences, and (iii) short-latency responses (minutes) versus longer-term consequences (hours or days). We shall examine here the first two distinctions, as well as a number of suggestions for intermediate events taking place during the course of NGF action. The third distinction, between short-latency and longer-term effects of NGF will be discussed in a separate section around a recently developed 'ionic hypothesis' of NGF action (Section 4).

3.1. NGF receptors

The first encounter between NGF and the target cell must occur at the surface of the cell membrane. NGF-specific, high-affinity binding sites have been demonstrated on intact cells or with membrane preparations from sympathetic ganglia (Banerjee et al., 1973, 1975, 1976; Frazier et al., 1974a; Costrini and Bradshaw, 1979), dorsal root ganglia (Herrup and Shooter, 1973; Frazier et al., 1973a, 1974a; Sutter et al., 1979), and PC12 cells (Herrup and Thoenen, 1979; Calissano and Shelanski, 1980; Landreth and Shooter, 1980). Binding sites for NGF have also been reported on nonresponsive tumoral cells such as C1300 neuroblastoma (Revoltella et al., 1974; Levi-Montalcini et al., 1974) and melanoma (Fabricant et al., 1977; Sherwin et al., 1979), and in brain (Frazier et al., 1974b; Szutowicz et al., 1976a,b). The NGF-binding constituent has been solubilized and purified from rabbit SCG membranes (Banerjee et al., 1976; Andres et al., 1977; Costrini and Bradshaw, 1979).

A much debated question has been whether *more than one subset* of high-affinity NGF binding sites occur on the surface of the same cell. Sutter et al. (1979) reported that chick embryo DRG cells display two such subsets (sites I and II), having apparent dissociation constants (K_D) of 2×10^{-11} and 2×10^{-9} M, respectively. Similarly, two sets of binding sites ($K_D = 10^{-11}$ and 10^{-9} M) were reported for E11 chick SG cells. With PC12 cells, only one set of sites ($K_D = 3 \times 10^{-9}$ M) was present at 0–4 °C (Herrup and Thoenen, 1979). At 37 °C, however, Landreth et al. (1980) reported that NGF-untreated PC12 cells first exhibit only lower-affinity binding sites (as seen at 4 °C), but that high-affinity binding (apparent $K_D = 10^{-10}$ M) appears within seconds of NGF presentation and rapidly increases in parallel with a decline of the first set and an increase in trypsin-resistance.

Designation of surface binding sites as *true NGF receptors* requires the demonstration that they are involved in the mode of action of NGF in some critical manner. Oxidation of the NGF tryptophan residues progressively reduces the binding competence (to both sites I and II), as well as the biological activity of NGF with regard to DRG cells (cf., Greene and Shooter, 1980). In PC12 cells, both the survival response and the neurite response display nearly identical dose-dependences, proposing the involvement of a common receptor site (Greene, 1978). In rat SCG neurons, on the other hand, the increase in transmitter-synthesizing enzymes requires 100-fold higher NGF concentrations than do survival, hypertrophy and neurite extension, paralleling the 100-fold difference in apparent K_D of the two bindings subsets (Chun and Patterson, 1977).

3.2. NGF internalization

Once bound to its surface binding sites, NGF can also be internalized by an endocytotic process resulting in intracellular vesicles which still carry the bound NGF inside them. Internalization and retrograde axonal transport of NGF has been reported for sympathetic neurons (Hendry et al., 1974; Hendry, 1977); and DRG neurons (Stoeckel et al., 1975; Stoeckel and Thoenen, 1975; Johnson et al., 1978; Brunso-Bechthold and Hamburger, 1979; Dumas et al., 1979), even in adult animals. Such accumulation in vivo of NGF, however, has also been reported for neurons that are not traditional NGF targets such as ciliary ganglionic cells (Max et al., 1978) and selected brain cells (Ebbott and Hendry, 1978). Evidence for NGF internalization has also been provided in vitro, where DRG cells were shown capable of both endocytotic and exocytotic transfers of NGF, involving specific NGF binding to surface sites (Burnham and Varon, 1973; Norr and Varon, 1975). PC12 cells are also capable of NGF internalization (Yanker and Shooter, 1979; Calissano and Shelanski, 1980; Marchisio et al., 1980).

The fate of the internalized NGF remains a matter of controversy. The NGF could be internalized only for translocation from one to another region of the neuronal plasma membrane, or merely for degradation purposes (cf., Varon, 1975a; Harper and Thoenen, 1980). On the other hand, an ability of internalized NGF to elicit functional consequences may be interpreted in two alternative ways. In one, the NGF continues to occupy its plasma membrane receptor (now located on the inner surface of the endocytotic vesicle), and can continue to trigger there the same events it can elicit when occupying the same receptor on the outer cell surface (Varon, 1975a; Varon and Adler, 1980, 1981; Harper and Thoenen, 1980). Alternatively, the internalized NGF becomes available in a free form to interact with its true site of action somewhere else within the target cell. Andres et al. (1977) using chick DRG cells, and Yanker and Shooter (1979) using PC12 cells observed that a portion of the NGF bound to their respective cell populations was Triton-insoluble and associated with a nuclear fraction (chromatin and membrane, respectively) displaying high-affinity binding sites. Marchisio et al. (1980) using immunofluorescence and autoradiography on PC12 cells at the light microscopy level described a progressive internalization of NGF into the cytoplasm, with discrete dots around the nucleus and on the nucleoplasm. On the other hand, Thoenen and collaborators using electron microscopy found no evidence of intact and functional NGF in cytoplasmic or nuclear compartments of ganglionic neurons (Schwab, 1977; Schwab and Theonen, 1977; Thoenen et al., 1979) or of PC12 cells (Heumann et al., 1981; Thoenen et al., 1981).

3.3. RNA and protein synthesis

We have seen (Sections 2.1 and 2.2) that NGF appears responsible both for general trophic effects (sustained survival, hypertrophy, prolonged neurite growth) and for selected cell behaviors (neurite and transmitter expressions, priming of PC12 cells). General or, respectively, specific synthesis of RNA and/or protein may be potentially involved at several different levels. This is not to say that (i) NGF must act *directly* on the RNA or protein synthesizing machineries, or (ii) internalized rather than surface-bound NGF must be responsible for such putative actions.

3.3.1. NGF effects on general synthesis of RNA and/or protein

Early studies (e.g., Angeletti et al., 1965) had reported that E8 chick DRG, incubated in suspension for a few hours with radiolabeled-precursors, accumulated more labeled RNA and protein in the presence that the absence of NGF. The results were taken to indicate that NGF stimulates RNA synthesis and, secondarily to it, protein synthesis. Subsequent studies with both chick SG (Partlow and Larrabee, 1971) and DRG (Burnham and Varon, 1974) have clarified that the difference in labeled RNA reflected a progressive decline in RNA labeling by the NGF-deprived ganglia, and that NGF did not stimulate, but only sustained the rate of RNA and protein labeling.

Using dissociated cell suspensions from E8 chick DRG, Horii and Varon (1975) showed that two sets of events could be recognized under NGF deprivation. First, there was an early and progressive reduction (0 to 6 h) in RNA labeling (but not in protein labeling), which was entirely reversible on delayed presentation of NGF. These reversible changes reflected a decline in the uptake of radio-uridine, and not a diminished rate of RNA synthesis (Horii and Varon, 1975, 1977). Beyond this 6 h period, continued NGF deprivation led to a further decline of RNA labeling, the progressive reduction of protein labeling and an increasing degradation of both RNA and protein – all of which developed in temporal coincidence and could not be reversed, but only interrupted, by delayed NGF administration. The time frame of this second set of events fits that of increasing neuronal death observed in explant (e.g., Levi-Montalcini et al., 1978) and dissociated cell cultures (e.g., Levi-Montalcini and Angeletti, 1963; Burnham et al., 1972). It is likely, therefore, that prolonged NGF deprivation leads to a decline in RNA and protein synthesis, eventually culminating in the death of the cell. Conversely, neurons that would display a hypertrophy response to NGF must be undergoing increased protein and presumably RNA synthesis and, therefore, a stimulation by NGF of these biosynthetic activities. Indeed, RNA polymerase activity in neonatal rat SCG increases both in vivo and in vitro within 4 to 12 h after administration of NGF (Huff et al., 1978). On the other hand, SDS gel analyses of labeled proteins have revealed no qualitative, and only few quantitative differences between naive and NGF-primed PC12 cells (McGuire et al., 1978).

Whether or not NGF is concerned with the stimulation of RNA or protein synthesis, ongoing synthesis may still be necessary either for the stimulation by NGF of selected cell responses or for their execution beyond the level of NGF instruction. Ganglionic neurons can survive for a brief period in vitro not only independently of NGF, but also independently of ongoing RNA synthesis for at least 6 h (Horii and Varon, 1975), and of protein synthesis for at least 12 h (Mizel and Bamburg, 1976). For PC12 cells cultured with NGF but no serum, survival is not affected for even 1 week by the presence of RNA synthesis inhibitors (Greene 1978). The neurite response to NGF in chick SG and DRG cultures takes place over 15 to 24 h even in the presence of RNA synthesis inhibitors, but fails to occur if protein synthesis is blocked (Partlow and Larrabee, 1971; Burnham and Varon, 1974; Mizel and Bamburg, 1976). With PC12 cells, blocking RNA synthesis prevents NGF from eliciting neurite extensions from 'naïve' cells but not from 'primed' cells, demonstrating that RNA synthesis is needed for the priming action of NGF but not for the execution of its neuritic consequence (e.g., Greene, 1978; Greene et al., 1980). Thus, once priming has taken place (in vivo, or respectively, in vitro), the action of NGF on neurite extension does not appear to occur via transcription – or translation-dependent events – although, in the longer term, protein and possibly RNA synthesis may still be needed to replenish or increase the necessary cell materials.

3.4. NGF intermediate effects

Our understanding of how NGF causes its traditional long-term consequences will ultimately depend on the recognition of *intermediate* events specifically elicited by NGF. Several such intermediate NGF effects have been reported.

3.4.1. Selected proteins
Nerve Growth Factor causes a marked increase in PC12 cells in at least three *glycoprotein* species (cf., Greene and Shooter, 1980). The NGF-induced increase of all three species is RNA-dependent, suggesting a possible involvement in the priming mechanism or its execution.

Ornithine decarboxylase (ODC) controls the synthesis of polyamines, which have been frequently implicated in regulation of cell growth and/or differentiation (e.g., Russell and Stambrook, 1975; Janne et al., 1978). Nerve Growth Factor has been reported to cause a rapid (4 to 6 h), transcription- and translation-dependent rise of ODC in young and adult rat SCG both in vitro and in vivo (MacDonnell et al., 1977; Huff and Guroff, 1979; Lakshmanan 1979; Hendry and Bonyhady, 1980), in mouse DRG (Lakshmanan 1979; Guroff and Yu, 1979), in PC12 cells (Hatanaka et al., 1978; Greene and McGuire, 1978), and even major brain areas in vivo (e.g., Lewis et al., 1978; Nagaiah et al., 1978; Ikeno et al., 1978). Blocking either the synthesis or the activity of ODC, however, did not block the effects of NGF on cell survival and hypertrophy or on neurite extension in PC12 or chick SG cell cultures (Greene and McGuire, 1978). Increases in ODC in PC12 cells could also be elicited with dibutyryl cyclic AMP (Hatanaka et al., 1978) or Epidermal Growth Factor (EGF) (Huff et al., 1981), which do not cause priming.

3.4.2. Membrane properties
Besides its role in growth cone activity and neurite growth, *adhesion* is also intricately related with several other cell properties, including cytoskeletal structures, membrane fluidity, cyclic nucleotide regulation and permeation to small molecules and ions (cf., Varon, 1979). Nerve Growth Factor has been reported to increase adhesiveness of ganglionic neurons (Weston, 1972; Varon, 1975a) and PC12 cells (Schubert and Whitlock, 1977; Shubert et al., 1978; Greene et al., 1979; Chandler and Herschmann, 1980). Adhesion of PC12 to a substratum is also promoted by EGF (Chandler and Herschmann, 1980) and cyclic AMP elevating agents (Schubert et al., 1978).

Connelly et al. (1979a,b) have described *morphological membrane alterations* in PC12 cells (detected by scanning electron microscopy), which occur in a rapid, synchronized sequence on presentation of NGF. There was a vast time difference between these surface events (minutes) and the elicitation by NGF of neurite formation (days) in PC12 cells. NGF has more recently been reported to have very rapid effects on growth cone activities of PC12 cells (Greene, 1981). When NGF is briefly withdrawn from primed PC12 cell cultures, *growth cone* activity ceases: representation of NGF calls forth a transcription-independent resumption of microspike formation within 5 min and their elongation within 10 min.

3.4.3. Cytoskeletal elements
Several observations suggest that microtubules and microfilaments are involved in NGF effects and, possibly, in its very mode of action. Nerve Growth Factor does not appear to stimulate the production of tubulin in ganglionic neurons, at least over the first 24 to 72 h in vitro (Yamada and Wessells, 1971; Mizel and Bamburg, 1976) or in vivo (Stoeckel et al., 1974): it appears, therefore, that tubulin synthesis does not mediate the neurite-stimulation effect of NGF when enough tubulin is already available in the cell. The increase in actin and tubulin observed in longer term ganglionic cultures (Fine and Bray, 1971) is, therefore, likely to reflect merely the increase in neurite mass and the overall trophic role of NGF.

On the other hand, NGF is likely to promote the assembly of tubulin into microtubules and of actin into microfilaments, as well as the subsequent stability of such structures. In cell-free systems (Calissano and Cozzari, 1974; Levi et al., 1975; Calissano et al., 1976, 1978; Monaco et al., 1977), NGF binds to soluble tubulin and G-actin and causes them to polymerize. Moreover, pre-formed microtubules presented with NGF form insoluble complexes and acquire resistance to vinblastine. Calissano et al. (1976,1978) have suggested that internalized NGF enhances formation and stabilization of microtubules and microfilaments, thereby altering membrane mobility and eliciting the membrane-related, transcription-independent effects of NGF. However, note (cf., Greene and Shooter, 1980) that (i) the NGF effects on cell-free tubulin and actin required 1 to 10 μM concentrations (three orders of magnitude higher than those involved in the usual biological effects), (ii) other basic proteins also could elicit them (Lee et al., 1978), and (iii) the opportunity for NGF to bind to cytoskeletal structures within or at the surface of intact cells remains to be demonstrated.

There may be indirect effects of NGF on cytoskeletal structures, or the involvement of the latter in NGF-induced consequences. A specific increase in tyrosyl-tubulin ligase activity has been reported in NGF-treated PC12 cells (Levi et al., 1978). Primed PC12 cells display a marked increase in the proportion of their microtubules that is colchicine-resistant (Greene, 1981). A greater protection against vinblastine also appears to be provided by NGF in vivo to rat SCG microtubules (Menesini-Chen et al., 1977; Johnson 1978).

3.4.4. Cyclic AMP involvements

A rapid, transient elevation of intracellular *cyclic AMP levels* has been reported to follow the presentation of NGF to several, though not all, target cells (Narumi and Fujita, 1978; Nikodijevic et al., 1975; Otten et al., 1978; Schubert and Whitlock, 1977; Schubert et al., 1978; Skaper et al., 1979b). However, other investigators have failed to detect such cyclic AMP responses (Frazier et al., 1973b; Hatanaka et al., 1978; Hier et al., 1973; Lakshmanan, 1978b; Otten et al., 1978). Given the known susceptibility of cyclic AMP to a variety of extrinsic and intrinsic influences, one should remain cautious about the negative results: nevertheless, it does not appear likely that cyclic AMP is *the* 'second message' for all NGF effects.

The record with cyclic AMP-elevating agents is also equivocal. Such agents promote neither survival nor neurites in chick SG cultures (Frazier et al., 1973b), but do elicit neurite outgrowth in chick DRG (Roisen et al., 1972; Hier et al., 1973; Narumi and Fujita, 1978) and rat DRG (Haas et al., 1972). In PC12 cells, cyclic AMP stimulants do not cause priming but can elicit neurite extension from NGF-primed cells for a short time and in a RNA-dependent manner (cf., Greene and Shooter, 1980): however, Schubert and Whitlock (1977) reported neurite responses with unprimed PC12 cells as well.

3.4.5. Phosphorylation effects

In vivo or in vitro treatment with NGF of rat SCG has been found to increase nearly 2-fold the phosphorylation of two specific *nuclear proteins* associated with the nuclear chromatin (Yu et al., 1978, 1980). This effect of NGF is detectable within 2 h and maximal within 4 h of NGF presentation, occurs with NGF concentrations as low as 10 ng/ml, and is independent of both transcription and translation. Cyclic AMP-elevating agents but not EGF elicited the response. Similar responses were detectable in the nuclei of E12 chick or E16 rat DRG, but not with NGF-unresponsive tissues. PC12 cells also respond to NGF, but with increased phosphorylation of a single nuclear protein. In PC12, however, not only dibutyryl cyclic AMP but also EGF mimicked the phosphorylation effect of NGF.

Several tissues display, under the influence of various physiological or pharmacological agents, an increased turnover of phosphatidylinositol (PI) – the so-called *'PI effect'* (e.g., Michell, 1975). The PI effect occurs only in intact cells (with the one exception of synaptosomes), requires ligand–receptor interaction, and is often accompanied by an increase of Ca^{2+} in the cytosol. Involvement of the PI effect has been suggested in ion transport at a synapse, macromolecular translocations across a membrane, and amplification of receptor-mediated signals including mitogenic ones. NGF (10^{-10} to 10^{-7} M) has been reported to increase PI turnover in rat SCG organ cultures (Lakshmanan, 1978a–c). Dibutyryl cyclic AMP also elicits a PI effect, which is additive to (hence distinct from) that of NGF.

4. The ionic hypothesis for NGF

In a recent review Greene and Shooter (1980) chose to define as '*actions* of NGF' the final results of NGF treatment, as '*responses* to NGF' the causal steps in the entire sequence of events leading to the final results, and as *mechanism of action* of NGF this entire sequence. These definitions overlook the possibility of distinguishing between (i) events that are strictly dependent on NGF (NGF responses), and (ii) cellular processes that require the NGF responses for their activation but then proceed on their own (or under additional extrinsic influences). For example, the same ganglionic neurons can be made to survive by NGF or by different GNTFs. The manners by which each factor provides the cell with the survival stimulus may well be different, but the 'survival-prone' neurons will probably use the same anabolic and energetic processes to achieve survival in both cases. Similarly, the assembly and elongation of neurites occurs in all neurons regardless of their relatedness to NGF, and very likely reflect the execution of neuronal programs for which NGF only represents one of several possible promotors: neurite enhancement by other agents, e.g., PNPF may again involve triggering events different from those precipitated by NGF but use, beyond such events, the same cellular machineries and processes for neurite extension as those used under NGF. Thus, it appears profitable to distinguish between two portions of the entire sequence of events observable under NGF treatment: (1) those earlier events that are exquisitely and uniquely dependent on this factor, and (2) those subsequent events that merely reflect the capabilities of neuronal machineries once they are brought into action by whatever means. Such a distinction, furthermore, raises the possibility that different consequences of NGF could be explained by different locations of the same molecular site of action, or by different cellular machineries involved beyond a common sequence of triggering events – rather than having to invoke different sites or different mechanisms of action in each case.

4.1. Concepts and strategies

We have adopted the working hypothesis that NGF must interact with a single molecular target in such a way as to trigger a series of rapid and early events (*short-latency responses*), the influence of which will in turn invest a veriety of cellular machineries to cause eventually the *longer-term composite consequences* for which NGF is traditionally known. Figure 1 illustrates this *hypothetical model*. The

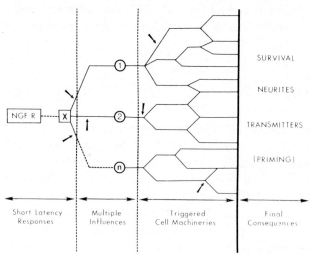

Fig. 1. Hypothetical model for NGF mode of action. See text for details.

first sequence of events, from NGF-receptor association to the 'X' response, represents the true mechanism of action of NGF, and is entirely unknown at present. The branching sequences between the X response and critical cellular functions (1, 2 etc.) are the keys to the multiplicity of the NGF consequences. The last sequences, beyond those critical cell functions, constitute the execution events making those consequences manifest, and they may proceed independently from or interacting with one another and/or with additional extrinsic cues (arrows).

In the absence of any starting information, a *temporal criterion* was adopted to seek short-latency responses, namely their occurrence within *minutes* of NGF presentation — as contrasted with the 16 to 40 *hours* required for the survival and neurite behaviors elicited by NGF in ganglionic neurons, and the even longer time frame of *days* involved in PC12 cell priming. One additional advantage of this choice was to avoid the pitfall inherent to any study of a 'survival-promoting' factor, namely that untreated cells are no longer viable to serve as control at the time when survival effects can be determined (with the exception of PC12 cells).

The experimental system sought for such investigations had to meet several requirements: (i) measurable deficits should develop within a few hours of NGF deprivation (before the start of neuronal death), (ii) delayed presentation of NGF should fully reverse such deficits (verifying the viable state of the cells), and (iii) the NGF-induced recovery should occur within minutes (the postulated feature of a short-latency response). The use of cell suspensions, freshly dissociated from NGF-sensitive ganglia (particularly E8 chick DRG), proved to meet the above requirements (cf., Varon and Skaper, 1980). As discussed earlier (Section 3.3), E8 chick DRG cells develop a reversible deficit in their RNA-labeling capabilities over the first 6 h of NGF deprivation. Development and reversal of an RNA-labeling deficit was the consequence of similar events concerning the transport of uridine rather than the rate of RNA synthesis (Horii and Varon, 1975, 1977). Subsequent findings that the *transport* of glucose (2-deoxyglucose) and γ-aminoisobutyric acid (AIB) were also dependent on NGF in the chick DRG cells led to the recognition that the NGF-regulated transport systems shared the property of being Na^+-coupled transports, i.e., requiring Na^+ gradients across the neuronal membrane as their energy source (Skaper and Varon, 1979a,b). This observation, in turn, brought about the realization that NGF regulates Na^+ and K^+ control mechanisms in its target neurons (Skaper and Varon, 1979c; 1980a,b; 1981a).

4.2. The ionic responses to NGF

4.2.1. Features of the ionic responses

The typical expressions of ionic behaviors related to NGF are illustrated in Fig. 2. DRG cells incubated over *6 h without NGF* (Fig. 2A) accumulate Na^+ (traced by $^{22}Na^+$) and lose K^+ (pre-equilibrated with $^{86}Rb^+$) until the intracellular levels of these ions have equilibrated with those of the surrounding medium. Both the Na^+ and the K^+ changes are passive processes, i.e., diffusion down their respective concentration gradients independently from each other — suggesting the progressive establishment of a cell deficit in Na^+ and K^+ control mechanisms. On *delayed NGF* presentation at 6 h (Fig. 2B), the excess Na^+ is extruded and the lost K^+ retrieved within minutes. This ionic recovery involves active processes, since the net movement of each ion is against its concentration gradient. It also involves coupled processes, since Na^+ extrusion will not take place unless K^+ is available outside the cells and, conversely, K^+ re-uptake will not occur unless Na^+ is present inside the cells (Skaper and Varon, 1981a). Thus, the presentation

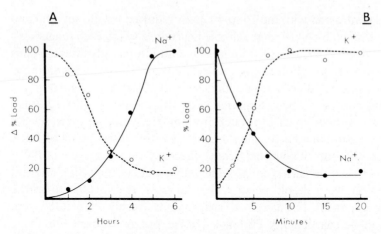

Fig. 2. Ionic changes in chick DRG cells in the absence and presence of NGF. (A) Cells were incubated over 6 h without NGF in the presence of ^{22}Na$^+$ (●) or ^{86}Rb$^+$ (○) to trace Na$^+$ and K$^+$ movements, respectively. (B) Delayed presentation of NGF at 6 h, followed by measurement of extrusion of excess Na$^+$ (●) and retrieval of lost K$^+$ (○).

of NGF turns on in a very short timeframe a *Na/K$^+$-pump mechanism*, just as the absence of NGF turns it off (albeit on a longer time scale).

Reactivation of the ionic pump by NGF depends on the NGF concentrations for both its magnitude and its speed (Skaper and Varon 1980a, 1981a). Low concentrations of NGF (e.g., <1 ng/ml) reverse the ionic defect only partially, and relatively slowly (30 min); high NGF concentrations not only cause complete recovery, they achieve it in shorter times – e.g., 10 min at 10 ng/ml (1 biological unit/ml, or the optimal NGF concentration in DRG culture assays) and less than 1 min at 500 to 1000 ng/ml. These dose–response relationships are typical of binding reactions, where adequate ligand concentrations are needed for complete saturation of all available binding sites, and even higher concentrations will achieve the same saturation in shorter and shorter times. Thus, the ionic responses to NGF appear to reflect thermodynamics and kinetics of the binding of NGF to its cell surface receptors.

The large differential in ^{22}Na$^+$ accumulation between NGF-deprived and NGF-treated cells offers an excellent basis for an objective NGF bioassay, sensitive to 10^{-11} M NGF and achievable within a single workday (cf., Skaper and Varon, 1982). More importantly, the discovery of an NGF impact on ionic behaviors opened up several new questions about its mode of action.

4.2.2. Correlations between ionic and traditional effects of NGF
The NGF-related ionic behaviors just described were first observed in dissociated cell suspensions from E8 chick DRG. The ionic responses of these cells appear to be specific for the NGF molecules: they are neither mimicked nor hindered by

the defined N1 supplement (see Section 2.1.2), or fetal and human sera (Skaper and Varon, 1980a; Skaper and Varon, 1982). Thus far, all primary ganglionic neurons examined have exhibited the ionic responses if they exhibit one of the more traditional consequences of NGF on survival and/or neurite outgrowth and, conversely, have failed to display them if their responsiveness to NGF has not otherwise been demonstrated (Skaper and Varon, 1980b). E8 chick DRG and E11 chick SG, two of the most studied NGF targets, show the ionic response in both intact ganglia and dissociated cell suspensions. E8 chick ciliary ganglia, known not to require NGF for survival or neurite extension (e.g., Manthorpe et al., 1981a), did not show NGF-related ionic differences. Neonatal mouse DRG display neurite outgrowth in explant cultures even without NGF, and their ionic behavior is not affected by the presence or absence of the factor. In contrast, dissociated cell cultures from these same mouse DRG demonstrate an NGF requirement for neuronal survival (Varon et al., 1973, 1974a), and ionic behaviors are regulated by NGF in the corresponding cell suspensions.

As already noted (Section 1.4.2), the responsiveness to NGF of chick DRG neurons is known to vary with embryonic age: in particular, between approximately E11 and E18, intact ganglia in explant culture display less and less of a neurite halo in the presence of NGF, and the NGF support of neuronal survival in dissociated cell cultures gradually disappears. A recent study (Skaper et al., 1982) has demonstrated that developmental changes in NGF responsiveness also hold true for the ionic responses: the differential Na^+ accumulation between NGF-deprived and NGF-treated *intact* chick DRG increases slightly between E6 and E10, but decreases to practically nought between E12 and E16. One important feature revealed by this study was that the reduction in differential Na^+ accumulation was not due to a progressive failure by NGF to restrict Na^+ accumulation, but to an increasing competence of the ganglia to do so independently from the availability of exogenous NGF: in other words, exogenous NGF becomes less and less *needed* for ionic control in the intact DRG (rather than less and less effective on them), as embryonic age progresses. Clearly, ionic control continues to be equally important beyond the NGF-dependent age span as it is during it, even though the means by which such control is achieved no longer involves exogenous NGF. Among the several possibilities yet to be investigated (by use of DRG dissociates, in suspension and in culture) are: (i) the neurons continue to require NGF, but now find it available from within the ganglion itself, (ii) the neurons switch their requirement to trophic factors other than NGF (e.g., the ganglionic neuronotrophic factors discussed in Section 2.1.2), which they can draw from indigenous sources, or, (iii) the neurons have matured to a stage where ionic control requires no extra-neuronal factors.

One other correlation between ionic and survival effects of NGF has been revealed, with intriguing new implications, by an investigation of *purified neurons* from E11 chick SG (Varon and Skaper, 1981b). These neurons can be collected practically free of ganglionic nonneural cells by differential cell attachment (Varon and Raiborn, 1972a). When reseeded on a collagen substratum with NGF-containing medium, they will attach but fail to survive for even 24 h, despite the availability of NGF: NGF, however, will again support their survival if an adequate number of ganglionic nonneurons are provided. When examined for sodium behaviors, purified SG neuronal suspensions accumulated Na^+ over 6 h of incubation even when NGF was present, thereby displaying the same failure to respond to NGF in the absence of nonneurons as they did for survival in culture. Conversely, and with equal consistency, re-addition of nonneuronal cells restored the capability of NGF to prevent, or reverse, the Na^+ accumulation. What nonneurons contribute to thse SG neuronal preparations to restore both their ionic and their survival responsiveness to NGF is not yet known, but new opportunities are offered to probe further into (i) the mechanisms involved in the control by NGF of ionic behaviors, and (ii) specific aspects of neuronal interactions with other cells. One point of caution must be stressed. The requirement for nonneuronal cell contributions seen in these SG preparations need not be the rule for all ganglionic systems: in different circumstances, cultured SG neurons have been shown to survive with NGF even in the absence of nonneuronal cells (Mains and Patterson, 1973; Edgar et al., 1981), and so have purified DRG neurons (Barde et al., 1980). However, what is contributed by nonneurons in the system described here may already be available in some other manner in the nonneuron-independent systems, and play there an equally essential role in the interactions between neurons and NGF.

4.3. Molecular aspects of the ionic effects of NGF

The very close correlation observed, thus far, between the ionic effects of NGF and the survival and neurite consequences of NGF in primary neurons strongly proposes that regulation of a Na^+/K^+-pump be viewed as one of the critical cell functions on which NGF exerts its action (see Fig. 1). This, then, defines two domains for future investigations (see Fig. 1): the sequence of events leading from NGF-receptor binding to control of the ionic pump, and the relevance of ionic control for longer-term, NGF-influenced neuronal behaviors.

4.3.1. From receptor binding to ionic pump control
One may start by considering two alternative eventualities concerning the sequence of events linking the NGF-receptor association to the control of ionic pump performance: (i) that all such events are contained within the plasma membrane of the neuron, or (ii) that extra-membrane cell constituents also intervene in the process. The first eventuality is particularly attractive, since one might be able to detect pump responses to NGF in a cell-free membrane

preparation: thus, one would open the way to experimentally reconstituted, membrane-like systems as the means to further resolve and identify the individual membrane constituents involved in this NGF action. The evidence collected to date, however, does not appear to support this eventuality (Skaper and Varon, 1981b). Microsomal fractions, prepared from NGF-treated and NGF-deprived E8 chick DRG, were examined in the presence or absence of NGF for their $(Na^+ + K^+)$-ATPase activity (the enzymatic equivalent of the Na^+/K^+-pump). No NGF-related differences could be detected in total levels of enzyme activity, or in the affinities of the enzyme for its active cations (Na^+ and K^+) and substrate (ATP).

Thus, it appears at present more likely that (i) cell constituents outside the plasma membrane are involved in either the inhibition of the pump in NGF-deprived cells or the stimulation of the pump in NGF-supported ones, and that (ii) the interactions between such cell constituents and the ionic pump in the intact cell do not survive the cell disruption required for direct $(Na^+ + K^+)$-ATPase studies. What such extra-membrane constituents may be, and how they may be altered by NGF in the intact neuron, remain to be determined. One may, however, speculate about a number of putative candidates, either suggested by already reported, transcription- and translation-independent, intermediate effects of NGF (Section 3.4), or believed to play regulatory roles in other cellular mechanisms affected by ligand–receptor interactions. Such candidates fall into one or both of two major categories, depending on whether they would modify the pump molecule itself or the membrane environment in which the pump must operate.

The list of putative agents or processes whereby intracellular regulation of the ionic pump could take place may be a long one, and the following are but few examples of them. *Cyclic AMP* has been ruled out as an NGF-regulated mediator (Skaper and Varon, 1981c): the cyclic AMP response to NGF has not been observed in some ganglia where the ionic response is evident, can be elicited regardless of the ionic situation of the ganglia (hence, does not appear to be itself a consequence of the pump regulation of NGF) and, most explicitly, fails by itself to preserve or re-instate ionic control. *Phosphorylation* by protein kinases of either the pump (Spector et al., 1980) or relevant membrane constituents (e.g., the so-called PI effect) could inhibit or disinhibit pump action, and be prevented or reversed by phosphatases upon tissue homogenization. A similar consideration might be given to *methylation* and demethylation processes, which (like the PI effect) have been implicated in the control of several receptor-dependent cell behaviors, possibly via changes in membrane fluidity (Hirata and Axelrod, 1980). *Calcium* shifts, either across the plasma membrane or among intracellular compartments, may be associated with (or even causal to) phosphorylation or methylation events, or may have independent effects of their own. Ca^{2+} changes under NGF have reported in PC12 by Schubert et al. (1978) but denied by others

(Landreth et al., 1980). Serious consideration should be given to an involvement of *cytoskeletal elements*, on which plasma membrane conformation and fluidity may depend. *Membrane fluidity* could also be altered from the outside of the cell via cell–cell or cell–substratum *adhesive interactions* (a possible basis for the nonneuronal requirements in ionic and survival responses to NGF by purified SG neurons). Finally, one cannot dismiss the possibility that NGF acts by permitting or hindering access to the pump of other *cytosol constituents* (e.g., vanadium – Cantley and Aisen, 1979), including ATP itself.

4.3.2. From ionic pump control to neuronal survival
A modified version of Fig. 1 is presented in Fig. 3 to facilitate the discussion of

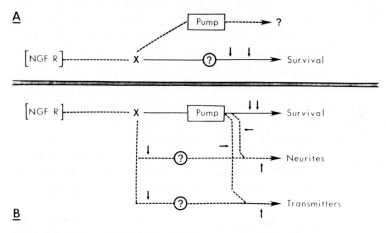

Fig. 3. Modified version of Fig. 1, illustrating the possible relationships of the ionic response to long-term NGF consequences. (A) Neuronal survival as a consequence of key changes independent of the ionic response. (B) Neuronal survival as a necessary consequence of ionic pump maintenance. See text for details.

the next question: do the ionic responses to NGF represent but one *epiphenomenon* in the overall spectrum of NGF responses or, could the ionic responses be *causal* to the survival (and, possibly, other) consequences of NGF action?

One may contend that neuronal survival is the consequence of other key changes in the cells which are entirely independent of the ionic response even though they, too, reflect changes imposed by NGF on the common X function (Fig. 3A). If so, the ionic defect developed under NGF deprivation would be only a *symptom* of neuronal illness, rather than a causal link to cell death, and the ionic recovery perceived on NGF re-presentation would be only an indirect sign of the resumption of a healthy state by the neurons. In other words, regulation of the ionic pump by NGF is on a 'sideline' rather than a 'mainline' with regard to the sequence leading to survival. The consistent occurrence of the ionic responses, however, would imply that disruption of ionic control be a *necessary* byproduct of neuronal illness. In addition, the ionic disruption *precedes* neuronal death as demonstrated by its full reversibility. At the very least, then, the ionic response to NGF would still remain the earliest measurable event presently known, which detects and faithfully monitors neuronal damage or recovery before the cell is irreversibly committed to its ultimate fate. Furthermore, tracking down the events linking NGF-receptor binding to ionic pump regulation (see previous subsection) should also illuminate early events that are causal to the survival outcome, since at least part of the sequence would be common to both pathways.

Given the obvious importance to a cell of intracellular Na^+ and K^+ levels and their transmembrane gradients, it is difficult to view the dependence of ionic pump control on NGF in the ganglionic neurons as having *no* critical role on other important cell functions. A viable alternative, until proven otherwise, would be that in fact the ionic defect developed on NGF deprivation be the main *cause* of subsequent neuronal death and that its prevention by NGF be the mechanism through which NGF also controls survival. This is not to say that (i) activation or disinhibition of the ionic pump is the direct, immediate effect of NGF, since several other events must intervene along the way, or that (ii) the ionic effect is the single mediator of NGF action, since one or more of the intervening events (e.g., X) may well give rise to independently relevant sequences (Fig. 3B). Also not to be implied is the notion that ionic control, while necessary, need be sufficient for ultimate survival, since additional extrinsic signals may be required to activate survival-supportive cellular activities (Fig. 3B: arrows).

The speculation that ionic control also controls survival can be put to several in vitro tests, e.g., by imposing changes in Na^+ or K^+ intracellular levels and/or transmembrane gradients in the continued presence of NGF and determining whether neuronal survival will be prevented as effectively as if NGF were absent. Potassium ions (and Mg^{2+}) in the medium are reportedly required by E9 chick DRG for their maintenance in explant cultures (Stach et al., 1979). A strong linkage between monovalent cations and neuronal survival in vitro has been demonstrated by the increased survival of cultured ganglionic neurons in the presence of high external K^+ concentrations (cf., Scott and Fisher, 1971; Phillipson and Sandler, 1975; Bennett and White, 1979; Chalazonitis and Fischbach, 1980), even in the absence of their corresponding neuronotrophic factors. Since external K^+ is a principal stimulus for the Na^+/K^+-pump, these findings are at least compatible with the ionic pump mediation hypothesized here for the survival consequence of NGF. Also compatible with this ionic hypothesis would be a functional role of internalized NGF: by acting on the receptors of the inner face of the endocytotic vesicle, NGF could continue to promote transfer of Na^+ and K^+ between cytoplasm and the intravesicular fluid (initially equivalent to extracellular fluid).

There are several possible ways through which ionic control by NGF could lead to survival control. One, for example, is the control of *nutrient uptake* (e.g., Schwartz et al., 1972). We have already noted (Section 4.1) that E8 chick DRG require a Na^+ gradient for their uptake of glucose, AIB (presumably representing the alanine, glycine and glutamine transport systems – Oxender and Christensen, 1963), and several nucleosides (Horii and Varon, 1975; Skaper and Varon, 1979a,b). This need not be true for all ganglionic target neurons, or at any developmental age: in E11 chick SG and neonatal mouse DRG, hexose and nucleoside transports are independent of Na^+ gradients and NGF (Skaper and Varon, unpublished data). In other studies (Levi and Lattes, 1968), NGF increased the uptake by chick DRG of acidic and not other amino acids (with no information provided with regard to Na^+-coupling features). Transport of AIB is also enhanced by NGF in PC12 cells – as it is by serum which, like NGF, promotes their survival and general growth (McGuire and Greene, 1979). Other cell properties affected by Na^+ and K^+ gradients, beside nutrient uptake, are membrane potentials and intracellular osmotic pressure (e.g., Baker, 1972). *Intracellular Na^+ and K^+* levels are equally important on their own terms (cf., Kaplan, 1978), for example, by regulating cation-dependent enzymes involved in biosynthetic pathways (e.g., Scholnick et al., 1973) and energy metabolism (Jimenez de Asua, et al., 1970). Macromolecular synthesis is known to be inhibited in various cells by a loss of intracellular K^+ or a reduction in K^+/Na^+ ratio (Lubin, 1967; McDonald et al., 1972; Lamb and McCall, 1972; Ledbetter and Lubin, 1977; Kaplan, 1978; among others). Monovalent cation ionophores alter Golgi complexes and inhibit glycoprotein secretion in several cells, (Smilowitz, 1979, 1980; Tartakoff and Vasselli, 1977, 1978; Uchida et al., 1979). Cytoplasmic Na^+ may play a role in the release of Ca^{2+} from mitochondria (Carafoli and Crompton, 1978; Haworth et al., 1980) and thus change Ca^{2+} levels in the cytosol and Ca^{2+}-dependent activities (e.g., Schneider et al., 1978; Cheung, 1980). Na^+/H^+-exchange systems have been described (e.g., Christensen, 1975; Johnson et al., 1976) and intracellular pH may be affected.

4.4. General aspects of the ionic hypothesis

The newly discovered ionic effects of NGF on its target neurons have potential, though admittedly even more speculative, implications beyond those already discussed. One speculation remains concerned with the mode of action of NGF and the multiplicity of its ultimate consequences: could the ionic response serve as a unitary basis for all or most of the NGF consequences, if one assumes that the same ionic response will affect different cell machineries *when it occurs in different* parts of the same cells? The other speculation concerns a more general question beyond the NGF phenomenon itself: can one view the regulation of ionic mechanisms as a widespread mode for the regulation, by a variety of extrinsic agents, of maintenance, growth and/or differentiation of cells other than neurons?

4.4.1. Multiple consequences of NGF action

Neurite extension and neuronal survival have been increasingly seen as reflecting a common mode of action of NGF in primary neurons and primed PC12 cells (cf., Greene, 1978; Harper and Thoenen, 1980; Greene and Shooter, 1980; see also Section 2.2). Campenot (1977) has shown that neurite maintenance and elongation require the local presence of NGF. Neurite extension itself is a *local* event. The NGF receptors present on the distal portion of a neurite or its growth cone need not be different from those present on the neuronal soma. Thus, the same ionic effects to NGF which could promote survival activities when applied to the cell soma might regulate local properties or machineries concerned with neurite behaviors when applied to such a specialized local organelle as the growth cone. Neurite extension involves adhesive interactions between growth cone and terrain and the participation of cytoskeletal elements – two features which have been associated with ionic changes (cf., Varon and Adler, 1981), and may be altered by the action of NGF (Section 3.4). Indeed, neurite extension has been elicited, in the absence of NGF, by application of electric or electromagnetic fields (e.g., Marsh and Beams, 1946; Sisken and Smith, 1975; Jaffe and Nuccitelli, 1977; Jaffe and Poo, 1979) or by minute injections of current (Sisken and Lafferty, 1978; see also Becker, 1981) – both stimulations involving ionic movements across the neuronal membrane. Calcium ions are important for cytoskeletal stability in neurites (e.g., Daniels, 1972; Schlaepfer and Bunge, 1973; Lasek and Hoffman, 1976), and calcium changes may be elicited by monovalent ion changes (e.g. Crompton et al., 1978; Schneider et al., 1978; Cheung, 1980; among others). Membrane lipid fluidity, also conceivably involved in the ionic responses (see Section 4.3.1) has been reported to increase in neuroblastoma cells in the course of their neuritic expression (DeLaat et al., 1978).

The case for an ionic role in the *transmitter promotion* consequence of NGF is more tenuous. Macromolecular synthesis may be regulated by intracellular monovalent cations (Section 4.3.2), and effects on specific macromolecular products have also been reported in various cells (e.g., Shinohara and Piatigorsky, 1977). Hawrot (1980) has reported that cell–substratum adhesion is one of the many stimuli causing rat SCG neurons to shift to a cholinergic transmitter mode. A more explicit ionic involvement in transmitter production has been described by Walicke et al (1977) in rat SCG neurons: depolarization by electrical stimulation, increased external K^+, or veratridine led to increased synthesis of noradrenaline and a prevention of the exogenously inducible shift to acetylcholine production.

The *priming* of PC12 cells is the only indication, thus far, for a possible role of NGF preceding the acquisition by the target cell of a bonafide neuronal status. Greene (1978) has pointed out that, in PC12 cells, NGF is likely to act on a single class of receptors, the results of which would branch into RNA-dependent (e.g., priming) and RNA-independent (survival, neurite expression) consequences. It would be at least plausible to speculate that NGF might use the same ionic mechanism to act as a 'phenotype-specifier' as well as a neuronotrophic agent. The key feature for such a speculation would be the assumption that intracellular monovalent cation levels have very pervasive consequences on proliferation control as well as on differentiation control – an assumption strongly encouraged by a variety of independent reports in the literature (see next subsection). One could thus hypothesize that (i) NGF binds to existing receptors on naive PC12 cells, and causes a moderate reduction in intracellular Na^+, (ii) this in turn turns off mitotic activities of the cell (cf., Cone and Cone, 1976, 1978) and diverts cell anabolism to more differentiated products, possibly including an increased production of ionic pump molecules, and (iii) NGF promotes further changes in the K^+/Na^+ intracellular ratio and elicits the trophic responses typical of a primary neuron. Cultured primary neurons can be caused to resume DNA synthesis and undergo nuclear, or even cell replication by experimentally imposing increases in their intracellular Na^+ (Cone and Cone, 1976, 1978) – a converse effect to that of NGF on PC12 cell proliferation. Serum-stimulated proliferation in neuroblastoma cells is accompanied by increased pump-activity which is mediated by amiloride-sensitive increases in Na^+ entry (Moolenaar et al., 1979 – see also Section 4.4.2). In neuroblastoma–glioma hybrid clonal cultures, an increase in membrane potential can be caused by the Na^+ ionophore monensin – presumably by eliciting increased pump activity via the higher entry of Na^+ (Lichtshtein et al., 1979). More directly, Boonstra et al. (1981) have reported that naive PC12 cells in serum-free cultures do respond to NGF with an early, rapid, and transient increase in Na^+/K^+-pump activity: in this case, too, the pump stimulation appears secondary to an increase in Na^+ influx. Note that changes in Na^+ permeability may lead to different effects in different cells with regard to intracellular levels of Na^+ (as distinct from rates of Na^+ influx) and membrane potential levels and, therefore, cause different consequences on cellular function (e.g., Rozengurt and Mendoza, 1980).

4.4.2. Ionic involvements in cell regulations
Wilson and Lin (1980) have pointed out that differentiation appears to parallel the development of specialized mechanisms addressing ionic control at both the phylogenetic and ontogenetic levels. Indeed, brain development in several animal species is accompanied by an increase in total content and specific activity of the $(Na^+ + K^+)$-ATPase (Abdel-Latiff et al., 1967; Formby, 1975; Peter et al., 1975; Bertoni and Siegel, 1978, 1979; among others). Maturation of newly grown neurites in vitro is also accompanied by an increase in intramembranous particles, presumed to represent Na^+/K^+-pump sites (cf., Pfenninger and Reese, 1976). Growth and repair in several tissues may be regulated by ionic signals across the cell membrane (for a survey, see Becker, 1981). Stimulation of Na^+/K^+-pump activity has been reported to be an early component in the action of several hormones – e.g., insulin on skeletal muscle in frog (Moore, 1973; Moore and

Rabovsky, 1979) or rat (Flatman and Clausen, 1979) and on frog skin (Grinstein and Erlij, 1974), or thyroxin on neuroblastoma and other cells (Draves and Timiras, 1980).

The greatest attention, thus far, has been directed to the involvement of monovalent cations in the *regulation of cell proliferation* (for reviews, see Kaplan, 1978; Rozengurt 1979; Rozengurt and Mendoza, 1980). Rozengurt and co-workers have proposed that mitogenic stimulation of fibroblasts and other non-excitable cells is mediated by early changes in Na^+ and K^+ fluxes. Specifically: (1) presentation to quiescent cells of serum or other known mitogens increases the rate of Na^+ entry, which can be shown to result in Na^+ accumulation if the Na^+/K^+-pump is blocked by ouabain, but not otherwise; a large portion of this Na^+ entry is through Na^+-specific channels that can be blocked by amiloride, while the remaining portion is subject to pump control (Smith and Rozengurt, 1978a,b); (2) in these cells, the Na^+/K^+-pump is particularly sensitive to intracellular Na^+ concentrations and, thus, is stimulated by the increased influx of Na^+ caused by the mitogens, or by Na^+ ionophores (Rozengurt and Heppel, 1975). It is this increased pump activity (and the consequent increase in K^+ influx) which would be responsible for the mitogen-induced stimulation of DNA-synthesis. The mitogenic response is blocked by ouabain, low external concentrations of either K^+ or Na^+, K^+ ionophores or amiloride. Conversely, DNA synthesis can be stimulated by amphotericin B (another ionophore), vasopressin (a stimulant of Na^+ transport) or a potent tumor-promoting agent (TPA) acting synergistically with other hormones or growth factors (insulin, EGF, MSA, platelet- or fibroblast-derived growth factors) (cf., Rozengurt and Mendoza, 1980; Lubin, 1980).

These several reports on ionic involvement in cell regulation, of course, cannot be taken as evidence for the ionic hypothesis for NGF action. At the very least, however, they point out the importance of ions and the plausibility of ionic mediations in the mode of action of various extrinsic agents. Of particular interest are the points of difference as well as of similarity between, for example, ionic behaviors in response to mitogens and to NGF: thus far, the effects of NGF on the ionic pump do not appear to be mediated by an NGF-induced increase in Na^+ entry (Skaper and Varon, 1980a). Clearly, much work remains to be done on ionic roles both with regard to NGF action and in the more general field of cell regulation.

Acknowledgement

This work was supported by USPHS grant NS-07606 from the National Institute of Neurological and Communicative Disorders and Stroke.

References

Abdel-Latiff, A.A., Brody, J. and Ramahi, H. (1967) J. Neurochem. *14*, 1133–1141.
Adler, R. and Varon, S. (1980) Brain Res. *188*, 437–448.
Adler, R. and Varon S., (1981a) Dev. Biol. *81*, 1–11.
Adler, R. and Varon, S. (1981b) Dev. Biol. *86*, 69–80.
Adler, R., Landa, K.B., Manthorpe, M. and Varon, S. (1979) Science *204*, 1434–1436.
Adler, R., Manthorpe, M. and Varon, S. (1980) Trans. Am. Soc. Neurochem. *11*, 230.
Adler, R., Manthorpe, M., Skaper, S.D. and Varon, S. (1981) Brain Res. *206*, 129–144.
Aloe, L. and Levi-Montalcini, R. (1979) Proc. Natl. Acad. Sci. USA *76*, 1246–1250.
Andres, R., Jeng, I., and Bradshaw, R. (1977) Proc. Natl. Acad. Sci. USA *74*, 2785–2789.
Angeletti R.A.H. and Bradshaw, R.A. (1979) in Handbook of Experimental Pharmacology (Lee, C.-Y., ed.) Vol. 52, pp. 276–294, Springer Verlag, Berlin.
Angeletti, P. and Vigneti, E. (1971) Brain Res. *33*, 601–604.
Angeletti, P.U., Gandini-Attardi, D., Toschi, G., Salvi, M.L. and Lewi-Montalcini, R. (1965) Biochim. Biophys. Acta *25*, 111–120.
Baker, P.F. (1972) in Metabolic Pathways (Hokin, L.E., ed.) Vol. 6, pp. 243–268, Academic, New York.
Banerjee, S.P., Snyder, S.H., Cuatrecasas, P. and Greene, L.A. (1973) Proc. Natl. Acad. Sci. USA *70*, 2519–2523.
Banerjee, S.P., Cuatrecasas, P. and Snyder, S.H. (1975) J. Biol. Chem. *250*, 1427–1433.
Banerjee, S., Cuatrecasas, P., and Snyder, S. (1976) J. Biol. Chem. *251*, 5680–5685.
Banker, G.A. (1980) Science *209*, 809–810.
Banks, B.E.C. and Walker, S.J. (1977) J. Neurocytol. *6*, 287–297.
Barde, Y.-H., Edgar, D., Thoenen, H. (1980) Proc. Natl. Acad. Sci. USA *77*, 1199–1203.
Becker, R.O. (1981) Mechanisms of Growth Control, Thomas, New York.
Bennett, M.R. and White, W. (1979) Brain Res. *173*, 549–553.
Bertoni, J.M. and Siegel, G.J. (1978) J. Neurochem. *31*, 1501–1511.
Bertoni, J.M. and Siegel, G.J. (1979) J. Neurochem. *32*, 573–580.
Black, I.B., Coughlin, M.D. and Cochard, P. (1979) Soc. Neurosci. Symp. *4*, 184–204.
Bloom, E.M. and Black, I.B. (1979) Dev. Biol. *68*, 568–578.
Boonstra, J., van der Saag, P.T., Moolenaar, W.H. and de Laat, S.W. (1981) Exp. Cell Res. *131*, 452–455.
Bottenstein, J.E., Skaper, S.D., Varon, S.S. and Sato, G.H. (1980) Exp. Cell Res. *125*, 183–190.
Bradshaw, R.A. and Young, M. (1976) Biochem. Pharmacol. *25*, 1445–1449.
Brunso-Bechthold, J.K. and Hamburger, V. (1979) Proc. Natl. Acad. Sci. USA *76*, 1494–1496.
Burnham, P.A. and Varon, S. (1973) Neurobiology *3*, 232–245.
Burnham, P.A. and Varon, S. (1974) Neurobiology *4*, 57–70.
Burnham, P., Raiborn, C. and Varon, S. (1972) Proc. Natl. Acad. Sci. USA *69*, 3556–3560.
Burstein, D.E. and Greene, L.A. (1978) Proc. Natl. Acad. Sci. USA *75*, 6059–6063.
Calissano, P. and Cozzari, C. (1974) Proc. Natl. Acad. Sci. USA *71*, 2131–2135.
Calissano, P. and Shelanski, M.L. (1980) Neuroscience *5*, 1033–1039.
Calissano, P., Levi, A., Alema, S., Chen, J.S. and Levi-Montalcini, R. (1976) in 26th Colloquim-Mosbach 1975. Molecular Basis of Motility (Heilmeyer, L., ed.) pp. 186–202, Springer-Verlag, Berlin.
Calissano, P., Monaco, G., Castellani, L., Mercanti, D. and Levi, A. (1978) Proc. Natl. Acad. Sci. USA *75*, 2210–2214.
Campenot, R.B. (1977) Proc. Natl. Acad. Sci. USA *74*, 4516–4519.
Cantley, L.C. and Aisen, P. (1979) J. Biol. Chem. *254*, 1781–1784.
Carafoli, E. and Crompton, M. (1978) Ann. N.Y. Acad. Sci. *307*, 269–284.

Chalazonitis, A. and Fischbach, G.D. (1980) Dev. Biol. 78, 173–183.
Chandler, C.E. and Herschman, H.R. (1980) J. Supramol. Struc. Supp. 4, 59.
Cheung, W.Y. (1980) Science 207, 19–27.
Christensen, H.N. (1975) Biological Transport, p. 373, Benjamin, London.
Chun, L. and Patterson, P. (1976) Soc. Neurosci. Abst. 2, 191.
Chun, L. and Patterson, P. (1977) J. Cell Biol. 75, 694–718.
Cochard, P., Goldstein, M. and Black, I.B. (1978) Proc. Natl. Acad. Sci. USA 75, 2986–2990.
Cochard, P., Goldstein, M. and Black, I.B. (1979) Dev. Biol. 71, 100–114.
Cohen, A.I., Nicol, E.C. and Richter, W. (1964) Proc. Soc. Exp. Biol. Med. 116, 784–789.
Collins, F. (1978a) Dev. Biol. 65, 50–57.
Collins, F. (1978b) Proc. Natl. Acad. Sci. USA 75, 5210–5213.
Collins, F. (1980) Dev. Biol. 79, 247–252.
Cone, C.D. and Cone, C.M. (1976) Science 192, 155–158.
Cone, C.D. and Cone, C.M. (1978) Exp. Neurol. 60, 41–55.
Connolly, J.L., Greene, L.A., Viscarello, R. (1979a) Fed. Proc. 38, 1430.
Connolly, J.L., Greene, L.A., Viscarello, R.R. and Riley, W.D. (1979b) J. Cell Biol. 82, 820–827.
Costrini, N.V. and Bradshaw, R.A. (1979) Proc. Natl. Acad. Sci. USA 76, 3242–3245.
Coughlin, M., Boyer, D. and Black, I. (1977) Proc. Natl. Acad. Sci. USA 74, 3438–3442.
Couglin, M.D., Dibner, M.D., Boyer, D.M. and Black, I.B. (1978) Dev. Biol. 66, 513–528.
Cowan, W.M. (1973) in Development and Aging in the Nervous System (Rockstein, M. and Sussman, M.C., eds.) pp. 19–41, Academic, New York.
Crompton, M., Moser, R., Lüdi, H. and Carafoli, E. (1978) Eur. J. Biochem. 82, 25–31.
Daniels, M. (1972) J. Cell Biol. 53, 164–176.
DeLaat, S.W., van de Saag, P.T., Nelemans, S. and Shinitzky, M. (1978) Biochim. Biophys. Acta 509, 188–193.
DeVellis, J. (1982) in Proteins in the Nervous System: Structure and Function (Haber, B., Perez-Polo, R. and Coulter, J.D., eds.), pp. 243–269, Liss, New York.
Dibner, M.D. and Black, I.B. (1976) Brain Res. 103, 93–102.
Dibner, M.D., Mytilineous, C. and Black, I.B. (1977) Brain Res. 123, 301–310.
Dichter, M., Tischler, A. and Greene, L.A. (1977) Nature (London) 268, 501–504.
Draves, D.J. and Timiras, P.S. (1980) in Tissue Culture in Neurobiology (Giacobini, E., Vernadakis, A. and Shahar A., eds.) pp. 291–302, Raven, New York.
Dumas, M., Schwab, M.E. and Thoenen, H. (1979) J. Neurobiol. 10, 179–198.
Ebbott, S. and Hendry, I. (1978) Brain Res. 139, 160–163.
Ebendal, T. (1979) Dev. Biol. 72, 276–290.
Ebendal, T. and Jacobson, C.-O. (1977) Exp. Cell Res. 105, 379–387.
Ebendal, T., Olson, L., Seiger, A. and Hedlund, K.-O. (1980) Nature (London) 286, 25–28.
Edgar, D., Barde, Y. and Thoenen, H. (1979a) Soc. Neurosci. Abstr. 5, 158.
Edgar, D., Barde, Y.-A., Thoenen, H. (1979b) Exp. Cell Res. 121, 353–361.
Edgar, D., Barde, Y.-A., Thoenen, H. (1981) Nature (London) 289, 294–295.
Fabricant, R.N., DeLarco, J.E. and Todaro, G.J. (1977) Proc. Natl. Acad. Sci. USA 74, 565–569.
Fedoroff, S. and Hertz, L. (1977) Cell, Tissue, and Organ Cultures in Neurobiology, Academic, New York.
Fine, R.E., and Bray, D. (1971) Nature (London) New Biol. 234, 115–118.
Flatman, J.A. and Clausen, T. (1979) Nature (London) 281, 580–581.
Formby, B. (1975) Experientia 31, 315–316.
Frazier, W., Boyd, L. and Bradshaw, R. (1973a) Proc. Natl. Acad. Sci. USA 70, 2931–2935.
Frazier, W., Ohlendorf, C., Boyd, L., Aloe, L., Johnson, E., Ferrendelli, J. and Bradshaw, R. (1973b) Proc. Natl. Acad. Sci. USA 70, 2448–2452.
Frazier, W.A., Boyd, L.F. and Bradshaw, R.A. (1974a) J. Biol. Chem. 249, 5513–5519.

Frazier, W., Boyd, L., Pullman, M., Szutowicz, A. and Bradshaw, R. (1974b) J. Biol. Chem. 249, 5918–5923.
Freed, W.J. (1976) Brain Res. Bull. 1, 393–412.
Giacobini, E., Vernadakis, A. Shahar, A. (1980) Tissue Culture in Neurobiology, Raven, New York.
Goldstein, L., Reynolds, C. and Perez-Polo, J.R. (1978) Neurochem. Res. 3, 175–184.
Gorin, P.D. and Johnson, E.M. (1979) Proc. Natl. Acad. Sci. USA 76, 5382–5386.
Greene, L.A. (1977a) Brain Res. 133, 350–353.
Greene, L.A. (1977b) Dev. Biol. 58, 106–113.
Greene, L.A. (1978) in Advances in Pharmacology and Therapeutics (Adolphe, M., ed.) Vol. 10, pp. 197–206, Pergamon, New York.
Greene, L.A. (1981) Trans. Am. Soc. Neurochem. 12, 129.
Greene, L.A. and McGuire, J.C. (1978) Nature (London) 276, 191–194.
Greene, L.A. and Rein, G. (1977a) Brain Res. 138, 521–528.
Greene, L.A. and Rein, G. (1977b) Nature (London) 268, 349–351.
Greene, L.A. and Shooter, E.M. (1980) Ann. Rev. Neurosci. 3, 353–402.
Greene, L.A. and Tischler, A.S. (1976) Proc. Natl. Acad. Sci. USA 73, 2424–2428.
Greene, L.A., Burstein, D.E., McGuire, J.C. and Black, M.M. (1979) Soc. Neurosci. Symp. 4, 153–171.
Greene, L.A., Burstein, D.E. and Black, M.M. (1980) in Tissue Culture in Neurobiology (Giacobini, E., Vernadakis, A. and Shahar, A., eds.) pp. 313–320, Raven, New York.
Grinstein, S. and Erlij, D. (1974) Nature (London) 251, 57–58.
Gundersen, R.W. and Barrett, J.N. (1979) Science 206, 1079–1080.
Haas, D.C., Hier, D.B. Arnason, B., and Young, M. (1972) Proc. Soc. Exp. Biol. Med. 140, 45–47.
Hamburger, V. and Levi-Montalcini, R. (1949) J. Exp. Zool. 111, 457–501.
Hamburger, V., Brunso-Bechtold, J.K. and Yip, J.W. (1981) J. Neurosci. 1, 60–71.
Harper, G.P. and Thoenen, H. (1980) J. Neurochem. 34, 5–16.
Harper, G.P., Barde, Y.A., Burnstock, G., Carstairs, J.R., Dennison, M.E., Suda, K. and Vernon, C.A. (1979) Nature (London) 279, 160–162.
Harper, G.P., Al-Saffar, A.M., Pearce, F.L. and Vernon, C.A. (1980a) Dev. Biol. 77, 379–390.
Harper, G.P., Pearce, F.L. and Vernon, C.A. (1980b) Dev. Biol. 77, 391–402.
Hatanaka, H., Thoenen, H. and Otten, U. (1978) FEBS Lett. 92, 313–316.
Haworth, R.A., Hunter, D.R. and Berkoff, H.A. (1980) FEBS Lett. 10, 216–218.
Hawrot, E. (1980) Dev. Biol. 74, 136–151.
Helfand, S., Smith, G. and Wessels, N. (1976) Dev. Biol. 50, 541–547.
Hendry, I.A. (1975) in Neurotransmission (Ahtee, L., ed.) Vol. 2 of Proceedings of the 6th International Congress Pharmacology, pp. 249–258, Forssen Kirjapamo Oy, Forssa (Finland).
Hendry, I.A. (1976) in Review of Neuroscience (Ephrenpreis, S. and Kopin, I.J., eds.) Vol. 2, pp. 149–194, Raven, New York.
Hendry, I.A. (1977) Brain Res. 134, 213–223.
Hendry, I.A. and Bonyhady, R. (1980) Brain Res. 200, 39–45.
Hendry, I.A. and Campbell, J. (1976) J. Neurocytol. 5, 351–360.
Hendry, I. and Iversen, L. (1973) Nature (London) 243, 500–504.
Hendry, I.A., Stoeckel, K., Thoenen, H. and Iversen, L.L. (1974) Brain Res. 68, 103–121.
Herrup, K. and Shooter, E.M. (1973) Proc. Natl. Acad. Sci. USA 70, 3884–3888.
Herrup, K. and Shooter, E.M. (1975) J. Cell Biol. 67, 118–125.
Herrup, K. and Thoenen, H. (1979) Exp. Cell Res. 121, 71–78.
Heumann, R., Schwab, M. and Thoenen, H. (1981) Nature (London) 292, 838–840.
Hier, D., Arnason, B. and Young, M. (1973) Science 182, 78–81.
Hirata, F. and Axelrod, J. (1980) Science 209, 1082–1090.
Hollyday, M. and Hamburger, V. (1976) J. Comp. Neurol. 170, 311–320.

Horii, Z.-I. and Varon, S. (1975) J. Neurosci. Res. *1*, 361–375.
Horii, Z. and Varon, S. (1977) Brain Res. *124*, 121–133.
Huff, K.R. and Guroff, G. (1979) Biochem. Biophys. Res. Comm. *89*, 175–180.
Huff, K., Lakshmanan, J., and Guroff, G. (1978) J. Neurochem. *31*, 599–606.
Huff, K., End, D. and Guroff, G. (1981) J. Cell Biol. *88*, 189–198.
Ikeno, T. and Guroff, G. (1979) Molec. Cell. Biochem. *28*, 67–91.
Ikeno, T., MacDonnell, P.C., Nagaiah, K. and Guroff, G. (1978) Biochem. Biophys. Res. Comm. *82*, 957–963.
Jaffe, L. and Nuccitelli, R. (1977) Ann. Rev. Biophys. Bioengin. *6*, 445–476.
Jaffe, L.F. and Poo, M.-M. (1979) J. Exp. Zool. *209*, 115–128.
Jänne, J., Poso, H. and Raina, A. (1978) Biochim. Biophys. Acta *473*, 241–293.
Jimenez de Asua, L., Rosengurt, E. and Carminatti, H. (1970) J. Biol. Chem. *245*, 3902–3905.
Johnson, D.G., Gordon, P. and Kopin, I.J. (1971) J. Neurochem. *18*, 2355–2362.
Johnson, E.M. (1978) Brain Res. *141*, 105–118.
Johnson, E.M., Andres, R.Y. and Bradshaw, R.A. (1978) Brain Res. *150*, 319–331.
Johnson, E.M., Gorin, P.D., Brandeis, L.D. and Pearson, J. (1980a) Science *210*, 916–918.
Johnson, J.D., Epel, D. and Paul, M. (1976) Nature (London) *262*, 661–664.
Johnson, M.I., Ross, C.D. and Bunge, R.P. (1980b) J. Cell Biol. *84*, 692–704.
Johnson, M.I., Iacovitti, L., Bunge, R.P., Higgins, D. and Burton, H. (1981) in Development of the Autonomic Nervous System (Milner, R.D.G. and Burnstock, G., eds.), pp. 108–122, CIBA Foundation Symposium No. 83, Pitman Medical, London.
Johnston, R.N. and Wessells, N.K. (1980) in Current Tropics in Developmental Biology (Moscona, A.A., Monroy, A. and Hunt, R.K., eds.) Vol. 16, pp. 165–206, Academic, New York.
Kaplan, J.G. (1978) Ann. Rev. Physiol. *40*, 19–41.
Kessler, J.A. and Black, I.B. (1980) Proc. Natl. Acad. Sci. USA *77*, 649–652.
Lakshmanan, J. (1978a) Biochem. Biophys. Res. Comm. *82*, 767–775.
Lakshmanan, J. (1978b) Brain Res. *157*, 173–177.
Lakshmanan, J. (1978c) FEBS Lett. *95*, 161–164.
Lakshmanan, J. (1979) Biochem. J. *178*, 245–248.
Lamb, J.F. and McCall, D. (1972) J. Physiol. *225*, 599–617.
Landa, K.B., Adler, R., Manthorpe, M. and Varon, S. (1980) Dev. Biol. *74*, 401–408.
Landmesser, L. and Pilar, G. (1978) Fed. Proc. *37*, 2016–2022.
Landreth, G.E. and Shooter, E.M. (1980) Proc. Natl. Acad. Sci. USA *77*, 4751–4755.
Landreth, G., Cohen, P. and Shooter, E.M. (1980) Nature (London) *283*, 202–204.
Lasek, R. and Hoffman, P. (1976) Cold Spring Harbor Conference on Cell Proliferation, *3*, 1021–1049.
Lasher, R. and Zagon, I.S. (1972) Brain Res. *41*, 482–488.
Lazarus, K.J., Bradshaw, R.A., West, N.R. and Bunge, R.P. (1976) Brain Res. *113*, 159–164.
Ledbetter, M.L.S. and Lubin, M. (1977) Exp. Cell Res. *105*, 223–236.
Lee, J.C., Tweedy, N. and Timasheff, S. (1978) Biochem. *17*, 2783–2790.
Letourneau, P.C. (1975) Dev. Biol. *44*, 92–101.
Letourneau, P.C. (1978) Dev. Biol. *66*, 183–196.
Letourneau, P.C. (1979) Exp. Cell Res. *124*, 127–138.
Levi, A., Cimino, M., Mercanti, D., Chen, J. and Calissano, P. (1975) Biochim. Biophys. Acta *399*, 50–60.
Levi, A., Castellani, L., Calissano, P., Deanin, G.G. and Gordon, M.W. (1978) Bull. Molec. Biol. Med. *1* (Suppl. 3), 425–515.
Levi, G. and Lattes, M.G. (1968) Life Sci. *7*, 827–834.
Levi-Montalcini, R. (1958) in A Symposium on the Chemical Basis of Development (McElroy, W.D. and Glass, B., eds.) pp. 646–664, John Hopkins, Baltimore.

Levi-Montalcini, R. (1963) in The Nature of Biological Diversity (Allen, J.M., ed.) pp. 261–295, McGraw-Hill, New York.
Levi-Montalcini, R. (1966) Harvey Lect. *60*, 217–259.
Levi-Montalcini, R. (1975) in The Neurociences: Paths of Discovery (Worden, F.G., Swazey, J.P. and Adelman, G. eds.) pp. 243–265, MIT, Cambridge.
Levi-Montalcini, R. (1976) in Progress in Brain Research (Corner, M.A. and Swaab, D.F., eds.) Vol. 45, pp. 235–256, Elsevier, Amsterdam.
Levi-Montalcini, R. (1979) Differentiation *13*, 51–53.
Levi-Montalcini, R. and Angeletti, P. (1963) Dev. Biol. *7*, 653–659.
Levi-Montalcini, R., and Angeletti, P. (1968) Physiol. Rev. *48*, 534–569.
Levi-Montalcini, R. and Hamburger, V. (1951) J. Exp. Zool. *116*, 321–362.
Levi-Montalcini, R., Meyer, H. and Hamburger, V. (1954) Cancer Res. *14*, 49–57.
Levi-Montalcini, R., Revoltella, R. and Calissano, P. (1974) in Recent Progress in Hormone Research (Greep, R.O., ed.) Vol. 30, pp. 635–669, Academic, New York.
Levi-Montalcini, R., Aloe, L., Mugnaini, E., Oesch, F. and Thoenen, H. (1975) Proc. Natl. Acad. Sci. USA *72*, 595–599.
Lewis, M., Lakshmanan, J., Nagaiah, K., MacDonnell, P. and Guroff, G. (1978) Proc. Natl. Acad. Sci. USA *75*, 1021–1023.
Lichtshtein, D., Dunlop, K., Kaback, H.R. and Blume, A.J. (1979) Proc. Natl. Acad. Sci. USA *76*, 2580–2584.
Longo, A.M. (1978) Dev. Biol. *65*, 260–270.
Longo, A.M. and Penhoet, E. (1974) Proc. Natl. Acad. Sci. USA *71*, 2347–2349.
Longo, F., Manthorpe, M. and Varon, S. (1982) Dev. Brain Res. *3*, 277–294.
Lubin, M. (1967) Nature (London) *213*, 451–453.
Lubin, M. (1980) Biochem. Biophys. Res. Comm. *97*, 1060–1067.
Luckenbill-Edds, J.L., Van Horn, C. and Greene, L.A. (1979) J. Neurocytol. *8*, 493–511.
MacDonnell, P., Nagaiah, K., Lakshmanan, J. and Guroff, G. (1977) Proc. Natl. Acad. Sci. USA *74*, 4681–4684.
Mains, R.E. and Patterson, P.H. (1973) J. Cell Biol. *59*, 329–345.
Manthorpe, M., Skaper, S., Adler, R., Landa, K. and Varon, S. (1980) J. Neurochem. *34*, 69–75.
Manthorpe, M., Adler, R. and Varon, S. (1981a) Dev. Biol. *85*, 156–163.
Manthorpe, M., Skaper, S.D. and Varon, S. (1981b) Brain Res. *230*, 295–306.
Marchisio, P.C., Cirillo, D., Naldini, L. and Calissano, P. (1980) in Control Mechanisms in Animal Cells: Specific Growth Factors (Jiminez de Asua, L., Shields, R., Levi-Montalcini, R. and Iacobell, S., eds.) pp. 53–60, Raven, New York.
Marsh, G. and Beams, H.W. (1946) J. Cell. Comp. Physiol. *27*, 139–157.
Max, S.R., Schwab, M., Dumas, M. and Thoenen, H. (1978) Brain Res. *159*, 411–415.
McCarthy, K.D. and Partlow, L.M. (1976) Brain Res. *114*, 391–414.
McDonald, T.F., Sachs, H.G., Orr, C.W. and Ebert, J.D. (1972) Dev. Biol. *28*, 290–303.
McGuire, J.C. and Greene, L.A. (1979) J. Biol. Chem. *254*, 3362–3367.
McGuire, J.C., Greene, L.A. and Furano, A.V. (1978) Cell *15*, 357–365.
Menesini-Chen, M.C., Chen, J.S., Calissano, P. and Levi-Montalcini, R. (1977) Proc. Natl. Acad. Sci. USA *74*, 5559–5563.
Menesini-Chen, M., Chen, J.S. and Levi-Montalcini, R. (1978) Arch. Ital. Biol. *116*, 53–84.
Michell, R.H. (1975) Biochem. Biophys. Acta *415*, 81–147.
Mizel, S.B. and Bamburg, J.R. (1976) Dev. Biol. *49*, 20–28.
Mobley, W.C., Server, A.C., Ishii, D.N., Riopelle, R.J. and Shooter, E.M. (1977) N. Engl. J. Med. *297*, 1096–1104, 1149–1158, 1211–1218.
Monaco, G., Calissano, P. and Mercanti, D. (1977) Brain Res. *129*, 265–274.
Moolenaar, W.H., deLaat, S.W. and van der Saag, P.T. (1979) Nature (London) *279*, 721–723.

Moore, R.D. (1973) J. Physiol. (London) *232*, 23–45.
Moore, R.D. and Rabovsky, J.L. (1979) Am. J. Physiol. C249–C254.
Murphy, R.A., Singer, R.H., Pantazis, N.J., Saide, J.D., Arnason, B.G.W. and Young, M. (1979) in Musscle Regeneration Conference on Muscle Regeneration, Rockefeller University (Mauro, A., ed.) pp. 443–452, Raven, New York.
Murray, M.A. (1965) in Cells and Tissues in Culture (Willmer, E.N., ed.) pp. 373–455, Academic, New York.
Nagaiah, K., Ikeno, T., Lakshmanan, J., MacDonnell, P. and Guroff, G. (1978) Proc. Natl. Acad. Sci. USA *75*, 2512–2515.
Narumi, S. and Fujita, T. (1978) Neuropharmacology *17*, 73–76.
Nikodijevic, B., Nikodijevic, O., Yu, M.-Y.W., Pollard, H. and Guroff, G. (1975) Proc. Natl. Acad. Sci. USA *72*, 4769–4771.
Nja, A. and Purves, D. (1978) J. Physiol. (London) *281*, 45–62.
Norr, S.C. and Varon, S. (1975) Neurobiology *5*, 101–118.
Oger, J., Arnason, B.G.W., Pantazis, N., Lehrich, J. and Young, M. (1974) Proc. Natl. Acad. Sci. USA *71*, 1554–1558.
O'Lague, P.H., Potter, D.D. and Furshpan, E.J. (1978) Dev. Biol. *67*, 424–443.
Otten, U., Hatanaka, H. and Thoenen, H. (1978) Brain Res. *140*, 385–389.
Oxender, D.L. and Christensen, H.N. (1963) J. Biol. Chem. *238*, 3686–3699.
Pannese, E. (1980) Adv. Anat. Embryol. and Cell Biol. *65*, 1–100.
Partlow, L. and Larrabee, M. (1971) J. Neurochem. *18*, 2101–2118.
Patterson, P. (1978) Ann. Rev. Neurosci. *1*, 1–19.
Perez-Polo, J., Hall, K., Livingston, K. and Westlund, K. (1977) Life Sci. *21*, 1535–1544.
Peter, H.W., Wiese, F. and Graszynski, K. (1975) Devel. Biol. *46*, 439–445.
Pfeiffer, S. (1982) Neuroscience Approached Through Cell Culture, CRC Press, Boca Raton, in press.
Pfenninger, K.H. and Reese, R.P. (1976) in Neuronal Recognition (Barondes, S.H., ed.) pp. 131–178, Plenum, New York.
Phillipson, O.T. and Sandler, M. (1975) Brain Res. *90*, 273–281.
Potter, D.D., Landis, S.C. and Furshpan, E.J. (1981) in Development of the Autonomic Nervous System (Milner, R.D.G. and Burnstock, G., eds.), pp. 123–138, CIBA Foundation Symposium No. 83, Pitman Medical, London.
Prestige, M.C. (1970) in The Neurosciences Second Study Program (Schmitt, F.O., ed.) pp. 73–82, Rockefeller University, New York.
Purves, D. (1976) J. Physiol. (London) *259*, 159–175.
Purves, D. and Nja, A. (1976) Nature (London) *260*, 535–536.
Revoltella, R., Bertolini, L. and Pediconi, M. (1974) Exp. Cell Res. *85*, 89–94.
Rohrer, H., Otten, U. and Thoenen, H. (1978) Brain Res. *159*, 436–439.
Roisen, F.J., Murphy, R.A., Pichicero, M.E. and Braden, W.G. (1972) Science *175*, 73–74.
Rozengurt, E. (1979) in Hormones and Cell Culture (Sato, G.H. and Ross, R., eds.) Vol. 6, pp. 773–788, ColdSpring Harbor Laboratory.
Rozengurt, E. and Heppel, L.A. (1975) Proc. Natl. Acad. Sci. USA *72*, 4492–4495.
Rozengurt, E. and Mendoza, S. (1980) Ann. New York Acad. Sci. *339*, 175–190.
Russell, D.H. and Stambrook, P.J. (1975) Proc. Natl. Acad. Sci. USA *72*, 1482–1486.
Sato, G. (1973) Tissue Culture of the Nervous System, Plenum, New York.
Schlaepfer, W.W. and Bunge, R. (1973) J. Cell Biol. *59*, 456–469.
Schneider, J.A., Diamond, I. and Rozengurt, E. (1978) J. Biol. Chem. *253*, 872–877.
Scholnick, P., Lang, D. and Racker, E. (1973) J. Biol. Chem. *248*, 5175–5182.
Schubert, D. and Whitlock, C. (1977) Proc. Natl. Acad. Sci. USA *74*, 4055–4058.
Schubert, D.S., Heinemann, S. and Kiddokoro, Y. (1977) Proc. Natl. Acad. Sci. USA *74*, 2579–2583.

Schubert, D., LaCorbiere, M., Whitlock, C. and Stallcup, W. (1978) Nature (London) *273*, 718–723.
Schwab, M.E. (1977) Brain Res. *130*, 190–196.
Schwab, M. and Thoenen, H. (1977) Brain Res. *122*, 459–474.
Schwartz, A., Lindenmayer, G.F. and Allen, J.C. (1972) in Current Topics in Membranes and Transport (Bronner, F. and Kleinzeller, A., eds.) pp. 1–82, Academic, New York.
Schwartz, J., Chuang, D.M. and Costa, E. (1977) Brain Res. *137*, 369–375.
Scott, B.S. (1971) Exp. Neurol. *30*, 297–308.
Scott, B.S. (1977) J. Cell Physiol. *91*, 305–315.
Scott, B.S. and Fisher, K.C. (1970) Exp. Neurol. *27*, 16–22.
Scott, B.S. and Fisher, K.C. (1971) Exp. Neurol. *31*, 183–188.
Sherwin, S.A., Sliski, A.H. and Todaro, G.J. (1979) Proc. Natl. Acad. Sci. USA *76*, 1288–1292.
Shinohara, T. and Piatigorsky, J. 1977) Nature (London) *270*, 406–411.
Sisken, B.F. and Lafferty, J.F. (1978) Biochem. Bioenerg. *5*, 459–472.
Sisken, B.F. and Smith, S.D. (1975) J. Embryol. Exp. Morph. *33*, 29–41.
Skaper, S.D. and Varon S. (1979a) Brain Res. *163*, 89–100.
Skaper, S.D. and Varon, S. (1979b) Brain Res. *172*, 303–313.
Skaper, S.D. and Varon, S. (1979c) Biochem. Biophys. Res. Comm. *88*, 563–568.
Skaper, S.D. and Varon, S. (1980a) J. Neurochem. *34*, 1654–1660.
Skaper, S.D. and Varon, S. (1980b) Brain Res. *187*, 379–389.
Skaper, S.D. and Varon, S. (1981a) Exp. Cell Res. *131*, 353–361.
Skaper, S.D. and Varon, S. (1981b) J. Neurosci. Res. *6*, 133–141.
Skaper, S.D. and Varon, S. (1981c) J. Neurochem. *37*, 222–228.
Skaper, S.D., Adler, R. and Varon, S. (1979a) Dev. Neurosci. *2*, 233–237.
Skaper, S.D., Bottenstein, J.E. and Varon, S. (1979b) J. Neurochem. *32*, 1845–1851.
Skaper, S.D., Selak, I. and Varon, S. (1982) Dev. Brain Res. *3*, 419–428.
Skaper, S.D. and Varon, S. (1982) Exp. Neurol. *76*, 655–665.
Smilowitz, H. (1979) Mol. Pharmacol. *16*, 202–214.
Smilowitz, H. (1980) Cell *19*, 237–244.
Smith, J.B. and Rozengurt, E. (1978a) Proc. Natl. Acad. Sci. USA *75*, 5560–5564.
Smith, J.B. and Rozengurt, E. (1978b) J. Cell. Physiol. *97*, 441–450.
Spector, M., O'Neal, S. and Racker, E. (1980) J. Biol. Chem. *255*, 8370–8373.
Stach, R.W., Stach, B.M. and West, N.R. (1979) J. Neurochem. *33*, 845–855.
Steiner, G. and Schönbaum, E. (1972) Immunosympathectomy, Elsevier, New York.
Stoeckel, R. and Thoenen, H. (1975) Brain Res. *85*, 337–341.
Stoeckel, K., Solomon, F., Paravicini, U. and Thoenen, H. (1974) Nature (London) *250*, 150–151.
Stoeckel, K., Schwab, M. and Thoenen, H. (1975) Brain Res. *89*, 1–14.
Sutter, A., Riopelle, R.J., Harris-Warwick, R.M. and Shooter, E.M. (1979) J. Biol. Chem. *254*, 5972–5982.
Szutowicz, A., Frazier, W.A. and Bradshaw, R.A. (1976a) J. Biol. Chem. *251*, 1516–1523.
Szutowicz, A., Frazier, W.A. and Bradshaw, R. (1976b) J. Biol. Chem. *251*, 1524–1528.
Tartakoff, A.M. and Vassalli, P. (1977) J. Exp. Med. *146*, 1332–1345.
Tartakoff, A. and Vassalli, P. (1978) J. Cell Biol. *79*, 694–707.
Thoenen, H., Angelletti, P.U., Levi-Montalcini, R. and Kettler, R. (1971) Proc. Natl. Acad. Sci. USA *68*, 1598–1602.
Thoenen, H., Barde, Y.A., Edgar, D., Hatanaka, H., Otten, U. and Schwab, M. (1979) in Progress in Brain Research (Cuenod, M., Kreutzberg, G.W. and Bloom, F.E., eds.) Vol. 51, pp. 94–108, Elsevier, Amsterdam.
Thoenen, H., Schäfer, T., Heumann, R. and Schwab, M. (1981) in Hormones and Cell Regulation (Dumont, J.E. and Nunez, J., eds.) Vol. 5., pp. 15–34, Elsevier, Amsterdam.
Turner, J.E., Delaney, R.K. and Johnson, J.E. (1980) Brain Res. *204*, 283–294.

Tuttle, J.B. (1977) Soc. Neurosci. Abstr. *3*, 529.
Uchida, N., Smilowitz, H. and Tanzer, M.L. (1979) Proc. Natl. Acad. Sci. USA *76*, 1868–1872.
Unsicker, K., Krisch, B., Otten, U. and Thoenen, H. (1978) Proc. Natl. Acad. Sci. USA *75*, 3498–3502.
Unsicker, K., Rieffert, B. and Ziegler, W. (1980) in Histochemistry and Cell Biology of Autonomic Neurons, SIF Cells, and Paraneurons (Franko, O., Soinila, S. and Paivarinta, H., eds.) pp. 51–60, Raven, New York.
Varon, S. (1970) in The Neurosciences Second Study Program (Schmitt, F.O., ed.) pp. 83–99, Rockefeller University, New York.
Varon, S. (1975a) Exp. Neurol. *48*, 75–92.
Varon, S. (1975b) Exp. Neurol. *48*, 93–134.
Varon, S. (1977) in Cell, Tissue and Organ Cultures in Neurobiology (Fedoroff, S. and Herz, L., eds.) pp. 93–103, Pergamon, New York.
Varon, S. (1979) Neurochem. Res. *4*, 155–173.
Varon, S. and Adler, R. (1980) Curr. Topics Dev. Biol. *16*, 207–252.
Varon, S. and Adler, R. (1981) Adv. Cell Neurobiol. *2*, 115–163.
Varon, S. and Bunge, R.P. (1978) Ann. Rev. Neurosci. *1*, 327–361.
Varon, S. and Manthorpe, M. (1982) Adv. Cell. Neurobiol. *3*, 35–95.
Varon, S. and Raiborn, C. (1972a) J. Neurocytol. *1*, 211–221.
Varon, S. and Raiborn, C. (1972b) Neurobiol. *2*, 183–196.
Varon, S. and Saier, M. (975) Exp. Neurol. *48*, 135–162.
Varon, S. and Skaper, S.D. (1980a) in Tissue Culture in Neurobiology (Giacobini, E., Vernadakis, A. and Shahar, A., eds.) pp. 333–347, Raven, New York.
Varon, S. and Skaper, S.D. (1980b) J. Supramol. Struct. Suppl. *4*, 59.
Varon, S. and Skaper, S.D. (1981a) in Chemisms of the Brain (Rodnight, R., Bachelard, H. and Stahl, W., eds.), pp. 151–176, Churchill-Livingstone, Edinburgh, Pitman Medical, London.
Varon, S. and Skaper, S. (1981b) in Development of the Autonomic Nervous System (Milner, R.D.G. and Burnstock, G., eds.) CIBA Foundation Symposium No. 83, in press.
Varon, S. and Somjen, G.G. (1979) Neurosci. Res. Prog. Bull. *17*, 1–239.
Varon, S., Nomura, J., Perez-Polo, J.R. and Shooter, E.M. (1972) in Methods and Techniques of Neurosciences (Fried, R., ed.) pp. 203–229, M. Dekker, New York.
Varon, S., Raiborn, C. and Tyszka, E. (1973) Brain Res. *54*, 51–63.
Varon, S., Raiborn, C. and Burnham, P.A. (1974a) J. Neurobiol. *5*, 255–371.
Varon, S., Raiborn, C. and Burnham, P.A. (1974b) Neurobiology *4*, 231–252.
Varon, S., Raiborn, C. and Burnham, P.A. (1974c) Neurobiology *4*, 317–327.
Varon, S., Raiborn, C. and Norr, S. (1974d) Exp. Cell Res. *88*, 247–256.
Varon, S., Manthorpe, M. and Adler, R. (1979) Brain Res. *173*, 29–45.
Varon, S., Skaper, S.D. and Manthorpe, M. (1981) Dev. Brain Res. *1*, 73–87.
Varon, S., Manthorpe, M., Skaper, S.D. and Adler, R. (1982a) in Proteins of the Nervous System: Structure and Function (Haber, B., Perez-Polo, R. and Coulter, J.D., eds.), pp. 225–242, Liss, New York.
Varon, S., Adler, R., Manthorpe, M. and Skaper, S.D. (1982b) in Neuroscience Approached Through Cell Culture (Pfeiffer, S.E., ed.) CRC Press, Boca Raton, in press.
Walicke, P.A., Campenot, R.B. and Patterson, P.H. (1977) Proc. Natl. Acad. Sci. USA *74*, 5767–5771.
Walker, P., Weichsel, M.E., Fisher, D.A., Guo, S.M. and Fisher, D.A. (1979) Science *204*, 427–429.
Weber, M.J. (1980) Proc. 1st Meeting Int. Soc. Dev. Neurosci., pp. 8–9.
Weston, J.A. (1972) in Cell Interactions (Silvestri, L.G., ed.) 3rd Lepetit Colloquium, pp. 286–292, Elsevier, New York.
Wilson, T.H. and Lin, E.C.C. (1980) J. Supramol. Struc. Suppl. *4*, 47.

Winiick, M. and Greenberg, R.E. (1965) Nature (London) 205, 180–181.
Yamada, K.M. and Wessells, N.K. (1971) Exp. Cell Res. 66, 346–352.
Yankner, B.A. and Shooter, E.M. (1979) Proc. Natl. Acad. Sci. USA 76, 1269–1273.
Young, M., Saide, J.D. and Murphy, R.A. (1976) J. Biol. Chem. 251, 459–464.
Yu, M.W., Hori, S., Tolson, N., Huff, K. and Guroff, G. (1978) Biochem. Biophys. Res. Comm. 81, 941–946.
Yu, M.W., Tolson, N.W. and Guroff, G. (1980) J. Biol. Chem. 255, 10481–10492.

Burnstock et al. (eds.) Somatic and Autonomic Nerve–Muscle Interactions
© 1983, Elsevier Science Publishers, B.V.

CHAPTER 8

The effects of autonomic nerves on smooth muscle properties

MARK D. DIBNER

*Neurobiology Group, Central Research and Development Department,
E.I. Du Pont de Nemours and Co., Du Pont Glenolden Laboratory,
Glenolden, PA 19036, USA*

1. Introduction

Smooth muscle is innervated by autonomic nerves in a highly complex fashion (Bevan, 1974; Holman and Hirst, 1977; Burnstock, 1979). There are not well-defined synapses between nerve and muscle as occur in skeletal muscle (Bevan, 1977). Moreover, smooth muscle is innervated by a variety of neuronal types which include adrenergic neurons of the sympathetic nervous system, cholinergic neurons of the parasympathetic nervous system and possibly some non-adrenergic/non-cholinergic neurons. Smooth muscle also responds to a large number of other hormones and neurotransmitters. For example, while in the gut the parasympathetics generally mediate smooth muscle contraction and the sympathetics often mediate smooth muscle dilatation (Dixon and Ransom, 1912; Bevan and Su, 1973; Shepherd, 1976; Nadel, 1980) receptors on the surface of other smooth muscle cells may respond to interaction with a certain compound by contraction while others respond to the same compound with relaxation. In other tissues, two compounds which usually produce opposite effects will induce the same response (Szurszewski, 1973). Additionally, responses of smooth muscle to autonomic nerve stimulation may be dependent on other cell types such as vascular endothelial cells (Furehgott and Zawadzki, 1980).

To compound the above-mentioned complexities, smooth muscles are not a homogenous group of organs. The amount and type of innervation varies from tissue to tissue. Receptors mediating smooth muscle contraction and relaxation are also variable between tissues. This is highlighted by the differences in response between the gut and the vasculature. Additionally, there is a large amount of species variability in the innervation and responses of smooth muscle (Bulbring, 1973b). Thus, many studies of the effects of autonomic nerves on smooth muscle properties are system-specific. In turn, only a few commonalities have been elucidated.

Numerous studies have focused on trophic mechanisms between nerve and skeletal muscle during development and in the adult (see Chapters 4, 5 and 9 in this volume). There is also much known about trophic influences between target organ and autonomic nerve during development (Bito et al., 1971; Dibner et al., 1977) (see Chapter 2). Relatively little is known about the trophic relationships between autonomic nerves and smooth muscle in mature animals (Chamley et al., 1973; Burnstock, 1971; Bevan et al., 1976; Hermsmeyer and Aprigliano, 1980). However, considerable evidence from studies addressing this subject have demonstrated supersensitivity and desensitization phenomena in smooth muscle. Also, a number of compounds which interact with autonomic nerves appear to affect smooth muscle as well.

This chapter attempts to encompass current concepts and knowledge of the relationships between smooth muscle and autonomic nerves. Focus will be made on desensitization and supersensitivity responses as well as the molecular pharmacology of smooth muscle. Despite the complexities of nerve/muscle interactions and the plethora of model systems available, a few basic properties have evolved and will be described.

2. Chemical interactions between nerve and muscle

Smooth muscle contains cell surface receptors for a number of agents. In general, parasympathetic nerves exert their influence via cholinergic receptors and sympathetic nerves exert their influence via adrenergic receptors. However, these relationships are not always clear cut and receptors for many different molecules exist on smooth muscle cells. The ones presently known include those for acetylcholine, norepinephrine, epinephrine, histamine, purines and various peptides and hormones. Some of these agents act directly following neuronal release (autonomic and others); others act indirectly by modifying autonomic input. While many pieces of this complex puzzle most likely remain to be discovered, a number of the components will be discussed below.

2.1. Cholinergic interactions

Agonist occupancy of cholinergic receptors generally results in smooth muscle contraction, especially in the gut (Table 1). Receptors for acetylcholine have been studied directly on smooth muscle cells employing various radiolabeled cholinergic ligands including benzilylcholine mustard (Fewtrell and Rang, 1973), quinuclidinyl benzilate (Hata et al., 1980), dibenamine (Takagi et al., 1965) and propylbenzilylcholine mustard (Burgen et al., 1974). Cholinergic receptors on smooth muscle

TABLE 1

Adrenergic and cholinergic receptors mediating smooth muscle response – a partial compilation

Tissue	Response	Source
1. α-Adrenergic receptors		
Rat blood vessel	Contraction	Vanmeel et al., 1981
Guinea pig intestine	Contraction	Kosterlitz et al., 1970
Guinea pig intestine	Contraction	Jenkinson and Morton, 1967
Rabbit ear artery	Contraction	Kreth and Vonderlage, 1979
Rabbit bladder	Contraction	Khanna et al., 1981
Rabbit colon	Contraction	Andersson, 1972
Cat nictitating membrane	Contraction	Pluchino and Trendelenburg, 1968
Bovine mesenteric artery	Contraction	Andersson, 1973
Human oviduct longitudinal muscle	Contraction	Lindblom et al., 1971
2. β-Adrenergic receptors		
Guinea pig intestine	Relaxation	Jenkinson and Morton, 1967; Kosterlitz et al., 1970
Chicken rectum	Relaxation	Komori et al., 1980
Rabbit colon	Relaxation	Andersson, 1972
Cat nictitating membrane	Relaxation	Pluchino and Trendelenburg, 1968
Dog bronchi	Relaxation	Vermeire and Vanhoutte, 1979
Bovine mesenteric artery	Relaxation	Andersson, 1973
Human oviduct circular muscle	Contraction	Lindblom et al., 1979
Rabbit coronary arteries[a]	Relaxation	Bevan and Purdy, 1973
Rabbit aorta[b]	Relaxation	Bevan and Purdy, 1973
Rabbit bladder[b]	Relaxation	Khanna et al., 1981
Human lung[b]	Relaxation	Takeyasu et al., 1981
3. Cholinergic receptors		
Rat ileum	Contraction	McPhillips, 1968
Guinea pig intestine	Contraction	Kosterlitz et al., 1970
Guinea pig vas deferens	Contraction	Takeyasu et al., 1981
Rabbit bladder	Contraction	Levin et al., 1980
Chicken esophagus	Contraction	Ohashi and Ohga, 1967
Cat iris	Contraction	Bito et al., 1971
Dog bronchi	Contraction	Vanhoutte, 1978

[a] β_1-Adrenergic
[b] β_2-Adrenergic

are generally of the muscarinic subtype (Heilbronn and Bartfai, 1978). These receptors appear similar in type to muscarinic receptors of the central nervous system (Beld et al., 1975). In addition to postsynaptic muscarinic receptors which are responsible for most of the actions of acetylcholine on smooth muscle, presynaptic cholinergic autoreceptors may provide a feedback inhibition of acetylcholine release from autonomic nerves (Kilbinger and Wagner, 1975).

The effects of stimulation of acetylcholine receptors on smooth muscle have been under study for many years (Heilbronn and Bartfai, 1978). In the guinea pig ileum, parasympathetic stimulation of muscarinic receptors produce a contractile response (Bolton, 1973; Primor, 1980). This appears to be due to an increased permeability to Na^+ and K^+, but not to Cl^- (Durbin and Jenkinson, 1961; Bevan and Su, 1973; Bolton, 1973). It has been suggested that two subtypes of muscarinic receptors exist: one at the neuronal endings mediating contraction and a second type which may mediate efflux of K^+ (Ohashi and Ohga, 1967; Burgen and Spero, 1968; Ito and Kuriyama, 1971; Primor, 1980). Cholinergic stimulation causes contraction in a number of tissues in various species. In addition to muscarinic receptor stimulation, nicotine has been demonstrated to produce excitatory stimulation in the guinea pig trachea (Jones et al., 1980). In rabbit arterial smooth muscle, cholinergic agents produce relaxation and this response may be mediated via receptors on associated endothelial cells (Furchgott and Zawadzki, 1980). Methacholine-induced contraction in canine airway smooth muscle is synergistic with histamine-induced contraction but not with norepinephrine-induced contraction (Leff and Munoz, 1981), indicating a separate mechanism for adrenergic and cholinergic effects in this system.

In general, three major changes have been shown to occur in smooth muscle following cholinergic receptor stimulation. The first is calcium release leading to contraction. The next is secondarily increased level of cyclic GMP in the cells (Heilbronn and Bartfai, 1978). Third, is an increase in the turnover of phosphatidylinositol (Pluchino and Trendelenburg, 1968). However, a number of endogenous and exogenous agents have been shown to effect all three changes and their role in smooth muscle function is yet to be determined.

2.2. Adrenergic interactions

Norepinephrine and epinephrine exert their influence on smooth muscle via a number of types of adrenergic receptors. The result may be smooth muscle contraction mostly mediated via α-adrenergic receptors (Table 1).

The synapse between adrenergic and smooth muscle cells varies greatly in size from tissue to tissue and from species to species (Bevan, 1977). Synaptic widths can vary from 20 to 1,900 nm. As such, the size of a smooth muscle response to adrenergic nerve stimulation can relate to nerve density and distribution (Heilbronn and Bartfai, 1978). It is interesting to note that both the response to nerve stimulation and the reuptake process are inversely proportional to synaptic cleft width (Bevan, 1979). Very narrow synaptic clefts look like and appear to behave similarly to a non-sympathetic specialized synapse with a 1 to 3 μm^2 area of contact and a small distance of neurotransmitter diffusion (Bevan and Su, 1973).

Adrenergic receptors on smooth muscles may respond both the norepinephrine released by sympathetic innervation and to circulating catecholamines: not all smooth muscle cells are directly innervated. For example, in the guinea pig trachea and lung there are inhibitory adrenergic receptors but only the trachea receives functional adrenergic innervation (Doidge and Satchell, 1980). In some tissues, the β-adrenergic receptors may not be innervated by sympathetic nerves whereas α-adrenergic receptors are. Thus, denervation supersensitivity in the dog saphenous vein and guinea pig vas deferens is not extended to the relaxant effects of β-adrenergic agonists (Pluchino and Trendelenburg, 1968; Guimarães, 1975). However, in the rabbit iris adrenergic denervation leads to an increased number of β-adrenergic receptors with no change in α-adrenergic receptors (Page and Neufeld, 1978).

An antagonistic relationship has been reported between α-adrenergic and β-adrenergic receptors in the same smooth muscle tissue. In guinea pig taenia coli, α-adrenergic receptors mediate contraction whereas β-adrenergic receptors mediate relaxation (Jenkinson and Morton, 1967; Bulbring and Tomita, 1969a,b). In the rabbit colon and bovine mesenteric artery, relaxation is mediated via β-adrenergic stimulation, whereas contraction in the bovine mesenteric artery is mediated via α-adrenergic receptors (Andersson, 1972; 1973). Other tissues where this phenomenon appears to occur include the guinea pig ileum (Kosterlitz et al., 1970), the human spleen (Davies and Withrington, 1973), the rabbit bladder (Khanna et al., 1981), and the chicken rectum (Komori et al., 1980). In a review of these phenomena, Bulbring (1973a) stated that α-adrenergic receptors directly affect K^+ conduction as well as Ca^{2+} and membrane permeability with different effects in different tissues. The β-adrenergic receptors, on the other hand, suppress slow depolarization and decrease spike generation to the same degree in all tissues (Bulbring, 1973a).

There are at least two subtypes of α-adrenergic receptors reported in smooth muscle systems. In the gut there is one type of α-adrenergic receptor directly on the smooth muscle membrane (α_1; post junctional) which mediates hyperpolarization of the muscle (Holm, 1967; Kosterlitz et al., 1970; Andersson, 1972; MacDonald and McGrath, 1980). This may be due to an increase in permeability to K^+ with an ensuing hyperpolarization (Bulbring, 1973b). The presynaptic α_2-adrenergic receptors serve to inhibit the effect of stimulating sympathetic neurons (Starke and Endo, 1976), perhaps by decreasing the release of norepinephrine (Majewski and Rand, 1981). There are different responses outside the gut. In studies on the guinea pig ear artery, Kajiwara et al. (1981) demonstrated that low doses of norepinephrine activate the prejunctional α-adrenergic receptors with no direct effect on the muscle itself. At higher doses, norepinephrine acts directly on the muscle to produce depolarization (Kajiwara et al., 1981). Norepinephrine-induced canine airway contraction is synergistically augmented by histamine, suggesting a separate mechanism for the two substances (Leff and Munoz, 1981). In the guinea pig aorta, α_1-adrenergic receptors mediate contraction whereas in the ileum and trachea, presynaptic α_2 receptors block contraction by inhibiting cholinergic neurotransmission (Grundström et al., 1981).

β-Adrenergic receptors generally mediate smooth muscle relaxation (Table 1). Two subtypes of β-adrenergic receptors have been identified (Lands et al., 1967) and appear to be under independent regulation (Minneman et al., 1979). Both subtypes of receptors have been demonstrated to be present in smooth muscle. In the rabbit, vasodilation is mediated via β_2-adrenergic receptors in the aorta and by β_1-adrenergic receptors in skeletal muscle vasculature (Bevan et al., 1978). The rabbit bladder contains β_2-adrenergic receptors which mediate relaxation but has no β_1-adrenergic receptors (Khanna et al., 1981). In human peripheral airway tissue β_2-adrenergic receptors also predominate and respond to β_2-adrenergic agonists by a generation of cyclic AMP (Davis et al., 1980).

Prejunctional β-adrenergic receptors may be present on sympathetic nerve terminals. In guinea pig atrium, neuronally released norepinephrine can activate α-adrenergic receptors and cause a secondary inhibition of further norepinephrine release (Majewski et al., 1980). Prejunctional β-adrenergic receptors can facilitate norepinephrine release but this is normally overshadowed by a prejunctional α-adrenergic receptor-mediated inhibition of further norepinephrine release (Majewski and Rand, 1981).

Similar to cholinergic interactions with smooth muscle (Bevan, 1979; Abdel-Latif and Luke, 1981), norepinephrine also causes a turnover of phosphatidylinositol. In rabbit iris, norepinephrine produces a sodium-dependent increase in ^{32}P labelling of phosphatidic acid and phosphatidylinositol (Abdel- Latif and Luke, 1981). This effect may be associated with Ca^{2+} influx and may be a critical component of autonomically-regulated smooth muscle function (Michell and Kirk, 1981).

2.3. Cross interactions

It has long been known that an antagonistic relationship appears to exist between adrenergic and cholinergic systems (Golla and Symes, 1913). In guinea pig ileum, catecholamines decrease the output of acetylcholine (Paton and Vizt, 1969). This is also seen in guinea pig stomach following stimulation of both the sympathetic nerve and vagus in vitro (Beani et al., 1971). Other studies which demonstrate this relationship have utilized the guinea pig stomach (Campbell, 1966) and the canine airway (Leff and Munoz, 1981).

Conversely, adrenergic neurotransmission (via norepinephrine release) is inhibited by acetylcholine in the vasculature (McGrath and Vanhoutte, 1978). Curiously, whereas adrenergic and cholinergic systems appear to have opposite effects on the same smooth muscle, both may have synergistic effects with histamine in augmenting histamine-induced contractions of canine airway smooth muscle (Leff and Munoz, 1981).

2.4. Non-adrenergic, non-cholinergic autonomic effects

Some effects of autonomic nerve stimulation may be mediated via a non-adrenergic and non-cholinergic receptor. For example, in the human oviduct a non-adrenergic, non-cholinergic inhibition of contraction was demonstrated (Lindblom et al., 1979). In the guinea pig trachea, nicotine produces non-cholinergic and non-adrenergic inhibition of contraction (Jones et al., 1980). Jager and Schevers (Jager and Schevers, 1980) postulated a purinergic inhibitory nerve system which inhibits acetylcholine-induced contraction in the taenia caecum. Adrenergic nerve stimulation of the rat basilar artery is not blocked by α-adrenergic blockers and this may be another site of novel innervation, perhaps by purines or peptides (Lee, 1977). From these examples and others (Duckles et al., 1977; Lee et al., 1978; Kalenberg and Satchell, 1979; Kanazawa et al., 1980) it is apparent that further studies are necessary to elucidate the anatomy and role of these putative non-adrenergic/non-cholinergic neurons.

2.5. Histamine effects

The actions of histamine on smooth muscle have been long established (Dale and Laidlaw, 1910). However, the sensitivity and response of smooth muscle to this compound varies not only for different tissues within the animal but also for the same tissue between species. Dale (1929) showed that histamine produces a contraction of arteriolar muscle in rats and cats and a dilatation in the same tissue in dog, monkey and in man. Most smooth muscles respond to histamine with contraction, but in a number of tissues a relaxation response occurs. A major source of this substance in peripheral tissues is the mast cell (Riley and West, 1953). In brain, histamine appears to be a neurotransmitter liberated by 'histaminergic' nerves (Schwartz, 1977) but the existence of similar neurons in the periphery is less clear. Histamine can affect autonomic neurotransmission by stimulating sympathetic ganglia (Trendelenburg, 1954), or perhaps by other mechanisms (Shepherd, 1978). Leff and Munoz (1981) demonstrated an augmentation of histamine-induced contraction in dog bronchial muscle by both adrenergic and cholinergic agents. In smooth muscle, contraction is produced via direct action with H_1 receptors and relaxation is mediated via H_2 receptors (Owen, 1977; Tenner and McCully, 1981).

2.6. Peptide and hormone effects

Peptides have recently been implicated as putative neurotransmitters for a number of neuronal systems both in the central nervous system and in the periphery. Some members of this growing list have been reported to affect smooth muscle. These include bradykinin, substance P, angiotensin II, neurotensin, thyrotrophin-releasing hormone, endorphins and vasoactive intestinal polypeptide.

Neurotensin, the tridecapeptide, causes a relaxation in a number of smooth muscle tissues. In the guinea pig colon, neurotensin is two times as potent as epinephrine and 50,000 times more potent than ATP in producing relaxation (Kitabgi and Vincent, 1981). With an IC_{50} of less than 0.5 nM, it is probable that there are high-affinity receptors for neurotensin on smooth muscle cells. In the intestine, neurotensin may interact with the cholinergic contraction mechanism (Kitabgi and Freychet, 1978).

Another peptide with smooth muscle action is prolactin. This hormone produces contraction in various tissues including myometrium and ileum (Horrobin et al., 1973; Pillai et al., 1981). Although contraction produced by acetylcholine is greater in magnitude, prolactin appears to exert its effect by a cholinergic mechanism (Pillai et al., 1981). Prolactin has an ED_{50} for contraction which is 50-fold less than that of acetylcholine and prolactin-induced contractile responses were blocked by atropine and potentiated by neostigmine (Burnstock, 1974). Evidence exists that other peptides such as thyrotophin-releasing hormone and vasoactive intestinal polypeptide also cause a contractile response in mammalian intestine through a cholinergic mechanism (Cohen and Landry, 1980), indicating a possible modulatory role for these peptides.

Opiates have both enhancing and suppressing functions of smooth muscle (Huidobro-Toro et al., 1981). In the guinea pig ileum, 3 different types of opiate receptors have been observed (Su et al., 1981). The exact function(s) of endogenous opiate peptides in smooth muscle needs to be elucidated.

The undecapeptide substance P has a potent effect in contracting smooth muscle. This has been demonstrated in many systems and many smooth muscles have substance P receptors (Hanley and Iversen, 1980). As with many other peptides, substance P-mediated smooth muscle contraction appears to occur via a cholinergic mechanism (Rosell et al., 1977). In the guinea pig ileum for example, substance P has been shown to be 40 times more potent than acetylcholine and 160 times more potent than histamine in inducing contraction (Rosell et al., 1977). Recent studies indicate the presence of substance P in the superior cervical sympathetic ganglion of the rat (Kessler et al., 1981). This peptide appears to be contained in the postganglionic sympathetic neurons and may reside in the same neurons as norepinephrine (Kessler et al., 1981). A trophic relationship between substance P levels and preganglionic sympathetic impulse activity appears to exist and the ontogenetic development of substance P in sensory ganglia appears to be under regulation by nerve growth factor (Kessler et al., 1981; Kessler and Black, 1981). However, adrenergic denervation of the submaxillary gland in the rat did not reduce substance P levels in the gland (Dibner, M.D., Hanley, M.R., Nagy, J. and Goedert, M., unpublished observation). Whether autonomic nerves directly affect smooth muscle by liberation of substance P is now under investigation.

Prostaglandins also function to modulate sympathetic neurotransmission in smooth muscle. Prostaglandin E_2 has been shown by Hedqvist (1970a) to decrease outflow of norepinephrine in cat spleen following sympathetic nerve stimulation. This effect does not appear to be due to an alteration in the reuptake of norepinephrine (Hedqvist, 1970a). Further studies employing cat spleen and guinea pig vas deferens show that prostaglandins have a dual function in inhibiting norepinephrine release and inhibiting smooth muscle response (Hedqvist, 1970b). High doses of prostaglandins E_1 and E_2 have the reverse effect and potentiate contractions due to sympathetic nerve stimulation on the administration of exogenous epinephrine (Hedqvist, 1972). Inhibition of prostaglandin synthesis increases norepinephrine release in rabbit heart caused by stimulation of sympathetic nerves (Samuelsson and Wennmalm, 1971). Thus, it appears that prostaglandins may effect a negative feedback mechanism on the activity of sympathetic nerves in smooth muscle.

There is also hormonal regulation of the smooth muscle response to autonomic nerve stimulation. Estrogen, injected into rabbits, causes a marked increase in response to α-adrenergic, muscarinic, and purinergic agonists (Levin et al., 1980). This is paralleled by an increase in the densities of α-adrenergic and muscarinic cholinergic receptors with no change in β-adrenergic receptor density or responsiveness (Levin et al., 1980). In rat anococcygeus muscle, 17-β-estradiol decreases sensitivity to K^+ with no effect on the norepinephrine or acetylcholine response (Gibson and Pollock, 1973). The relevance of hormonal regulation of autonomic response in smooth muscle cells needs to be further studied.

3. Changes in smooth muscle under autonomic influence

Autonomic nerves affect certain molecules in the smooth muscle. These include ions and cyclic nucleotides, and they, in turn, can directly affect smooth muscle responses. These interactions and other trophic responses will be outlined below.

3.1. Ion changes

Changes in Ca^{2+} concentration in smooth muscle accompany smooth muscle contraction and relaxation. In an early report, Bulbring (1973b) hypothesized that a primary action of catecholamines in smooth muscle was to control the levels of free Ca^{2+} uptake and to decrease free Ca^{2+} concentration. This has been demonstrated in a number of tissues including guinea pig taenia coli (Bulbring, 1978; Mueller and van Breemen, 1979). Stimulation of α-adrenergic receptors increases Ca^{2+} release and possibly its extrusion. This relationship also holds in that intracellular levels of free Ca^{2+} were demonstrated to be inversely proportional to cyclic AMP levels in rabbit intestinal smooth muscle which, in turn, is under adrenergic control (Andersson and Nilsson, 1972). Changes in the intracellular levels of free Ca^{2+} are believed to be the basis of depolarization and hyperpolarization in smooth muscle (Andersson and Nilsson, 1972; Bulbring, 1973b). Drugs which decrease intracellular Ca^{2+} concentrations also inhibit smooth muscle contraction (Sanner and Prusa, 1980). In a Ca^{2+}-free solution, norepinephrine is largely ineffective in producing contraction of the rabbit aorta (Karaki et al., 1979). This can be reversed by short-term Ca^{2+} loading of the calcium-depleted muscle and the magnitude of contraction can be correlated with Ca^{2+} concentration (Karaki et al., 1979). Cholinergic agents also appear to work by stimulating Ca^{2+} influx across the smooth muscle cell membrane as demonstrated by Durbin and Jenkinson (1961) in guinea pig taenia coli. In a recent review, Berridge (1980) suggested that Ca^{2+} signaling may indeed be an important event mediating neurotransmitter-induced muscle cell functioning. Intracellular Ca^{2+} concentration also appears to be an important modulator of smooth muscle responsiveness to autonomic nerve stimulation. A decrease in Ca^{2+} concentration inhibits release of norepinephrine from sympathetic neurons in cat spleen whereas increasing the Ca^{2+} concentration has the opposite effect (Kirpekor et al., 1980). Lastly, in the cat nictitating membrane, Ca^{2+} antagonizes the inhibition of sympathetically-induced contraction produced by the morphinomimetic fentanyl (Roquebert and Demichel, 1981).

The compound A23187, a divalent cation-selective ionophore, has been shown to produce contractions of the guinea pig ileum (Jim and Triggle, 1981). However, the effects of this drug could be greatly reduced by preincubation with a muscarinic agonist. This desensitization to the ionophore may not be caused by both drugs working at the muscarinic receptor but rather by competition for a post-receptor contractile event. Regardless of the mechanism, the ionophore may act as more than a simple Ca^{2+} carrier (Jim and Triggle, 1981).

Ca^{2+} antagonists such as verapamil and nifepidine have antihypertensive effects. It was recently hypothesized that these agents may decrease blood pressure by interfering with the vascular tone maintained by postsynaptic α_1-adrenergic receptors (Vanmeel et al., 1981). In pithed normotensive rats these drugs or EDTA acted to antagonize the effects of α_1 but not α_2-selective drugs. Also, constriction was antagonized by these agents (Vanmeel et al., 1981). Thus, in vivo vasoconstriction may be mediated by an influx of Ca^{2+} due to an α_1-adrenergic receptor stimulation. Inhibition of Ca^{2+}-dependent contractile responses can be mimicked by inhibiting Ca^{2+} influx across the cell membrane with verapamil (Turlapaty et al., 1981).

Of the ion changes observed in smooth muscle following autonomic nerve stimulation, perhaps the one most frequently reported is a change in muscle membrane permeability to K^+. In the guinea pig taenia coli, carbachol increases the membrane permeability for K^+ (Durbin and Jenkinson, 1961). In the same tissue, epinephrine also increased K^+ conductance (Bulbring and Tomita, 1969; Ohashi, 1971). The changes in K^+ permeability appear to be mediated by α- but not by β-adrenergic receptors in this tissue (Jenkinson and Morton, 1967; Bulbring and Tomita, 1969). Accompanying the K^+ flow, Cl^-, and to a lesser extent Na^+, also have increased permeability across the membrane (Bulbring and Tomita, 1969; Casteels et al., 1977). In addition, in vascular smooth muscle, K^+-induced relaxation occurs with nitroglycerine treatment (Karashima, 1980). K^+ may also act as a modulator of sympathetic nerve function (Vanhoutte et al., 1976). In the dog vascular system increasing K^+ concentration decreases the release of norepinephrine following nerve stimulation (Bevan, 1974). Thus, K^+ may have dual effects on both muscle and nerve.

Another ion which has recently reached importance in smooth muscle pharmacology is vanadium. This ion, which is found in many tissues, has been demonstrated to be an inhibitor of the sodium-potassium ATPase in many organs (Hopkins, 1974). In the rat vas deferens, a number of vanadium compounds were shown to cause dose-dependent contractions (Garcia et al., 1981). The mechanism of action for this phenomenon is uncertain but vanadium may be working at a number of possible sites including the membrane Ca^{2+} pump, adenylate cyclase or at the ATPase (Garcia et al., 1981). However, vanadium appears to act at different sites than does ouabain, another blocker of the sodium-potassium ATPase (Hopkins, 1974).

Cadmium is another modulator of norepinephrine response in smooth muscle. Rats given high dietary cadmium exhibited elevated resting blood pressure (Fadloun and Leah, 1980). Cadmium-induced hypertension may be caused by direct effects on sympathetic functioning (Fadloun and Leah, 1980).

3.2. Cyclic nucleotides

Cyclic nucleotide levels in smooth muscle appear to be under autonomic control. In rat uterus, β-adrenergic receptor stimulation can produce a large increase in cyclic AMP with no concurrent change in cyclic GMP. Cholinergic potentiation with theophylline causes an increase in cyclic GMP levels with a cyclic AMP response which is half as strong (Diamond and Hartle, 1974). Muscarinic cholinergic receptor stimulation appears to elevate cyclic GMP levels 2- to 5-fold in a number of different tissues from a number of different species (Heilbronn and Bartfai, 1978).

An increase in cyclic GMP has been demonstrated in guinea pig taenia coli following stimulation of contraction with carbachol or KCl (Janis and Diamond, 1981). However, the elevated cyclic GMP levels can be seen without muscle contraction after coincubation with verapamil or with low extracellular Ca^{2+}. Also, following stimulation with low levels of carbachol, maximum contraction is observed with little change in cyclic GMP levels (Janis and Diamond, 1981). Thus, in the tissue, elevated cyclic GMP levels are calcium independent and are not necessary for smooth muscle contraction. Fiscus and Dyer (1981) using sheep and human ubilical arteries, also found no consistent correlation between elevation of cyclic GMP levels and contraction. They were, however, able to find a temporal relationship between elevated cyclic GMP and smooth muscle relaxation (Fiscus and Dyer, 1981).

Cyclic AMP may act as a second messenger within the smooth muscle cell. One function may be to mediate the phosphorylation of myosin light chain kinase and to alter actin and myosin interactions. Kerrick and Hoar (1981) postulated that β-adrenergic elevation of cyclic AMP activates a catalytic subunit of protein kinase and inhibits Ca^{2+}-activated tension.

3.3. Phenotypic properties

Smooth muscle cells from a number of sources have been shown to proliferate when placed in cell culture (Chamley et al., 1979). These cells can also maintain or regain their differentiated morphology under certain specific conditions (Chamley et al., 1979). Autonomic nerves present with the explant may inhibit the dedifferentiation process (Chamley et al., 1974; Chamley and Campbell, 1976), and a similar inhibition of morphologic change occurs when cells are grown in the presence of extracts of homogenized sympathetic ganglia. Thus, a chemical mechanism for this trophic effect is indicated (Chamley and Campbell, 1976). Norepinephrine or acetylcholine do not block dedifferentiation, however. The effect of sympathetic nerves or sympathetic ganglion extracts can be mimicked by dibutyryl cyclic AMP or by theophylline, which is a cyclic AMP phosphodiesterase inhibitor (Chamley and Campbell, 1975). Since cyclic AMP is considered a 'second messenger' in the cell, the primary trophic factor from nerves is still to be identified.

In a recent report, Campbell et al. (1980) questioned whether autonomic nerves control specific smooth muscle phenotypic properties. The caudal artery of spontaneously hypertensive rats has an altered electrogenesis and response to norepinephrine when compared with that from Kyoto normotensive rats (Hermsmeyer, 1976). Pieces of this tissue from hypertensive rats were transplanted to the anterior eye chamber of normotensive rats where sympathetic reinnervation occured (Campbell et al., 1980). When tissue from a two-week old normotensive or hypertensive animal was transplanted into the arterior chamber of a host from the opposite group, phenotypic characteristics of the host developed (Campbell et al., 1980). However, when adult tissue was transplanted, the graft phenotype was maintained. In this system, tropic control from nerve to muscle was only possible during development and not in the adult.

4. Regulation of smooth muscle responsiveness

A major aspect of autonomic nerve/smooth muscle interaction is the plasticity which is inherent in the system. How muscle responds to neuronal stimulation is not constant but is subject to regulation. One type of regulation is a supersensitivity

TABLE 2
Supersensitivity responses in smooth muscle

Tissue	Treatment	Supersensitivity response	Source
Cat nictitating membrane	Denervation	To norepinephrine	Hampel, 1935
		To α-adrenergic agents	Pluchino and Trendelenburg, 1968
		To acetylcholine	Trendelenburg and Weiner, 1962
Dog saphenous vein	Denervation	To α-adrenergic agents	Guimarães, 1975
Guinea pig vas deferens	Denervation	To norepinephrine	Westfall et al., 1972
		Muscarinic receptors	Hata et al., 1980
Rabbit iris	Denervation	β-adrenergic receptors	Page and Neufeld, 1978
Rat portal vein	6-Hydroxydopamine	To norepinephrine	Aprigliano et al., 1976
Rat vas deferens	6-Hydroxydopamine	To norepinephrine	Kasuya et al., 1969
Rabbit basilar artery	Reserpine	To norepinephrine	Hudgins and Fleming, 1960; Bevan, 1980.
Rabbit aorta	Reserpine	To norepinephrine	Taylor and Green, 1970
		To acetylcholine	Taylor and Green, 1970
Guinea pig taenia coli	Stimulate inhibitory nerves	Firing rate	Bennett, 1963a,b
Avian gizzard	Stimulate inhibitory nerves	Firing rate	Bennett, 1969

which muscle develops under certain conditions. In skeletal muscle, a supersensitivity has been well defined in response to denervation (Thesleff, 1960; Fleming, 1978). This is characterized by a spread of receptors to areas outside the neuromuscular junction (Thesleff, 1960). In smooth muscle, supersensitivity also develops in response to denervation as well as to other stimuli (Table 2). In addition, a subsensitivity of smooth muscle can also occur when there is an increase in stimulation of the receptor (Table 3). Thus, a negative trophic relationship appears between the amount of agonist made available to the receptors and the responsiveness of smooth muscle via receptors.

The trophic relationship between innervation and response occurs on a local level in vivo (Fleming, 1976). Bevan and Purdy (1973) demonstrated that some portions of rabbit saphenous artery have less adrenergic innervation than others. Whereas more innervated areas have more response to sympathetic nerve stimulation, the less innervated areas are more sensitive to applied norepinephrine (Bevan and Purdy, 1973).

TABLE 3
Desensitization responses in smooth muscle

Tissue	Treatment	Desensitized response	Source
Rat ileum	Anticholinesterase	Cholinergic	McPhillips, 1968
Cat iris	Anticholinesterase	Cholinergic	Bito and Dawson, 1970
Guinea pig vas deferens	Incubation with acetylcholine	To acetylcholine	Takeyasu et al., 1981
Chick expansor secundarium	Incubation with acetylcholine	To acetylcholine	Gonoi, 1980
Guinea pig intestine	Nerve stimulation	Hyperpolarization	Kuriyama et al., 1967
Human lung	Incubation with isoproterenol	β-Adrenergic	Davis and Connolly, 1980
Mouse ileum	Acute morphine	Cholinergic	Ramaswamy et al., 1980
Rat vas deferens	Endorphins	Contraction	Huidobro-Toro et al., 1981

4.1. Supersensitivity

In a recent review, Westfall (1981) divided supersensitivity of smooth muscle into two classifications. One, termed deviation supersensitivity, is caused by a change in the amount of various drugs available to the receptor. This can occur through changes in a number of physiological processes including uptake into sympathetic neurons, metabolism of catecholamines or extra-neuronal uptake. In turn, drugs such as monoamine oxidase inhibitors, catechol-O-methyltransferase inhibitors or drugs which affect neurotransmitter uptake may effect a supersensitivity response in smooth muscle.

All of the drugs mentioned above require an intact neuronal input to the muscle. It is apparent that most types of deviation supersensitivity are caused by factors which increase the amount of neurotransmitter or drug available to the receptors on muscle. For example, blocking neuronal uptake of catecholamines by cocaine leads to increased amounts of neurotransmitter available to the muscle. This, in turn, leads to a supersensitive response of the muscle to a given neuronal signal (Westfall, 1981).

The second type, non-deviation supersensitivity, is due to an actual alteration in the physiology of the smooth muscle cells (Westfall, 1981). A number of manipulations can cause this phenomenon (Fleming et al., 1973). They include chronic receptor blockade, chronic release of neurotransmitter, surgical denervation, chemical denervation and chronic ganglionic blockade (Fleming et al., 1973; Westfall, 1981). These processes share the common property of reducing the amount of neurotransmitter available to the receptor and have been widely studied.

Perhaps the most studied type of supersensitivity in smooth muscle cells is denervation supersensitivity (Fleming, 1978). In an early study, Pluchino and Trendelenburg (1968) examined the development of denervation supersensitivity in the nictitating membrane of the cat. They demonstrated changes in responses mediated through α-adrenergic receptors while β-adrenergic receptor responses were unchanged. Similar results were obtained in the dog saphenous vein (Guimarães, 1975). Westfall and co-workers (1972) denervated the guinea pig vas deferens by cutting the hypogastric nerve. Within 1 week there was a large (63-fold) increase in sensitivity of the vas deferens to norepinephrine. Smaller supersensitivity responses were also produced to acetylcholine and histamine. One week after preganglionic denervation, there were less marked changes, however, with only a 30% to 50% increase in the sensitivity to norepinephrine. The observed supersensitivity was not accompanied by changes in maximal contraction (Westfall et al., 1972). Using the same model system, Hata and co-workers (1980) looked directly at changes in α-adrenergic and muscarinic cholinergic receptors. Employing radiolabelled ligand binding techniques, they demonstrated that the supersensitivity to norepinephrine and acetylcholine, observed 4 days after denervation, was accompanied by a 50% increase in binding sites for the muscarinic cholinergic antagonist [^{3}H]quinuclidinyl benzilate. There was no shift in the affinity of the muscarinic receptors for the ligand. In contrast, the number of binding sites and affinity for ^{3}H-labelled WB4101, an α-adrenergic receptor antagonist, was not altered with denervation. Thus, whereas the increase in acetylcholine sensitivity in the denervated vas deferens was accompanied by an increase in the number of receptors for acetylcholine this relationship does not appear to obtain for the adrenergic system (Hata et al., 1980). In contrast, in the rabbit iris, sympathetic denervation led to an increase in β-adrenergic receptor numbers with no change in α-adrenergic receptors (Page and Neufeld, 1978). Thus, any common mechanism behind the increased sensitivity to norepinephrine remains to be elucidated.

One model system commonly used in early studies of smooth muscle denervation supersensitivity is the cat nictitating membrane. In 1935, Hampel (1935) observed a supersensitive response to epinephrine in the nictitating membrane only two weeks after surgery. The response to denervation by cutting the postganglionic sympathetic nerve was always much greater in magnitude than was the response to decentralization. This model system was examined in much greater detail in the 1960s primarily by Trendelenburg (1955; 1963a; Trendelenberg and Weiner, 1962; Pluchino and Trendelenburg, 1968). A decrease in norepinephrine content or in sympathetic influence on the nictitating membrane produced by decentralization or denervation led to a supersensitivity to norepinephrine. This effect generally took 7 to 14 days to occur (Fleming, 1963; Rapoport and Bevan, 1979). Supersensitivity to acetylcholine was also demonstrated in the denervated or decentralized membrane and this response also occured after one week (Fleming, 1963; Rapoport and Bevan, 1979). Sympathetic denervation also led to a subsensitivity to tyramine which appears to be releasing norepinephrine from the neurons which degenerate after the surgical procedure (Trendelenburg and Weiner, 1962; Trendelenburg, 1963a).

Trendelenburg (1963a,b) compared the increase in sensitivity to norepinephrine in the denervated nictitating membrane to that seen in decentralized tissue. Two days after surgery, the denervated side had over ten times the sensitivity to norepinephrine when compared to the decentralized side. This ratio did not change significantly, even at 2 weeks when both responses were maximal (Trendelenburg, 1966). He thus divided denervation supersensitivity into two components. The decentralization-like supersensitivity required 2 weeks to fully occur. The other component, termed 'cocaine-like', results from direct loss of postsynaptic neuron functioning and occurs much more rapidly, after 1 to 2 days (Trendelenburg, 1963b). The first component may be due to a slow alteration in muscle sensitivity following decreases in neuronal input. The second may be due to a decrease in the nerve's ability to release or to take up norepinephrine.

Chemically-induced adrenergic denervation of the rat portal vein by the drug

6-hydroxydopamine leads to an increased sensitivity to norepinephrine as well as an increase in muscle membrane excitability (Aprigliano and Hermsmeyer, 1976; Aprigliano et al., 1976). Similar effects were observed when smooth muscle of various types are removed from its innervation and incubated in vitro (Kasuya et al., 1969; Abel et al., 1980). However, the supersensitivity responses could be prevented by incubation with norepinephrine and a trophic effect of adrenergic influence on smooth muscle function is apparent (Abel et al., 1980). Supersensitivity can also be induced by injection of reserpine, which leads to a supersensitivity to phenylephrine, acetylcholine, and K^+, but not to histamine in rabbit aortic strips (Taylor and Green, 1970). In vitro exposure of rabbit aortic strips to reserpine causes a supersensitivity to acetylcholine, norepinephrine and K^+ but not to serotonin or histamine (Hudgins and Fleming, 1966). The reserpine-induced supersensitivity is thus highly non-specific.

Gerthoffer et al. (1979) correlated an inhibition of the sodium-potassium pump with the development of denervation supersensitivity. They interrupted the excitatory innervation to the guinea pig vas deferens and noted a reduction of sodium-potassium ATPase activity. This in turn inhibited the sodium-potassium pump mechanism. Furthermore, it was demonstrated that supersensitive tissues are not affected by treatments which inhibit the pump directly whereas normal tissues are affected. They thus postulated that denervation led to a partial inhibition of the sodium-potassium pump which resulted in a partial depolarization of the cell membrane. It is this partial depolarization which may be responsible for supersensitive responses (Gerthoffer et al., 1979), since the smooth muscle cells are now in a resting state closer to the activation potential.

An alternative method of producing supersensitivity in smooth muscle is by stimulation of inhibitory neurons to the muscle. This may act to block release of excitatory substances and thus create a denervation-like effect. In 1966, Bennett and co-workers (1966) reported an increase in basal firing rate of guinea pig taenia coli following stimulation of its inhibitory nerve (Bennett, 1966). This might represent a denervation supersensitivity. Similar responses were observed in the avian gizzard (Bennet, 1969).

4.2. Desensitization

In contrast to supersensitivity responses, subsensitivity of muscarinic cholinergic receptors in smooth muscle has also been observed. For example, chronic administration of a cholinesterase inhibitor in rats led to a subsequent subsensitivity of the ileum to cholinomimetics (McPhillips, 1968). Bito and Dawson (1970) demonstrated that chronic topical treatment of the cat eye with an anticholinesterase led to a subsensitivity of the iris sphincter to cholinomimetics. In a later study, Bito et al. (1971) demonstrated that the sensitivity of cat iris to acetylcholine varies inversely with the intensity of stimulus background. Incubation of guinea pig vas deferens with acetylcholine led to a decreased sensitivity to that compound (Takeyasu et al., 1981). This effect was paralleled by a decrease in the number of muscarinic receptors as defined by radioligand binding. In contrast, the sensitivity to norepinephrine and K^+ and the density of α-adrenergic receptors were not altered by pre-incubation of tissue with acetylcholine (Takeyasu et al., 1981). When chick expansor secundarium muscle was placed in organ culture with acetylcholine, there was a subsensitive response to acetylcholine but not to norepinephrine (Gonoi, 1980). Acetylcholine-induced desensitization thus appears to be specific for the cholinergic response.

Repeated electrical stimulation of smooth muscle cells from rabbit colon led to a waning of the mechanical response (Gillespie, 1962a,b). Similarly, repetitive stimulation of nerve to guinea pig small intestine in vitro led to an initial hyperpolarization and then a gradual long-lasting depolarization (Kuriyama et al., 1967). This effect persisted even after 50 hours of rest. The cholinergic desensitization may have been due to a depletion of acetylcholine stores (Kuriyama et al., 1967).

Rapoport and Bevan (1979) studied the relationship between stress and sympathetic firing rate to the rabbit ear artery. After stress of various types, there was an increase in sympathetic discharge along with a decrease in responsiveness to norepinephrine. This behaviorally-induced desensitization can be mimicked in vitro by sympathetic nerve stimulation or by exposure to nerepinephrine. Desensitization of β-adrenergic response can also be induced by exposure to β-adrenergic agonists. In isolated strips of human lung, in vitro exposure to isoproterenol causes a reversible diminution of responsiveness to β-adrenergic agonists (Davis and Conolly, 1980).

Desensitization of smooth muscle to opiates has also been reported. In the mouse, acute morphine treatment causes a desensitization of the ileum to acetylcholine (Ramaswamy et al., 1980). In the same animals, there is a supersensitivity of the vas deferens to norepinephrine and both effects can be blocked by blocking the opiate receptors during morphine treatment (Ramaswamy et al., 1980). Additionally, morphine normally potentiates the twitch response evoked by electrical stimulation in the isolated rat vas deferens. The ability of morphine to cause this effect is greatly diminished in vasa deferentia from rats chronically treated with morphine (Huidobro-Toro et al., 1981). Endophins in contrast, cause a diminution of the electrically-stimulated contractile response in the vas deferens. The endorphin-produced effects were also decreased following chronic morphine treatment. It was suggested that opiates act at two different sites on the rat vas deferens (Huidobro-Toro et al., 1981).

5. Conclusions

Despite the complexities of the system, there are a number of general phenomena which arise from the study of smooth muscle responses to autonomic stimulation. Most of these phenomena have been discussed in this overview chapter. First of all, especially in the gut, sympathetic nerve stimulation often results in smooth muscle relaxation and parasympathetic stimulation results in smooth muscle contraction. The relaxation response appears to be mediated by β-adrenergic receptors and the contraction response is mostly mediated by muscarinic cholinergic receptors. However, α-adrenergic receptors on most smooth muscle cells also mediate contraction. In addition, numerous other receptors on smooth muscle cells affect autonomic response or directly affect the muscle. Also, the vascular endothelial cells appear to play an important role in cholinergic response. The exact nature of these interactions is only beginning to come to light.

As with skeletal muscle, responses in smooth muscle are regulated by neuronal input. The negative correlation between input and receptors/response appears to be maintained in this system. Thus, factors which reduce autonomic input to the smooth muscle, such as chemical and physical denervation, will increase responsiveness to the lost transmitter (Table 2). Conversely, increasing stimulation at receptor will decrease the responsiveness of that receptor (Table 3). Again, the exact nature of this trophic relationship remains to be elucidated.

In the coming years we should see an increase of our understanding of the actions of recently discovered peptides and hormones which interact with smooth muscle. Some of these may reside in neurons of the autonomic nervous system. We are just beginning to understand the possibility of coexistence of multiple neurotransmitters within a neuron. However, the interactions of these transmitters at the release site is not yet understood. Whether there is co-release and whether transmitters act directly on postsynaptic receptors or as neuromodulators are just some of the many questions which now face researchers.

As outlined in this review, many endogenous substances have effects on smooth muscle. It is likely that numerous substances yet to be discovered also affect smooth muscle properties. Since the in vivo state of smooth muscle at any time is determined by the sum total of responses to many different substances, an understanding of interactive as well as individual effects of these chemicals is necessary and should emerge in years to come.

References

Abdel-Latif, A.A. and Luke, B. (1981) Biochem. Biophys. Acta *173*, 64–74.
Abel, P.W., Trapani, A., Aprigliano, O. and Hermsmeyer, K. (1980) Circ. Res. *47*, 770–775.
Adelstein, R.S., Conti, M.A. and Hathaway, D.R. (1978) J. Biol Chem. *253*, 8347–8350.
Andersson, R. (1972) Acta Physiol. Scand. *85*, 312–322.

Andersson, R. (1973) Acta Physiol. Scand. *87*, 84–95.
Andersson, R. and Nilsson, K. (1972) Nature New Biol. *238*, 119–120.
Aprigliano, O. and Hermsmeyer, K. (1976) J. Pharmacol. Exp. Therap. *198*, 568–577.
Aprigliano, O. and Hermsmeyer, K. (1977) Circ. Res. *41*, 198–206.
Aprigliano, O., Rybarczyk, K.E., Hermsmeyer, K. and VanOrden III, L.S. (1976) J. Pharmacol. Exp. Therap. *198*, 578–588.
Beani, L., Bianchi, C. and Crema, A. (1971) J. Physiol. (London) *217*, 259–279.
Beld, A.J., van der Hosen, S., Wouterse, C.A., and Zegers, M.A.P. (1975) Eur. J. Pharmacol. *30*, 360–363.
Bennett, M.R. (1966a) Nature (London) *211*, 1149–1152.
Bennett, M.R. (1966b) J. Physiol. (London) *185*, 124–131.
Bennett, M.R., Burnstock, G. and Holman, M.E. (1966) J. Physiol. (London) *182*, 541–558.
Bennett, T. (1969) J. Physiol. (London) *204*, 669–686.
Berridge, M.J. (1980) Trends Pharmacol. Sci. *1*, 419–424.
Bevan, J.A. (1974) Circ. Res. *45*, 161–171.
Bevan, J.A. (1977) Fed. Proc. *36*, 2439–2443.
Bevan, J.A. (1979) Fed. Proc. *37*, 187–190.
Bevan, J.A. (1981) J. Pharmacol. Exp. Ther. *216*, 83–89.
Bevan, J.A., Bevan, R.D. and Duckles, S.P. (1976) Handb. Physiol. 516–566.
Bevan, J.A., Pegram, B.A., Prehn, J.L. and Winquist, R.J. (1978) Mechanisms of Vasodilatation, pp. 258–265, Satellite Symp. 27th Int. Cong. Physiol. Sci., Karger, Basel, 1977.
Bevan, J.A. and Purdy, R.E. (1973) Circ. Res. *32*, 746–751.
Bevan, J.A. and Su., C. (1973) Ann. Rev. Pharmacol. *13*, 269–286.
Bito, L.Z. and Dawson, M.J. (1970) J. Pharmacol. Exp. Ther. *175*, 673–684.
Bito, L.Z., Dawson, M.J. and Petrinovic, L. (1971) Science *172*, 583–585.
Black, I.B. (1978) Ann. Rev. Neurosci. *1*, 183–214.
Bolton, T.B. (1973) in Drug Receptors (Rang, H.P., ed.), pp. 87–104, University Park Press, Maryland.
Bulbring, E. (1973a) in Drug Receptors (Rang, H.P., ed.), pp. 1–14, University Park Press, Maryland.
Bulbring, E. (1973b) in Frontiers in Catecholamine Research (Usdin, E. and Snyder, S., eds.), pp. 389–391, Pergamon Press, London.
Bulbring, E. and Tomita, T. (1969a) Proc. Roy. Soc. (London) *172*, 89–102.
Bulbring, E. and Tomita, T. (1969b) Proc. Roy. Soc. (London) *172*, 103–119.
Burgen, A.S.V., Hiler, C.R. and Young, J.M. (1974) Br. J. Pharmacol. *50*, 145–151.
Burgen, A.S.V. and Spero, L. (1968) Br. J. Pharmacol. *34*, 99–115.
Burnstock, G. (1974) in Dynamics of Degeneration and Growth in Nerves (Fuxe, K., Olson, L. and Zotterman, Y., eds.), pp. 509–519, Pergamon Press, Oxford.
Burnstock, G. (1979) Br. Med. Bull. *35*, 255–262.
Campbell, G. (1966) J. Physiol. (London) *185*, 600–612.
Campbell, G.R., Chamley-Campbell, J., Robinson, R. and Hermsmeyer, K. (1980) in Vascular Neuroeffector Mechanisms (Bevan, J.A., ed.), pp. 107–113, Raven Press, New York.
Casteels, R., Kitamura, K., Kuriyama, H. and Suzuki, H. (1977) J. Physiol. (London) *271*, 41–61.
Chamley, J.H. and Campbell, G.R. (1975) Cell Tissue Res. *161*, 497–510.
Chamley, J.H. and Campbell, G.R. (1976) in Vascular Neuroeffector Mechanisms (Bevan, J., ed.), pp. 10–18, Karger, Basel.
Chamley, J.H., Campbell, G.R. and Burnstock, G. (1974) J. Embryol. Exp. Morphol. *32*, 297–373.
Chamley, J.H., Campbell, G.R. and Ross, R. (1979) Physiol. Rev. *59*, 1–61.
Chamley, J.H., Goller, I. and Burnstock, G. (1973) Dev. Biol. *31*, 362–379.
Cohen, M.L. and Landry, A.S. (1980) Life Sci. *26*, 811–819.
Dale, H.H. (1929) Lancet *1*, 1179–1183.

Dale, H.H. and Laidlaw, P.P. (1910) J. Physiol. (London) 41, 318–344.
Daniel, E.E. and Kwan, C.Y. (1981) Trends Pharmacol. Sci. 2, 220–223.
Davies, B.N. and Withrington, P.G. (1973) Pharmacol. Res. 25, 373–413.
Davis, C. and Conolly, M.E. (1980) Br. J. Clin. Pharmacol. 10, 417–423.
Davis, C., Conolly, M.E. and Greenacre, J.K. (1980) Br. J. Clin. Pharmacol. 10, 425–432.
Diamond, J. and Hartle, D.K. (1974) Can. J. Physiol. Pharmacol. 52, 763–767.
Dibner, M.D., Mytilineou, C. and Black, I.B. (1977) Brain Res. 123, 301–310.
Dixon, W.E. and Ransom, F. (1912) J. Physiol. (London) 45, 413–428.
Doidge, J.M. and Satchell, D.G. (1980) Comp. Biochem. Physiol. 65, 53–58.
Duckles, S.P., Lee, T.J.F. and Bevan, J.A. (1977) in Neurogenic Control of Brain Circulation (Owman, C. and Edvinsson, L., eds.), Proc. Int. Symp. No. 30, pp. 133–141, Wenner-Gren, Pergamon Press, Oxford.
Durbin, R.P. and Jenkinson, D.H. (1961) J. Physiol. (London) 157, 74–89.
Fadloun, Z. and Leah, G.D.H. (1980) Br. J. Pharmacol. 70, 167.
Fewtrell, C.M.S. and Rang, H.P. (1973) in Drug Receptors (Rang, H.P., ed.), pp. 211–224, University Park Press, Maryland.
Fiscus, R.R. and Dyer, D.C. (1981) Eur. J. Pharmacol. 73, 283–291.
Fleming, W.W. (1963) J. Pharmacol. Exp. Ther. 141, 173–179.
Fleming, W.W. (1976) Neuroscience 2, 43–90.
Fleming, W.W. (1978) Life Sci. 22, 1223–1228.
Fleming, W.W., McPhillips, J.J. and Westfall, D.P. (1973) Rev. Physiol. 68, 55–119.
Furchgott, R.F. and Zawadzki, G.V. (1980) Nature (London) 288, 373–376.
Garcia, A.G., Jurkiewicz, A. and Jurkiewicz, N.H. (1981) Eur. J. Pharmacol. 70, 17–23.
Gerthoffer, W.T., Fedan, J.S., Westfall, D.P., Goto, R. and Fleming, W.W. (1979) J. Pharmacol. Exp. Ther. 210, 27–36.
Gibson, A. and Pollock, D. (1973) Br. J. Pharmacol. 49, 506–513.
Gillespie, J.S. (1962a) J. Physiol. (London) 162, 54–75.
Gillespie, J.S. (1962b) J. Physiol. (London) 162, 76–92.
Golla, F.L. and Symes, W.L. (1913) J. Physiol. (London) 46, 33.
Gonoi, T. (1980) Eur. J. Pharmacol. 68, 287–293.
Grundström, N., Anderson, C.G. and Witberg, J.E.S. (1981) Life Sci. 28, 2981–2986.
Guimarães, S. (1975) Eur. J. Pharmacol. 34, 9–19.
Hampel, C.W. (1935) Am. J. Physiol. 111, 611–621.
Hanley, M.R. and Iversen, L.L. (1980) in Neurotransmitter Receptors (Enna, S.J. and Yamamura, H.I., eds.), pp. 73–103, Chapmen and Hall, London.
Hata, F., Takeyasu, K., Morikawa, Y., Tsanlai, R., Ishida, H. and Yoshida, H. (1980) J. Pharmacol. Exp. Ther. 215, 716–722.
Hedqvist, P. (1970a) Life Sci. 9, 269–278.
Hedqvist, P. (1970b) Acta. Physiol. Scand. Suppl. 345, 1–40.
Hedqvist, P. (1972) Acta. Physiol. Scand. 84, 506–511.
Heilbronn, E. and Bartfai, T. (1978) Prog. Neurobiol. 11, 171–188.
Hermsmeyer, K. (1976) Circ. Res. 38, 362–367.
Hermsmeyer, K. and Aprigliano, O. (1980) in Vascular Neuro Effector Mechanisms (Bevan, J.A., ed.), pp. 113–119, Raven Press, New York.
Holman, M.E. (1967) Circ. Res. 21, 71–82.
Holman, M.E. and Hirst, C.D.S. (1977) in Handbook of Physiology: The Nervous System (Brookhart, J.M. and Mountcastle, V.B., eds.), pp. 417–461.
Hopkins, L.L. (1974) in Trace Element Metabolism in Animals, Part 2, p. 397, University Park Press, Baltimore.
Horrobin, D.F., Lipton, A., Muiruri, K.L., Munku, M.S., Bramley, P.S. and Burstyn, P.G. (1973) Experientia 29, 109–116.

Hudgins, P.M. and Fleming, W.W. (1966) J. Pharmacol. Exp. Ther. *153*, 70–80.
Huidobro-Toro, J.P., Miranda, H. and Huidobro, F. (1981) Life Sci. *28*, 773–779.
Ito, Y. and Kuriyama, H. (1971) Jpn. J. Physiol. *21*, 277–294.
Jager, L.P. and Schevers, J.A.M. (1980) J. Physiol. (London) *299*, 75–83.
Jafferji, S.S. and Michell, R.H. (1976) Biochem. J. *154*, 653–657.
Janis, R.A. and Diamond, J. (1981) Eur. J. Pharmacol. *70*, 149–156.
Jenkinson, D.J. and Morton, I.K.M. (1967) J. Physiol. (London) *188*, 387–402.
Jim, K.F. and Triggle, D.J. (1981) Biochem. Pharmacol. *30*, 95–96.
Jones, T.R., Lefcoe, N.M. and Hamilton, J.T. (1980) Eur. J. Pharmacol. *67*, 53–64.
Kajiwara, M., Kitamura, K. and Kuriyama, H. (1981) J. Physiol. (London) *315*, 283–302.
Kalenberg, S. and Satchell, D.G. (1979) Clin. Exp. Pharmacol. Physiol. *6*, 549–559.
Kanazawa, T., Ohashi, H. and Takewaki, T. (1980) Br. J. Pharmacol. *71*, 519–524.
Karaki, H. Kuboto, H. and Urakawa, N. (1979) Eur. J. Pharmacol. *56*, 237–245.
Karashima, T. (1980) Br. J. Pharmacol. *71*, 489–497.
Kasuya, Y., Goto, K., Mashimoto, H., Watanabe, H., Munakata, H. and Watanabe, M. (1969) Eur. J. Pharmacol. *8*, 177–184.
Kerrick, W.G.L. and Hoar, P.E. (1981) Nature (London) *292*, 253–255.
Kessler, J.A., Adler, J.E., Bohn, M.C. and Black, I.B. (1981) Science *214*, 335–336.
Kessler, J.A. and Black, I.B. (1981) Brain Res. *208*, 235–245.
Khanna, O.P., Barbieri, E.J. and McMichael, R. (1981) J. Pharmacol. Exp. Ther. *216*, 95–100.
Kilbinger, H. and Wagner, P. (1975) Naunyn-Schmiedebergs Arch. Exp. Pathol. Pharmakol. *287*, 47–60.
Kirpekar, S.M., Garcia, A.A. and Prat, J.C. (1980) Biochem. Pharmacol. *29*, 3029–3031.
Kitabgi, P. and Freychet, P. (1978) Eur. J. Pharmacol. *50*, 349–361.
Kitabgi, P. and Vincent, J.P. (1981) Eur. J. Pharmacol. *74*, 311–318.
Komori, S., Ohashi, H. and Takewaki, T. (1980) Br. J. Pharmacol. *71*, 479–488.
Kosterlitz, H.W., Lydon, R.J. and Watt, A.J. (1970) Br. J. Pharmacol. *39*, 398–413.
Kreth, E. and Vonderlage, M. (1979) Pfleugers Arch. *377R*, 38.
Kuriyama, H., Osa, T. and Toida, N. (1967) J. Physiol. (London) *191*, 257–270.
Lands, A.M., Arnold, A., McAuliff, J.P., Luduena, F.P. and Brown, T.G. (1967) Nature (London) *214*, 597–598.
Lee, T.J.F. (1977) Fed. Proc. *36*, 1036.
Lee, T.J.F., Hume, W.R., Su, C. and Bevan, J.A. (1978) Circ. Res. *42*, 535–542.
Leff, A.R. and Munoz, N.M. (1981) J. Pharmacol. Exp. Ther. *218*, 582–587.
Levin, R.M., Shofer, F.S. and Wein, A.J. (1980) J. Pharmacol. Exp. Ther. *215*, 614–618.
Lindblom, B., Ljung, B. and Hamberger, L. (1979) Acta. Physiol. Scand. *106*, 215–220.
MacDonald, A. and McGrath, J.C. (1980) Br. J. Pharmacol. *71*, 445–458.
Majewski, H., McCulloch, M.W., Rand, M.J. and Story, D.F. (1980) Br. J. Pharmacol. *71*, 435–444.
Majewski, H. and Rand, M.J. (1981) Eur. J. Pharmacol. *69*, 493–498.
McGrath, M.A. and Vanhoutte, P.M. (1978) in Mechanisms of Vasodilation, pp. 248–257, Karger, Basel.
McPhillips, J.J. (1968) J. Pharmacol. Exp. Ther. *166*, 249–254.
Michell, R.H. and Kirk, C.J. (1981) Trends Pharmacol. Sci. *2*, 86–89.
Minneman, K.P., Dibner, M.D., Wolfe, B.B. and Molinoff, P.B. (1979) Science *252*, 297–299.
Mueller, T. and van Breemen, C. (1979) Nature (London) *281*, 682–683.
Nadel, J.A. (1980) Lung Biol. Health Dis. *15*, 217–257.
Ohashi, H. (1971) J. Physiol. (London) *212*, 561–575.
Ohashi, H. and Ohga, A. (1967) Nature (London) *216*, 291–292.
Owen, D.A.A. (1977) Gen Pharmacol. *8*, 141–156.
Page, E.D. and Neufeld, A.H. (1978) Biochem. Pharmacol. *27*, 953–964.

Paton, W.D.M. and Vizt, E.S. (1969) Br. J. Pharmacol. 35, 10–28.
Pillai, N.P., Ramaswamy, S., Gopalakrishnan, V. and Ghosh, M.N. (1981) Eur. J. Pharmacol. 72, 11–16.
Pluchino, S. and Trendelenburg, U. (1968) J. Pharmacol. Exp. Ther. 163, 257–265.
Primor, N. (1980) Eur. J. Pharmacol. 68, 497–500.
Ramaswamy, S., Pillai, N.P., Gopalakrishnan, V. and Ghosh, M. (1980) Eur. J. Pharmacol. 68, 205–208.
Rapoport, S. and Bevan, J.A. (1979) Experiental 35, 1609–1611.
Riley, J.F. and West, G.B. (1953) J. Physiol. (London) 120, 528–537.
Roquebert, J. and Demichel, P. (1981) Eur. J. Pharmacol. 70, 87–90.
Rosell, S., Bjokroth, V., Change, D., Yamaguchi, I., Wan, Y.P., Rackur, G., Fisher, G. and Folkers, K. (1977) in Substance P (von Euler, U.S. and Pernow, B., eds.), pp. 83–88, Raven Press, New York.
Samuelsson, B. and Wennmalm, A. (1971) Acta. Physiol. Scand. 83, 163–168.
Sanner, J.H. and Prusa, C.H. (1980) Life Sci. 27, 2565–2570.
Schwartz, J.-C. (1977) Ann. Rev. Pharmacol. Toxicol. 17, 325–331.
Shepherd, J.J. (1978) Fed. Proc. 37, 179–180.
Starke, J. and Endo, T. (1976) Comp. Gen. Pharmacol. 7, 307–312.
Su, T.P., Clements, T.H. and Gorodetzky, C.W. (1981) Life Sci. 28, 2519–2528.
Szurszewski, J.H. (1973) in Frontiers in Catecholamine Research (Usdin, E. and Snyder, S., eds.), pp. 383–388, Pergamon Press, London.
Takagi, K., Atao, M. and Takahashi, A. (1965) Life Sci. 4, 2165–2169.
Takeyasu, K., Uchida, S., Lai, R., Higuchi, J., Noguchi, Y. and Yoshida, H. (1981) Life Sci. 28, 527–540.
Taylor, J. and Green, R.O. (1970) J. Pharmacol. Exp. Ther. 177, 127–135.
Tenner, T.C. and McCully, J.P. (1981) Eur. J. Pharmacol. 73, 293–300.
Thesleff, S. (1960) Physiol. Rev. 40, 734–752.
Trendelenburg, U. (1954) Br. J. Pharmacol. 9, 481–487.
Trendelenburg, U. (1963a) Pharmacol. Rev. 15, 225–276.
Trendelenburg, U. (1963b) J. Pharmacol. Exp. Ther. 142, 335–342.
Trendelenburg, U. (1966) Pharmacol. Rev. 18, 629–640.
Trendelenburg, U. and Weiner, N. (1962) J. Pharmacol. Exp. Ther. 136, 152–161.
Turlapaty, P.D.M.W., Weiner, R. and Altora, B.M. (1981) Eur. J. Pharmacol. 74, 263–272.
Vanhoutte, P.M. (1978) Fed. Proc. 37, 181–186.
Vanhoutte, P.M., Verbeuren, T.J. and Lorenz, R.R. (1976) Inst. de la Santé et de la Recherche Médicine Colloque, Paris. 50, 425–441.
Vanmeel, J.C.A., Dejonge, A., Kalman, H.O., Wilffert, B., Timmermans, P. and Vanzwieten, P. (1981) Eur. J. Pharmacol. 69, 205–208.
Vermeire, P.A. and Vanhoutte, P.M. (1979) J. Appl. Physiol. 46, 787–791.
Westfall, D.P. (1981) in Smooth Muscle – An Assessment of Current Knowledge (Bulbring, E., Brading, J., Jones, A.W. and Tomita, T., eds.), pp. 285–320, Edward Arnold, London.
Westfall, D.P., McClure, D. and Fleming, W.W. (1972) J. Pharmacol. Exp. Ther. 181, 328–338.

CHAPTER 9

The influence of the motor nerve on the contractile properties of mammalian skeletal muscle

A.J. BULLER

University of Bristol, University Walk, Bristol, England

1. Introduction

Normally innervated mammalian skeletal muscle remains at rest until it is caused to contract by the application of an adequate stimulus. Under physiological conditions the stimuli leading to contraction are typically the arrival of motor nerve impulses at the neuromuscular junction. This characteristic quiescence of skeletal muscle until the arrival of an adequate stimulus differentiates it from both cardiac muscle and most smooth muscle. It also endows the innervating motoneurones with a degree of control over the contractile characteristics of skeletal muscle, a fact which was first suggested by the experiments of Buller et al. (1960).

It has been known since the time of Ranvier (1874) that the indirectly elicited isometric responses of the limb muscles of various adult animals differ in their contraction times. Because of a common, but not invariable, association between muscle colour and contraction time, mammalian limb muscles were for a time divided into two groups, fast-white and slow-red. The colour difference was later shown to be due to differences in the concentration of myoglobin (the more slowly contracting muscles containing a higher concentration), but the factors determining the speed of contraction were not to be elucidated for many years. However, in 1922 Banu showed that the differences in the speed of contraction which Ranvier had demonstrated between muscles of an adult animal were not apparent at the time of birth, but developed postnatally over a period of weeks, the duration of this period of differentiation varying from species to species.

During the monumental work on mammalian reflexes undertaken by Sherrington and his colleagues at Oxford important, but peripheral, studies were made on the isometric responses of various mammalian muscles in an effort to understand the development of fast-white and slow-red muscles in reflex activity (e.g., Cooper and Eccles, 1930). It was at this time that Denny-Brown (1929), also working in

Sherrington's department, stressed that the common association between muscle colour and speed was not invariant, and as a result the terms of fast-muscle and slow-muscle replaced the earlier nomenclature of fast-white and slow-red.

For a period of some thirty years from 1930 onwards the physiological investigation of skeletal muscle was dominated by the work of A.V. Hill and his colleagues at University College, London. Their preferred preparation was frog striated muscle, most often the sartorius muscle. Whilst occasional important observations on mammalian muscle were reported by others working elsewhere, for example Gordon and Phillips (1953) at Oxford who demonstrated that in the limbs of mammals slow muscle was usually placed deep to more superficial fast muscle, the opportunities and advantages offered by the study of striated muscle obtained from cold-blooded amphibia undoubtedly led to a rapid increase in the understanding of the mechanical characteristics of skeletal muscle.

Detailed examination of various frog striated muscles did not demonstrate marked differences between their contraction times as had been observed in mammals, but it became clear that there were at least two patterns of motor innervation. Some frog muscle fibres receive multiple innervations along their entire length, whilst others receive innervation from only one motoneurone, typically at a single neuromuscular junction. Unfortunately the former type of muscle (i.e., with multiple innervation) was dubbed 'slow' muscle (Kuffler and Vaughan Williams, 1953), and the latter type 'twitch' muscle. This led to an ambiguity over the use of the term 'slow muscle' which had a different connotation in amphibia and mammals. This confusion has now been minimised by recommending the term tonic muscle to describe muscle fibres which receive multiple innervation and twitch muscle to describe muscle fibres receiving innervation from a single neurone. In mammals tonic fibres have been demonstrated in some striated muscles innervated by cranial nerves (e.g., Hess and Pilar, 1963), but only twitch fibres have so far been observed in the limb musculature. Twitch fibres can be clearly divided into fast-twitch and slow-twitch.

In 1958 J.C. Eccles was preparing to undertake a series of experiments to study the plasticity of the monosynaptic connections within the lumbar spinal cord of the new-born kitten. His plan was to cross over the innervations of certain hind-limb muscles and, after an interval of 6 to 12 months, examine the central connections established by incoming 1A afferents upon the motoneurones which, by that time, had reinnervated 'foreign' muscles. Because of the attention which Eccles' experiments subsequently attracted, it is important to appreciate that this was not the first occasion upon which the technique of crossing nerves has been used. Sperry (1941) had employed this procedure in rats some 18 years earlier in order to study the effect upon motor performance. The significance of Eccles' experiments lay in the fact that, by chance rather than design, he produced cross-reinnervation of a fast-twitch muscle by a nerve which had previously innervated a slow-twitch muscle and vice versa. However, it was not chance which led him to observe the twitch times of the cross-reinnervated muscles but the need to establish that reinnervation had occurred.

A brief description of the first of Eccles' terminal experiments on a cross-reinnervated kitten has been given by Buller and Pope (1977). Suffice it to say that visual observation of the twitch contractions of the two cross-reinnervated muscles was sufficient to satisfy Eccles that he had unexpectedly discovered a powerful neuronal control of the contractile characteristics of mammalian skeletal muscle.

Since Eccles' original experiments much time and effort has been, and continues to be, spent on increasing our understanding of the two questions raised by his observations:

(1) What are the precise changes produced in the contractile machinery which bring about the altered mechanical responses of cross-reinnervated muscles?

(2) How are these changes induced by an alteration of the muscle's innervation?

A variety of techniques have been utilised in seeking answers to these two questions. No attempt will be made here to provide a full catalogue of these techniques. In particular, no reference will be made to the contribution which histochemistry has made to our present understanding. Whilst histochemistry can, and has, made important contributions to our knowledge of the biochemical differences between the individual fibres of skeletal muscle, it has not yet achieved a perfect correlation between the histochemical profile of an anatomical muscle and that muscle's contractile performance. Unfortunately some authors have attempted to substitute histochemistry for mechanical measurements. If this is practiced at the level of whole muscles the results can be misleading. For this reason the present review identifies the mechanical performance of muscle (both isotonic and isometric) as the dependent variable which is subjected to experimental modification.

2. Contractile responses

2.1. General

Studies of the contractile responses of mammalian hind limb muscles have been made on a number of species including the mouse, rat, rabbit and cat. There is increasing evidence that the more in-bred the stock, the smaller the animal to animal differences become (assuming that all members of the colony are kept under similar conditions). Whilst inbred strains of rat are common the same is not generally true of rabbits or, less still, cats. For this reason the variability in the results obtained from cats which has been reported in the literature is greater than reported for rats. In the results reported below only young adult cats obtained from a breeding farm over a period of a few years were used.

It is also important to appreciate that whilst mammalian hind limb muscles may be conveniently classified as either fast-twitch or slow-twitch, effectively all muscles contain a mixture of motor unit types (Bessou et al., 1963; Weurker et al., 1965; Burke et al., 1973). Indeed it is now generally recognised that there are more than two physiologically distinct types of motor unit, but the mixtures of units within individual muscles did not allow the differentiation of more than two distinct physiological types of whole muscle (fast-twitch and slow-twitch). Recent physiological experiments have confirmed the previously tacit assertion that the skeletal muscle fibres supplied by a single motoneurone have very similar, if not identical, contraction characteristics.

2.2. Isometric contractions

The isometric contraction characteristics of analogous skeletal muscles from different adult mammalian species having dimensional similarity and comparable life patterns vary with the size of the animal. The mouse muscle has a shorter contraction time than the comparable muscle of a rat which, in turn, has a shorter contraction time than that of the cat. However, in all species the contraction time of the maximum isometric twitch in slow-twitch muscle is some 3 to 4 times longer than in a fast-twitch.

In order to obtain reproducible results when recording isometric contractions from whole muscles indirectly stimulated via the distal stump of their cut motor nerves, it is essential to control certain characteristics such as the muscle's initial length and its temperature. These factors will not be discussed here, as they have been detailed elsewhere (Buller and Buller, 1980).

Under optimal recording conditions there are no significant differences between normal and self-reinnervated cat FDL muscles (flexor digitorum longus, the smaller of the two long flexors in the calf), or FHL muscles (flexor hallucis longus, the larger of the two long flexors), both fast-twitch muscles, or soleus muscles (a slow-twitch calf muscle) in respect of time to peak (the time from the commencement of the contraction to the instant of peak tension development) or half relaxation time (the time from the instant of peak tension development to the time at which tension has fallen to half its maximum value). The mean time to peak value for normal cat FDL muscles is 21.4 ms (S.D. 2.7, $n = 30$) and the mean half-relaxation time is 17.5 ms (S.D. 2.6, $n = 20$). For self-reinnervated FDL muscles the mean values are 18.7 ms (S.D. 2.5, $n = 10$) and 19.0 ms (S.D. 2.1, $n = 9$) respectively. For the larger FHL muscles comparable normal figures are 26.4 (S.D. 2.0, $n = 14$) and 20.9 ms (S.D. 3.2, $n = 14$) and for self-reinnervated muscles 27.7 (S.D. 3.4, $n = 12$) and 23.3 ms (S.D. 3.3, $n = 12$). For normal slow-twitch cat soleus muscles the mean time to peak is 76.9 ms (S.D. 8.5, $n = 29$) and the mean half relaxation time 93.4 ms (S.D. 12.1, $n = 18$). The comparable figures following self-reinnervation are 79.0 ms (S.D. 10.1, $n = 25$) and 90 ms (S.D. 17.6, $n = 25$).

This series of data illustrates the difference that may be seen between the times to peak of different fast-twitch muscles (i.e., a mean value of 21.4 ms for FDL and 26.4 ms for FHL). However, the differentiation between fast-twitch and slow-twitch muscle is not based solely on measurements of times to peak (T.P.) and half relaxation times (H.R.), but also on the muscle's post-tetanic characteristics (Brown and Euler, 1938) and the effect of change of temperature on the maximum twitch tension (Buller et al., 1968; Close and Hoy, 1968). For details see Buller and Buller (1980). In both these characteristics self-reinnervated cat muscles behave identically with their normal counterparts.

Following the cross-reinnervation of either the FDL muscle or the FHL muscle with the nerve that formerly innervated the soleus muscle, significant increases ($P < 0.001$) are seen in the mean values obtained for the times to peak and half relaxation of these muscles. For FDL muscles the T.P. lengthened to 62.9 ms (S.D. 4.7, $n = 6$) and H.R. to 73.2 ms (S.D. 5.9, $n = 6$). For FHL the comparable figures were 51.0 ms (S.D. 3.7, $n = 13$) and 55.6 ms (S.D. 10.0, $n = 13$).

Cross-reinnervation of soleus muscles with the nerve that had formerly innervated either FDL or FHL produced significant decreases ($P < 0.001$) in the mean values for T.P. and H.R. of the soleus muscles. Using FDL nerves the mean T.P. of soleus muscles fell to 42.4 ms (S.D. 3.8, $n = 9$) and the value for H.R. to 50.7 ms (S.D. 5.0, $n = 9$). With FHL nerves the figures were 36.5 ms (S.D. 8.0, $n = 12$) and 43.0 ms (S.D. 11.7, $n = 12$).

From these results, all obtained on whole muscles, it may be seen that unambiguous changes in T.P. and H.R. occur following cross-reinnervation, but not following self-reinnervation. However, transformations measured in cross-reinnervated muscle are always incomplete and clear differences in the extent of the conversion may be seen depending upon the nerve and muscle chosen. Cross-reinnervation of the (smaller) FDL muscle by the soleus nerve results in greater changes in T.P. and H.R. than are observed if the soleus nerve is used to cross-reinnervate the FHL muscle. Conversely if the nerve formerly supplying the FDL muscle is used to cross-reinnervate soleus it produces smaller changes in the T.P. and H.R. of that muscle than are observed if the nerve formerly innervating the FHL muscle is used for cross-reinnervation.

The significance of these differences is not yet clear. They do not appear to be time dependent, since cross-reinnervations left for up to 4 years produce (at the level of whole muscle recordings) terminal results similar to cross-reinnervations left for only 6 to 9 months. Perhaps the explanation of the differing results obtained using whole muscle preparations will come from the study of individual motor units in self-reinnervated and cross-reinnervated muscles. Certainly the observations recently published by Bagust et al. (1981) demonstrate that the characteristics of the motor units in cross-reinnervated cat FDL and soleus muscles differ markedly in size distribution from those of normal muscles. A brief report of later experiments (Lewis and Owens, 1979) suggests that, like the changes described above for whole muscles, the alterations at motor unit level are also not time-dependent, at least for intervals of up to 3 years between the cross-reinnervation and the final experiment.

There is a further characteristic of the isometric responses that is consistently altered following cross-reinnervation of soleus, namely the twitch/tetanus ratio (the ratio of the optimal twitch tension to the maximum tetanic tension). The twitch/tetanus ratio for normal fast-twitch muscles is approximately 0.2 and for normal slow-twitch muscle approximately 0.23. Following cross-reinnervation with the soleus nerve the twitch/tetanus ratio of both FDL and FHL remain in the normal range for fast-twitch muscles, but soleus muscle, cross-reinnervated by the nerve which formerly innervated either FDL or FHL produces a twitch/tetanus ratio of 0.16. Again this result is not time dependent, and, to the present time no satisfactory explanation can be given for the observed results.

Essentially all the more recent whole muscle experiments described above do no more than confirm the observations of Buller et al. (1960). Unfortunately the passage of time has led some others to successive approximations in their descriptions of what has actually been observed, so that one may now read that 'cross-reinnervation changes fast-twitch muscle to slow-twitch muscle and vice versa'. Such statements obscure many remaining unsolved problems, the solution of which would enlarge our understanding of nerve–muscle interactions. It is to such problems, some of which have been described above, that further isometric studies should be directed.

2.3. Isotonic studies

Whilst isometric recordings from mammalian skeletal muscles are technically easy to achieve, and have therefore been commonplace for many years, it is only comparatively recently that adequate systems for the accurate recording of isotonic contractions of mammalian muscles at 35 to 37 °C have been devised. It is true that the first unequivocal description of the curvilinear relationship between load and shortening velocity of skeletal muscle by Fenn and Marsh (1935) contained some preliminary observations on cat muscle, but it was not until the studies of rat skeletal muscle by Close (1964) that the stage was set for the full exploration of the isotonic responses of mammalian skeletal muscle. Close also introduced the practice of measuring shortening velocity in units of micrometers per second per sacromere (μm/s/sar), a device which allows comparison between muscles possessing both different shortening speeds and fibre lengths.

If the shortening velocities during the isotonic contractions of a muscle against various loads are fitted with Hill's equation (Hill, 1938) a plot of the rate of shortening (expressed in micrometers per second per sacromere) against load (expressed as a fraction of the muscle's maximum isometric tetanic tension) can provide two characteristics of that muscle. The first, the maximum shortening velocity at zero load (V_s^{max}), is obtained by extrapolating the Hill curve back to the point at which it intersects the velocity axis (typically the ordinate). The second is a measure of the curvature of the velocity/load relationship (a/P_0), which is derived from the Hill equation. It has been suggested by Woledge (1968) that the degree of curvature is related to the metabolic efficiency of the muscle, a higher value of a/P_0 indicating a low efficiency and vice versa.

The isotonic contraction characteristics of normal cat FDL and FHL muscles are remarkably similar when expressed in the manner described above. For FDL the mean value for V_s^{max} is 12.94 (S.D. 1.68, $n = 29$) and that for a/P_0 is 0.336 (S.D. 0.56, $n = 29$). For FHL the comparable mean values are 12.27 (S.D. 1.92, $n = 14$) and 0.278 (S.D. 0.081, $n = 14$). No significant changes are found in the values of either V_s^{max} or a/P_0 for either FDL or FHL following self-reinnervation, the figures for FDL being 12.55 (S.D. 2.55, $n = 6$) and 0.433 (S.D. 0.098, $n = 6$) and for FHL 12.37 (S.D. 1.87, $n = 12$) and 0.310 (S.D. 0.048, $n = 12$).

For normal soleus muscles the mean maximum shortening velocity of zero load is 5.61 (S.D. 1.18, $n = 26$) and the mean value for a/P_0 is 0.144 (S.D. 0.030, $n = 27$). Again no significant changes are seen following self-reinnervation, the figures being 5.73 (S.D. 0.95, $n = 21$) and 0.174 (S.D. 0.041 $n = 21$).

Following cross-reinnervation with the nerve which previously innervated the soleus muscle there are significant changes ($P < 0.001$) in the mean values for both V_s^{max} and a/P_0 obtained from both the FDL and FHL muscles. In the FDL muscles the mean values move to 4.09 (S.D. 0.51, $n = 4$) and 0.131 (S.D. 0.025, $n = 4$), and in the FHL muscles to 8.02 (S.D. 1.65, $n = 12$) and 0.172 (S.D. 0.038, $n = 12$).

For cross-reinnervated soleus muscles the results depend upon the nerve used for reinnervation. If soleus muscles are each cross-reinnervated using the nerve that formerly innervated the fast-twitch FHL muscle there is a significant increase in the muscles' maximum shortening velocities ($P < 0.001$), the mean value for such experiments being 10.56 (S.D. 1.63, $n = 11$). The mean value for a/P_0 becomes 0.259 (S.D. 0.130, $n = 11$), also a significant change ($P < 0.001$).

If the nerve which normally supplies FDL is used to cross-reinnervate soleus, the mean value for the maximum shortening velocity in such experiments is 6.23 (S.D. 2.38, $n = 7$) a result not significantly different from that found for normal or cross-reinnervated soleus muscles, though showing considerable variation from experiment to experiment. However, the mean value for a/P_0 of such cross-reinnervated muscles does change significantly ($P < 0.001$), becoming 0.286 (S.D. 0.078, $n = 7$).

As in the case for the isometric results the changes in contractile behaviour are not easy to interpret, and certainly lend no credence to the general statement that cross-reinnervation converts fast-twitch muscle to slow-twitch muscle and vice versa. It is noteworthy that cross-reinnervation of FDL muscles may result in slower maximum shortening velocities than are observed in normal, or self-reinnervated, soleus muscles.

3. Biochemical correlates

It is not the role of this chapter to describe the biochemical changes which occur in cross-reinnervated skeletal muscle. However, it is perhaps worth commenting upon the difficulty a physiologist may have in obtaining expert (practical!) biochemical help with his problems, and presumably vice versa.

Following the clear demonstration of the changes which can be induced by appropriate skeletal muscle cross-reinnervation in both the cat (Buller et al., 1960; Buller and Lewis, 1965) and the rat (Close, 1965, 1969) it was inevitable that the thoughts of those investigators turned to wondering what changes might be occurring in the contractile proteins and/or the sacroplasmic reticulum which might explain the mechanical findings. However, practical biochemical help was not easily forthcoming. For myself a chance meeting with Professor Wilfried Mommaerts at the International Congress of Physiology in Tokyo in 1965 provided my first opportunity to collaborate with a muscle biochemist (Buller et al., 1969) whilst Russel Close also had to collaborate at a great distance, in his case with Dr. Michael Barany (Barany and Close, 1971). Since that time further international collaborations have been set up by others, but it has never seemed to me an ideal method of working long-term.

There is an adage, which I have heard attributed to A.V. Hill, which states that if a scientist wishes to answer a biological problem he is wise to spend time selecting the biological preparation which is optimal for his experiments. Unfortunately, in the field of muscle biochemistry, the skeletal muscle used is often that which has been poorly studied by the physiologists, or is of such a heterogeneous nature (physiologically speaking) as to render the physiological significance of the biochemical results difficult to interpret. Today the terms 'red' and 'white' muscle are still used in some biochemical laboratories in the belief that the colour differentiation carries some precise indication of the physiological characteristics of the muscle. Again, much biochemical research is carried out on rabbit muscle, a species used infrequently by physiologists studying mammalian muscle. Are the two camps, physiology and biochemistry, so firmly entrenched that some coming together is not possible? Close collaboration certainly seems desirable if rapid progress is to be made with the remaining problems which confront researches in the field of the neural regulation of contraction.

4. Patterns of nerve activity

4.1. Introduction

Following the original observations on the changes in contractile characteristics which followed the reinnervation of fast-twitch muscle by motor nerve fibres that had previously innervated slow-twitch muscle, and vice versa, two hypotheses were advanced to explain these results. The first was that the motor nerve exerted a trophic influence on the skeletal muscle fibres it innervated. The second was that the pattern of nerve impulses reaching the muscle fibres determined the characteristics of the contractile response. It should be noted that these two postulates were not, and are not, mutually exclusive.

Whilst in a number of experiments in which it is claimed that motoneurones have remained electrically silent, changes have been produced in the contraction characteristics of the silent muscles (e.g., Eldridge and Mommaerts, 1980), it has to be admitted that the search for the trophic factor/s, or 'mysterine' as it has irreverently been called by Daniel Drachman, has not produced startling results.

In contrast there can no longer be any doubt that the pattern of nerve impulses (or the pattern of direct electrical stimulation) has a profound effect upon the biochemical and contractile characteristics of mammalian skeletal muscle.

4.2. Normal activity patterns

Considering the time that has elapsed since the pioneering work of Adrian and Bronk (1929) in the field now known as electromyography, it is surprising how little is known about the quantitative electrical activity of skeletal muscle in the execution of willed and reflex movements. Not that the electrical activity of muscle is invariably linearly related to either force generation or movement (shortening or lengthening). Nevertheless, what is known is essentially qualitative rather than quantitative (cf., Basmajian, 1974). It has to be admitted that the technical difficulties in quantitation are considerable, cross-talk from one muscle to another (from antagonist as often as from synergist) is a perennial problem which, to the present, has not been unequivocally overcome. It is therefore remarkable that much of the work that is described below, and is of undoubted importance, is based on the electromyographic evidence obtained by Gerta Vrbová from a small series of experiments on rabbits. In the earlier series of experiments aseptic operations were performed on one hind limb of each animal to cut the Achilles tendon. This produced a 'tenotomy' of the slow-twitch soleus muscle on that side, which silenced the normally continuous, approximately 10/s electrical activity of the soleus muscle fibres. However the tenotomised soleus muscle was reported to respond to postural alterations of the animal with short high frequency bursts of electrical activity (Vrbová, 1963a). Following tenotomy in the rabbit there is a rapid degeneration of the soleus muscle (McMinn and Vrbová, 1962) and it would be desirable to have confirmation that the high frequency bursts of electrical activity really did emanate from the remaining soleus motor units which, at least prior to tenotomy, had been of the slow-twitch type.

In a subsequent paper Vrbová (1963b) described a marked shortening of the isometric time to peak of the tenotomised soleus muscles, claiming, perhaps dubiously, that the operation had converted 'slow muscle to fast muscle'. Be that as it may, it was on such evidence that the first chronic stimulation experiments were undertaken by Gerta Vrbová with results that are now as well known as the cross-reinnervation experiments of Eccles.

4.3. Stimulation experiments

The first full description of the consequences of long-term stimulation on mammalian skeletal muscle were reported by Salmons and Vrbová (1969). They stimulated normally innervated muscle, and the presence of an intact innervation made it possible to distinguish between those changes which could be attributed solely to altered muscle activity and those which might be induced by neurotrophic factors resulting from increased motoneurone activity.

In an effort to dissociate these two possibilities Lømo and his colleagues (Lømo et al., 1974, 1980) devised an ingenious method for the long-term stimulation of chronically denervated rat soleus muscles. Considering the different mean contraction times of the rat and rabbit soleus muscles (approximately 34 ms for the rat and 64 ms for the rabbit) it is remarkable that Lømo chose to use stimulation rates similar to the discharge rates that had been reported for the soleus muscle of normal rabbits, and used by Vrbová in her rabbit stimulation experiments. Nevertheless the proof of the pudding was in the eating, and Lømo and his colleagues have produced convincing evidence that the pattern, as opposed to the total number, of nerve impulses is an important determinant of the isometric contractile properties of mammalian muscle.

Both Vrbová and Lømo reported isometric measurements in their results; more recently Buller and Pope (1977) have demonstrated that following prolonged stimulation at 10 Hz. the maximum rate of shortening at zero load of cat FDL muscle is greatly reduced, and the value of a/P_0 is also reduced.

Chronic stimulation experiments have also been shown to produce marked biochemical changes in muscle, and it is interesting to note that the biochemical correlates were again obtained by international collaboration, Salmons working with Sreter and Vrbová with Pette.

5. Conclusion

To the casual observer the original observations of Eccles may appear to be satisfactorily and totally explained, certainly as far as mechanical studies go, by two subsequent series of experiments. The first was the studies by Close on the isometric and isotonic responses of normal, self-reinnervated and cross-reinnervated muscle. The second was the observations by Lømo and his colleagues on the effects of different patterns of chronic stimulation on denervated soleus muscles. Both sets of experiments were undertaken on rats. Certainly these experiments should be carefully studied by any worker entering this field since they are models of well-designed and executed investigations.

However, certain problems remain, and I hope at least one of these has been highlighted by the presentation in Sections 2.2 and 2.3 of this paper of detailed data obtained from two different fast-twitch muscles cross-reinnervated by soleus nerve and vice versa. The results for the two muscles cross-reinnervated by the same nerve are not identical. To take but one example of the discrepancies. If the nerve previously supplying FDL is used to cross-reinnervate soleus no significant increase in maximum shortening velocity is obtained, whereas if the nerve previously supplying FHL is used a marked increase is observed. (A similar increase is observed in the cat if the nerve previously supplying extensor digitorum longus is used. This was the fast-twitch muscle chosen by Close for his experiments in the rat.)

In the chronic stimulation experiments by Lømo and his colleagues gross changes in the twitch/tetanus ratio were produced by some stimulation regimes, but these have not been commented upon, let alone explained. It is possible that by more detailed variation of the stimulation patterns a more precise match with normal muscle could be obtained, but this has yet to be achieved.

It is also surprising that since chronic stimulation experiments have provided such marked changes in contractile response, no attempt has been made to record the electromyograms of cross-reinnervated muscles.

Remembering the heterogeneity of the motor units of the muscles that have been used in the rat, rabbit and cat experiments one cannot avoid the conclusion that unambiguous answers will only come from the study of individual motor units. The isometric responses of single units have now been measured by a number of investigators notably, for self and cross-reinnervated muscle, by D.M. Lewis and his colleagues. More work needs to be done on isotonic studies before the formidable task of recording the in vitro discharge pattern of single units over long periods of time is undertaken, but only by such work will certain answers be obtained.

Surveying the progress that has been made in our understanding of the influence of the motor nerve on the contractile properties of skeletal muscle since the chance observation by Eccles twenty-two years ago, it is difficult to avoid the conclusion that we have reached our present position by a series of fortuitous steps. Whilst, for whole muscles, there is a broad similarity between self-reinnervated slow-twitch muscle, cross-reinnervated fast-twitch muscle and appropriately stimulated denervated slow-twitch muscle, there remain sufficient differences between these three to warrant more detailed examination of precisely what has, and has not been achieved and to view with scepticism any claim that slow-twitch muscle can be completely converted to fast-twitch muscle or vice versa.

References

Adrian, E.D. and Bronk, D.W. (1929) J. Physiol. (London) *67*, 119–151.
Banu, G. (1922) Récherches Physiologiques sur le Developpement Neuromusculaire Chez l'Homme et l'Animal. Maretheux, Paris.
Bagust, J., Lewis, D.M. and Westerman, R.A. (1981) J. Physiol. (London) *313*, 223–235.
Barany, M. and Close, R.I. (1971) J. Physiol. (London) *213*, 455–474.
Basmajian, J.V. (1974) Muscles Alive, Their Function Revealed by Electromyography. Williams and Wilkins, Baltimore.
Bessou, P., Emmet-Denand, F. and Laporte, Y. (1963) C.R. Acad. Sci. Paris, Ser. D. *256*, 5625–5627.
Brown, G.L. and Euler, U.S. (1938) J. Physiol. (London) *93*, 39–60.
Buller, A.J. and Buller, N.P. (1980) The Contractile Behaviour of Mammalian Skeletal Muscle, Carolina Biology Reader No. 36.
Buller, A.J., Eccles, J.C. and Eccles, R.M. (1960) J. Physiol. (London) *150*, 417–439.
Buller, A.J., and Lewis, D.M. (1965) J. Physiol. (London) *178*, 343–358.
Buller, A.J., Ranatunga and Smith (1968) J. Physiol. (London) *196*, 82 P.
Buller, A.J., Momaerts, W.F.H. and Seraydarian, K. (1969) J. Physiol. (London) *205*, 581–597.
Buller, A.J. and Pope, R. (1977) Phil. Trans. R. Soc. London Ser. B: *278*, 295–305.
Burke, R.E., Levine, D.N., Tsairis, P. and Zajac III, F.E. (1973) J. Physiol. (London) *234*, 723–748.
Close, R. (1964) J. Physiol. (London) *173*, 74–95.
Close, R. (1965) Nature (London) *206*, 831–832.
Close, R. (1969) J. Physiol. (London) *204*, 331–346.
Close, R. and Hoy, J.F.Y. (1968) Nature (London) *217*, 1179.
Cooper, S. and Eccles, J.C. (1930) J. Physiol. (London) *69*, 377–385.
Denny-Brown, D. (1929) Proc. R. Soc. London Ser. B: *104*, 371–411.
Eldridge, L. and Mommaerts, W.M. (1980) in Plasticity of Muscle (Pette, D., ed.) de Gruyter.
Fenn, W.O. and Marsh, B.S. (1935) J. Physiol. (London) *85*, 277–297.
Gordon, G. and Phillips, C.G. (1953) Quart. J. Exp. Physiol. *38*, 35–45.
Hess, A. and Pilar, G. (1963) J. Physiol. (London) *169*, 780–798.
Hill, A.V. (1938) Proc. R. Soc. London Ser. B: *126*, 136–195.
Kuffler, S.W. and Vaughan Williams, E.M. (1953) J. Physiol. (London) *121*, 289–317.
Lømo, T., Westgaard, R.H. and Dahl, H.A. (1974) Proc. R. Soc. London Ser. B: *187*, 99–103.
Lømo, T., Westgaard, R.H. and Engebretsen, L. (1980) in Plasticity of Muscle (Pette, D., ed.) de Gruyter.
Lewis, D.M. and Owens, R. (1979) J. Physiol. (London) *296*, 111P.
McMinn, R.M.H. and Vrbová, G. (1962) Nature (London) *195*, 509.
Ranvier, L. (1874) Arch. Physiol. Norm. Pathol. (2e ser.) *6*, 1–15.
Salmons, S. and Vrbová, G. (1969) J. Physiol. (London) *201*, 535–549.
Sperry, R.W. (1941) J. Comp. Neurol. *75*, 1–19.
Vrbová, G. (1963a) J. Physiol. (London) *166*, 241–250.
Vrbová, G. (1963b) J. Physiol. (London) *169*, 513–526.
Weurker, R.B., McPhedran, A.M. and Henneman, E. (1965) J. Neurophysiol. *28*, 71–84.
Woledge, R.C. (1968) J. Physiol. (London) *197*, 685–707.

CHAPTER 10

Dependence of peripheral nerves on their target organs

TESSA GORDON

Department of Pharmacology, University of Alberta, Edmonton, Alberta T6G 2H7, Canada

1. Introduction

During embryogenesis, the development of peripheral nerves and of skeletal muscle are independent of each other in the initial stages of differentiation. While the alignment and fusion of myoblasts to form multinucleated cross-striated myotubes occurs in the absence of innervation, maturation and further development of muscle cells is critically dependent on the formation of functional nerve–muscle contacts. Similarly, the early stages of neuronal development with proliferation of neuro-epithelial germinal cells and early differentiation, are unaffected by ablation of peripheral target organs. Once motoneurones have sent growth cones into the periphery, however, further maturation of the axons and their terminals is dependent on the formation of functional connections. Interaction between nerve and muscle is not only necessary in this stage of development but critical for the survival of the two tissues. The developmental aspects of the interdependence of somatic nerves and their target organs have been discussed in earlier chapters and will not be considered further here. Rather the mutual interaction of nerve and muscle will be considered in the adult animal. The influence of target organs on peripheral nerve, muscle in particular, will be considered by describing the response of neurones to injury or axotomy and changes that occur during regeneration and reformation of functional connections.

Changes in axotomized motoneurones may be viewed as dedifferentiation from the mature chemically transmitting cell to an actively growing cell. Evidence will be presented which indicates that the early stages of regeneration and axon growth are determined primarily by inherent properties of the neurones but that, as in the embryonic neurone, further differentiation and maturation will not proceed without the formation of functional connections with peripheral target organs. The growth of sprouts and the elongation of axons recapitulates axon growth during embryogenesis but, in the adult, the mature cell body can maintain the axons in an immature state for long periods of time without peripheral connections. Final maturity, however, is totally dependent on functional contact with the periphery during regeneration in the adult, as it is during embryogenesis.

2. Effects of axotomy

2.1. Wallerian degeneration in the distal nerve stump

Axons have little or no capacity for synthesis of protein or lipids, containing few if any ribosomes (Zelená, 1972). It is now widely recognized that the cell body of neurones is the main site of macromolecular synthesis and that axons depend on their transport systems for supply of their required materials which are synthesized in the soma (Droz, 1969, Droz and Koenig, 1970; Lasek 1970; Lasek et al., 1977). Axons which are physically separated from the cell body after nerve injury, cannot survive. Degeneration of these isolated axons, or *Wallerian degeneration* was described in 1850 (Waller, 1850) and showed that the cell body was essential for the survival of axons and that connection with end-organs cannot support axons in the absence of the cell body. Cajal (1928) suggested that the trophic support of the cell body was of 'a dynamic and not of a material nature' because degeneration appeared to occur simultaneously along the entire isolated stump but, more recent evidence has shown that degeneration progresses in a proximo-distal direction. Isolated axons in the distal stump continue to function for up to seven days depending on the length of the distal nerve stump (reviewed by Lubínska, 1975; Sunderland, 1978). In rat nerves for example, neuromuscular failure is delayed by 2 h for every additional centimeter of distal nerve stump (Miledi and Slater, 1968, 1970). This dependence of survival time on length suggested but did not prove that degeneration spread longitudinally along the nerve stump. Lubínska and Waryszewska provided more direct evidence for longitudinal spread of degeneration in histological studies of the distal nerve stump (quoted by Lubínska, 1975). Numbers of degenerating axons were highest closest to the injury site and decreased along the length of the stump as degeneration spread longitudinally with time until the proximo-distal gradient was finally lost with time (Fig. 1).

The short-term functioning of the axons in the distal stump was ascribed to their nutrition by continuing axonal transport in the isolated stump. The eventual deterioration with ultimate degeneration was ascribed to depletion of nutrients in axons without replenishment from the cell body (Lubínska, 1964, 1975). There is a minor source of axonal protein from the glial cells surrounding axons (Lasek

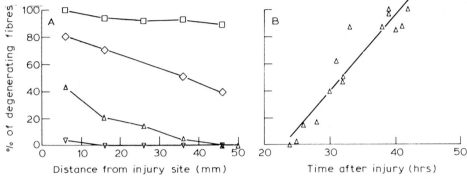

Fig. 1. Numbers of degenerating fibres in the distal nerve stump of sectioned rat phrenic nerves, are expressed as a % of the total fibre content and plotted (A) as a function of distance from the injury, 22 (▽), 26 (△), 30 (◇), and 38 (□) hours after nerve section; and (B) as a function of time after section at a distance of 5 mm from the injury site (replotted from Lubínska and Waryszewska, 1975). Note that degeneration decays over distance and increases as a function of time at any fixed distance from the injury site.

et al., 1977) but this source is insufficient to maintain the mammalian axon in the absence of the cell body and of axonal transport. In invertebrate axons, the glial cells maintain isolated axons somewhat longer (Bittner, 1973). The bulk of axoplasm is normally supplied by slow transport or axoplasmic flow and includes the cytoskeletal proteins: actin, tubulin and subunits of neurofilaments (Hoffman et al., 1975; (Willard et al., 1974). The rapid phase of axonal transport is two orders of magnitude faster than slow transport, 400 mm/day as compared with 1 to 5 mm/day in mammalian peripheral axons (Ochs, 1972; Lasek, 1968), and transports particulate material. This consists primarily of membranous organelles for normal membrane turnover, for transmitter synthesis and secretion, and axonal metabolism (McEwen and Grafstein, 1968; Cuénod et al., 1972; Droz, 1975; Grafstein, 1977).

Fast but not slow transport continues in isolated nerve segments (Dahlström, 1967; Lubínska, 1964, Lasek, 1970). Neuromuscular transmission continues until supply of transmitter is depleted (MacIntosch, 1938; Feldberg, 1943) and fails more rapidly if the isolated terminals are activated (Cook and Gerard, 1931; Gerard, 1932). Chemical transmission usually fails before impulse conduction (Lubínska, 1964). Materials will become depleted in the isolated nerves in a proximo-distal direction at a rate which will vary with transport rate and usage of materials. This depletion would account for anterograde spread of degeneration from the injury site to the terminals (Lubínska, 1975).

The final stages of Wallerian degeneration involve dissolution and disorganization of neurotubules and neurofilaments with final disintegration of axons, their phagocytosis and intense proliferation of Schwann cells (Speidel, 1935; reviewed by Gutmann, 1958; Sunderland, 1978). The final degeneration of the isolated axon is probably due to depletion of materials required for normal turnover on the one hand and on the other, to death of the mitochondria which are not replaced but accumulate at the injured end (Webster, 1962). Failure of oxidative phosphorylation would lead to loss of membrane potential and disruption of chemical and electrical gradients. Calcium ion accumulation in the axolemma causes rapid depolymerization of microtubules and may also be involved in activation of hydrolytic enzymes. In cell cultures in vitro where material is artificially supplied to axons, peripheral stumps severed from growing neuroblasts do not die but sometimes even grow and fuse again with the growing stump (Boeke, 1950).

2.2. Metabolic response of axotomized neurones

Axotomized cells maintain their synthetic ability in the neuronal soma and can survive and regenerate, unlike the isolated axons in the distal nerve stump. Observations that outgrowth of growth cones from the proximal stump of transected nerves led to axonal growth and regeneration, while growth of sprouts in severed distal nerve stumps was shortlived, indicated that continuity with the cell body was a prerequisite for regeneration (Ranson, 1912; Cajal, 1928). Axotomized neurones undergo a number of structural, metabolic and functional changes which have been viewed as an anabolic response appropriate for repair of cellular damage and regeneration of the lost axon (Grafstein, 1977). The changes may also be considered as a process of dedifferentiation to a growing state from a fully differentiated secretory cell.

2.2.1. Chromatolysis and the metabolic response

The cell body of axotomized neurones typically undergoes a number of characteristic morphological alterations, which were first described by Nissl (1982) and have since been extensively studied (reviewed by Hydén, 1960; Liebermann, 1971, 1974). These changes are collectively termed 'chromatolytic changes' and include swelling and migration of the nucleus to an eccentric position, and an increase in nuclear and nucleolar size. The changes refer particularly to the apparent disappearance of the prominent basophilic staining Nissl granules. These granules are ribosome clusters and ordered arrays of rough endoplasmic reticulum which can no longer be visualized when they become disorganized, freeing polyribosomes and even ribonucleotides into the cytoplasm. Disorganization of the ribosome clusters indicated that the morphological changes in the cell body were associated with increased protein synthesis, which suggested that the chromatolytis was an anabolic response of the injured neuron.

With few exceptions, this interpretation is supported by evidence from studies of RNA and protein content in axotomized neurones. These show reduced DNA repression, an increase in RNA synthesis with transfer of RNA from nucleus to cytoplasm, and increased cellular protein content (Brattgard et al., 1957, 1958; Gutmann et al., 1962; Murray, 1973; Watson 1965, 1968, 1970, 1974). Concentrations of enzymes of the oxidative pentose phosphate shunt, which are required for RNA synthesis, are also raised (Harkonen and Kauffman, 1974; Sinicropi and Kauffman, 1979). Changes in RNA and protein synthesis often coincided with the maximal chromatolytic response observed histologically (Cragg, 1970). Chromatolysis was more severe and was observed earlier in neurones where axonal lesions were close to the cell body (Marinesco, 1898). This indicated that chromatolysis could be a metabolic response to reconstitute the length of axon lost. Chromatolysis and RNA and protein synthesis are however not always mutually exclusive. There are injured neurones which neither exhibit a chromatolytic response nor show significant changes in cellular levels of total RNA or protein but still show significant changes in RNA metabolism, and are capable of regeneration. Examples include the rat nodose ganglion and the central process of sensory fibres in the dorsal root ganglion (Anderson, 1902; Cajal, 1928; Carmel and Stein, 1969; Lieberman, 1971).

Recent biochemical data have indicated that the changes in RNA and protein content which have been supposed to underlie the chromatolytic changes, are later manifestations of early changes in RNA metabolism. These early changes occur within hours and precede chromatolytic changes that take place within a few days of injury (Austin and Langford, 1980). Within a few hours of axotomy, the pattern of migration on polyacrylamide gels of rapidly labelled RNA in axotomized vagus nerves differed from ganglionic RNA from intact vagi (Gunning et al., 1977). This changed migration pattern and the marked increase in the uridine incorporation into messenger RNA, indicated that the pattern of production of messenger RNA changed within hours of axonal injury (Burrel et al., 1978). Incorporation of labelled precursors into ribosomal and transfer RNAs increased significantly whether or not there were significant increases in total RNA content and often before the peak of chromatolysis (Kaye et al., 1977). Thus, RNA turnover increases in axotomized neurones within hours of injury, and this turnover is not necessarily associated with an increase of total cellular content of DNA, RNA or protein, or other chromatolytic changes.

The signal for a change in RNA synthesis is still a matter of speculation, but it is clear that the injury alters gene expression, resulting in a change in the composition and amount of transported materials in regenerating as compared with normal adult axons (see Section 2.2.3). The signalling process to the cell body is extremely rapid. Changes that take place in phospholipid synthesis occur within 30 min of peripheral nerve injury at distances of more than 10 cm from the cell body. The rapidity of these changes suggests that there is a retrograde signal moving at velocities of at least 140 mm/h, which is very much faster than any previously reported velocity of axonal transport either anterograde or retrograde (Dziegielewska et al., 1980).

It has been argued that chromatolysis is not a regenerative response but rather a response to the injury of the cell (Cragg, 1970; Engh and Schofield, 1972). The chromatolytic response involves, in addition to changes in nuclear and ribosomal function, sequestration of intracellular organelles in membrane bound vacuoles with ultimate degradation in secondary lysosomes (Lieberman, 1971; Mathews and Raisman 1972; Mathews and Nelson, 1975; Sumner, 1975). The lysosomal response in axotomized neurones resembles in many respects, the changes that occur in secretory cells which follow physiological curtailment of hormonal secretion (Watson, 1976) and the response may be involved in the cessation of neurotransmitter synthesis and release. In the anterior pituitary gland, a characteristic lysosomal response accompanies the decline in synthesis of luteinizing hormone releasing factors by the mammotrophic cells when lactation ceases. The primary lysosomes from autophagic vacuoles and multivesicular bodies within which intracellular organelles are sequestrated and finally destroyed.

2.2.2. Two state model of the motoneurone; growing or transmitting
Neurones exhibit a burst of RNA synthesis in response to nerve section, and also when the regenerating axons make functional nerve–muscle connections. This was shown in experiments in which rat hypoglossal nerves were cut and sutured to an already innervated muscle, where they do not make functional connections. The response to axotomy in this situation was maximal at about 10 days after the operation and could be induced again by a crush injury of the same nerve a few months later. The severed hypoglossal nerve responded a third time with a burst of RNA synthesis when the axons were encouraged to make functional nerve–muscle contacts by cutting the original muscle nerve (Watson, 1970). The hypoglossal motoneurones therefore responded to axotomy, to a second stimulus to grow and to the formation of functional nerve–muscle contacts. Watson interpreted the second burst of RNA synthesis to nerve crush of the already axotomized hypoglossal nerves, as the transformation of the nerve's metabolism from the 'resting nontransmitting mode' to a second 'seeking mode' of the recently axotomized neurone. This burst of RNA synthesis after a crush of the axotomized nerve may be responsible for the enhanced regeneration after two consecutive injuries (McQuarrie and Grafstein, 1973; McQuarrie et al., 1977) and for earlier findings of Gutmann (1942) that the success of regeneration increased after double crush of a peripheral nerve. The third RNA burst when nerves made functional connections indicated that the nerves increased their synthetic activity possibly to supply transmitters and their precursors to developing nerve terminals.

Thus axons appear to exist in essentially two states, that of the *growing state or seeking mode*, or in the *resting or transmitting state* and neurones respond to change of state (Watson, 1976). Neurones showed chromatolytic changes when axons with intact peripheral contacts were induced to grow and sprout after an intramuscular injection of botulinum toxin (Watson, 1969). Sprouting induced by botulinum toxin occurs during the course of muscle paralysis which follows a rapid and potent block of acetylcholine release from the nerve terminals (Duchen and Strich, 1968). The chromatolytic response may reflect the axonal growth and/or reestablishment of transmission in the poisoned axons because, the axonal sprouts rapidly form functional nerve–muscle contacts, which restore transmission and muscle function. Recovery of neuromuscular transmission after botulinum toxin poisoning has a similar time course and resembles the development of functional transmission between regenerating axons and denervated muscle (Tonge, 1974a,b). This suggests that the new functional contacts formed by sprouts replace the toxin blocked terminals, and that the new terminals restore neuromuscular transmission.

Neuroglial cells, such as astrocytes and oligodendrocytes, hypertrophy after axotomy of motoneurones and also show changes in RNA synthesis. These glial cells show a biphasic response similar to motoneurones, with an increase following axotomy and a second peak during reinnervation. They also respond to botulinum poisoning of the motoneurones that they surround (Watson, 1972). Watson suggested that these glial responses may be related to changes in the dendrites of the axotomized motoneurones because glial responses occurred simultaneously with detachment of boutons from the axotomized neurones (Section 2.3.1) and did not occur a second time in response to a second injury. When the axotomized cells reinnervate muscle, there is an expansion of the dendritic field which may be associated with a second phase of astrocytic response.

2.2.3. Fast axonal transport in axotomized neurones
There are many similarities between glandular secretion and the release of neurotransmitters from neurones. Much of the neurone's metabolism, as in secretory cells, is normally committed to secretion with elaboration of membranes for storage and transport of neurotransmitters (reviewed by Dahlström, 1971; Heslop, 1975; Saunders, 1975). Neurosecretion involves many or all of the steps of synthesis, packaging and release described for glandular secretion with the specialization of the axonal transport systems in neurones appropriate for the axonal length (Droz, 1969, 1975; Droz et al., 1975; Palade, 1975; Holtzman, 1977). The rough endoplasmic reticulum in the neurone cell body and dendritic shafts is the major site of macromolecular synthesis, as in gland cells, and material destined for the axolemma, the axon and its terminals is transported by axonal transport systems (Droz, 1975; Grafstein and Forman, 1980). Most proteins and lipids, including transmitters and their percursors, are associated with membranes and are transported by fast axonal transport. Transport of membrane associated materials has been considered to be primarily secretory in function in the anterograde direction and lysosomal in the retrograde direction (Droz, 1975; Schwartz, 1979).

When the peripheral nerve is cut, transmitters and their precursors initially accumulate at the site of injury. This accumulation has been used to determine rates of transport (Lubínska, 1964; Dahlström, 1965, 1967; Brimijoin, 1975; Heslop, 1975; Saunders, 1975). However, the amounts of the transmitters and their precursors which are transported, fall within days after axotomy until they are barely measurable some time after axotomy (e.g., Boyle and Gillespie, 1970; Cheah and Geffen, 1973; Frizell and Sjöstrand, 1974a; Reis and Ross, 1973). Transport rates were unaltered suggesting that depression of synthesis is responsible for reduced transport of the neurosecretory materials. The extent and duration of the depressed synthesis appears to be determined by the type of injury. For example, depressed synthesis and transport of amine storage granules in adrenergic neurones was less marked in axotomized axons following nerve crush than after nerve section or ligation. The duration of depression was also shorter (Karlstrom and Dahlström, 1973). Transport of acetylcholinesterase declined in cholinergic neurones and did not recover if regeneration was impeded by nerve section (O'Brien, 1978; Heiwell et al., 1980). Although reinnervation was not followed in these experiments, reinnervation normally occurs earlier and is more successful after nerve crush than following nerve section (Gutmann and Sanders, 1943; Davis et al., 1978). This may account for the less severe effects of nerve crush compared to nerve section on synthesis of transmitter-related proteins. Synthesis and transport may be expected to recover more quickly in axons which make functional connections earlier and more readily after nerve crush injury as compared to after nerve section. The findings suggest that metabolic direction of the regenerating cell moves away from secretion with reduction in synthesis and transport of transmitters until connections are remade.

The anabolic response of the axotomized cell has been presumed to reflect an increased synthesis of materials required for axonal growth. However, direct evidence for alteration in amounts and composition of materials transported by injured neurones has been difficult to obtain. Fast transport conveys relatively small amounts of a large number of different proteins, few of which have been identified other than transmitters and related proteins. Increased incorporation of radioactivity into fast transported proteins has been documented in pulse labelled axotomized neurones in a number of different systems (e.g., Grafstein and Murray, 1969; Frizell and Sjöstrand, 1974a,b; Griffen et al., 1976; Elam and Maxwell, 1978; Bisby, 1978). But these experiments cannot determine whether this increase is due to increases in precursor uptake, changes in the rate of protein synthesis or alterations of the quantities of protein transported by fast axoplasmic flow during regeneration. There is now evidence of specific changes in the spectrum of rapidly transported proteins in goldfish optic nerve after axotomy (Benowitz et al., 1981). The spectrum of labelled protein shifted continuously during axon regeneration, which indicated that the pattern of synthesis and/or degradation of fast-transported proteins changes during the course of regeneration. Some of these changes are probably related to axonal growth and some to reformation of connections, since reinnervation of target cells would have occurred in the time course of 62 days of the experiments. Bisby (1980) has recently detected some changes in composition of transported proteins in regenerating rat motor nerves, particularly in the ratio of 2 readily identifiable low molecular weight polypeptides. These changes were maximal about 14 days after crush injury and returned to normal by day 42 when there was clear evidence of reinnervation (Bisby, 1980). If reinnervation was prevented by ligating the proximal nerve stump, the ratio of the 2 polypeptides did not return to normal values (Bisby, 1981).

These changes in proteins transported are consistent with histological demonstrations of polyribosomal dispersion and the changes in RNA metabolism described (Section 2.2.1). Putrescine is transported by regenerating as well as developing axons but not by normal adult axons (Fisher and Schmakolla, 1972; Ingoglia et al., 1977). Putrescine is one of the polyamines which are widespread in different cell types and is generally associated with actively growing and proliferating cells (Russell and Snyder, 1968). Polyamines are found in high concentrations in developing brain (Sieler, 1973). Although their function in regeneration is not understood, their presence is another indicator of cellular growth. Amounts of phospholipids transported by fast transport also change after axotomy, almost doubling in quantity within 10 h of nerve crush injury (Dziegielewska et al., 1980). There appeared to be a burst in synthesis which raised the levels of phospholipids in the axons above normal for up to 20 h. Almost 90% of the transported phospholipid was phosphotidylcholine, a major constituent of cell membranes (Bretcher and Raff, 1975). The phospholipid would be required for immediate sealing of damaged axons and for subsequent axon regeneration.

Although fast axonal transport has been considered primarily as secretory in function and the analogy has been made between neurosecretion and glandular secretion, transport systems are not necessarily linked with secretion. Different materials are directed along different processes in the same neurone. In adrenergic neurones, for example, noradrenaline containing vesicles and associated dopamine-β-hydroxylase, chromagranin A and ATP are transported down the axon to the nerve terminals but they are not transported in the dendrites, which have their own intrinsic transport system (Schubert et al., 1972). Transport in the central process of sensory neurones in the dorsal root ganglia but not in the peripheral axon is related to neurosecretion (Anderson and McClure, 1973).

Many of the changes of the axotomized cell can be mimicked by blocking axonal transport with colchicine in intact cells (Purves, 1976). These changes may be related to the isolation of the terminal connections by chemical means, in the way that axotomy does physically. Synaptic transmission is critically dependent on fast axoplasmic transport. When axoplasmic transport is blocked by colchicine or a reduction in the temperature, chemical transmission fails before impulse conduction (Schwartz, 1979). The changes are unlikely to be due to loss of trophic signals or material from the target organs (cf., Cragg, 1970) because changes in RNA, proteins, and phospholipid synthesis are triggered by a second injury of axotomized neurones (Watson, 1969; Dziegielewska et al., 1980). The changes may be linked to loss of transmitting terminals and/or of the specialized nerve endings and to the new requirements of the growing axons.

2.2.4. Cytoskeleton: slow axoplasmic flow in axotomized neurones
A shift in metabolic direction of the axotomised cell also involves a shift in proportion of the fibrillar proteins which make up the cytoskeleton. The fibrillar proteins can be recognised and distinguished by electron microscopy as microfilaments or actin containing filaments of approximately 7 nm in diameter; microtubules which are tubulin polymers and appear as hollow cylinders 20 to 30 nm in diameter; and neurofilaments which are seen as rods 10 nm in diameter and are composed of neurofilament triplet protein (Hoffman and Lasek, 1975; Schwartz, 1979; Bray and Gilbert, 1981). The microfilaments and neurotubules are orientated longitudinally along the proximo-distal axis of nerve processes with neurofilaments being the most prominent – even visible with the light microscope (Droz et al., 1973). In the mature axon, the number of neurofilaments far exceeds the number of neurotubules (Friede and Samorajski, 1970). Their number may determine axon size since the cross-sectional area of large axons is directly proportional to numbers of axonal neurofilaments (Forman and Borenberg, 1978; Willard et al., 1979).

When a mature axon is cut, the amount of neurofilament triplet protein transported down the axon is decreased so that the microtubular and microfilamental proteins increase relative to the neurofilaments (Hoffman and Lasek, 1980). This change may be regarded as a process of dedifferentiation to an immature state akin to the embryonic neurone where the cytoskeleton of the growth cone and newly formed axon is composed almost entirely of microtubules and microfilaments (Peters and Vaughn, 1967; Yamada et al., 1971). It is only when the embryonic axon reaches a peripheral end organ that neurofilaments appear in the axons for the first time. In the adult the proximal end of the cut axon is converted into a growth cone, which is like the embryonic growth cone. The new growth cone is characterised by an extensive network of smooth endoplasmic reticulum in its central core and the presence of actin containing filaments and microtrabeculae. Microtubules and neurofilaments are noticably absent or few in number (Del Cerro, 1974; Kuczmarski and Rosenbaum, 1979; Jockusch and Jockusch, 1981).

Since growth cones in regenerating axons very closely resembled the cones in developing neurones, Cajal (1928) considered the growth of axons in both situations as the same phenomenon. Lasek (1981) suggested that the growth cone in regenerating axons arises specifically as a result of reorganisation of the fibrillar proteins which accumulate in the cut end of the axons during the latent period before regenerative axonal elongation. He also suggested that regeneration is supported by increased transport of tubulin and actin relative to neurofilamental protein.

A combination of pulse-labelling experiments and two-dimensional gel electrophoretic analysis of the radiolabelled fibrillar proteins have shown that the fibrillar proteins are transported distally in normal and regenerating nerves in two distinguishable waves of radioactivity, slow component a and b (SCa and SCb) (Black and Lasek, 1980; Lasek and Hoffman, 1976). Tubulin and the neurofilament triplet proteins represent the bulk of the protein moving in SCa at a rate of about 1 mm/day in mammalian neurones (Hoffman and Lasek, 1975). SCb contains at least 20 major proteins including actin (Black and Lasek, 1979; Hoffman and Lasek, 1976), clathrin (Garner and Lasek, 1980), myosin-like protein (Willard, 1977), matrix proteins (Black and Lasek, 1980) and tubulin (Lasek and Hoffman, 1976). These also move as a coherent peak at an average velocity of 2 to 5 mm/day (Lasek and Hoffman, 1976); Black and Lasek, 1980). Rate of slow transport does not change in regenerating axons. Rather, neurones alter their synthesis of fibrillar proteins with increased incorporation of SCb proteins and decrease in supply of neurofilament protein in SCa. Transport in regenerating adult axons then resembles more closely the transport in developing axons, with the proportion of fibrillar proteins biased toward actin and tubulin with less neurofilamentous protein (Hoffman and Lasek, 1980).

Protein which is required immediately for regeneration, could not arrive at the cut end of the axon at the transport rate of 1 to 5 mm/day. Consequently, it appears that there is sufficient material present locally in the axon for the outgrowth of the cones. Elongation of the axon, however, would require supply of material at the rate of axonal growth. The rate of regeneration in mammals of 3 to 4 mm/day, established by the pinch test or by advancement of axonally transported radioactivity within the nerve, closely corresponds with the rate of transport of SCb proteins (Gutmann et al., 1942; Lasek and Hoffman, 1976; Black and Lasek, 1979; Hoffman and Lasek, 1980). In sensory neurones different rates of regeneration of the central and peripheral processes correspond with different rates of SCb transport (Wujck and Lasek, 1980). The rate of regeneration of axons declines from birth to adulthood and Black and Lasek (1979) have recently found a corresponding decrease in transport rate of SCb proteins. Therefore, cytoskeletal structures located near the tips of the regenerating sprouts appear to be moving at the same velocity as the rate of axonal elongation (Forman and Berenberg, 1978; Black and Lasek, 1980).

Hoffman and Lasek (1980) suggested that axons grew after peripheral nerve section because the axotomized cells could not break down cytoskeletal proteins and the proteins continue to advance forward. The neurones regenerate at the rate of advancement of the proteins. Weiss and Mayr (1971) had proposed earlier that the cytoskeletal proteins were constantly synthesized in the cell body and degraded in the nerve terminals of intact neurones. Waves of radioactivity of the slowly transported proteins advance as sharp coherent peaks even months after pulse labelling, suggesting that little protein is deposited en route along the axon and that the proteins could be transported as organized structural complexes to the nerve terminals (Lasek, 1981; Black and Lasek, 1980). The level of radioactivity in SCa proteins declines sharply in the terminals showing that tubulin and neurofilament protein are degraded rapidly once they arrive at the terminals (Lasek and Hoffman, 1976; Lasek and Black, 1977). This is consistent with the presence of fewer microtubules and neurofilaments in nerve terminals than in preterminal axons (Peters et al., 1976). When axons are cut, SCa proteins accumulate at the cut end presumably because they are not degraded (Lasek and Black, 1977). SCb proteins in contrast to SCa proteins appear to be degraded very slowly since they normally remain in the nerve terminals for long periods of time. The SCb proteins may constitute the fine filamentous lattice of microfilaments and trabeculae in the nerve terminals where they, particularly actin, have been implicated in normal neurosecretion.

Lasek and Hoffman (1976) extended Weiss and Mayr's (1971) original suggestion that the axonal cytoskeleton is dissembled in the axon terminals and put forward a model in which calcium ions play a central role in regulating depolymerization of microtubules in the terminals. Calcium also activates an endogenous protease which cleaves neurofilament protein. This protease is present in invertebrate giant axons and may dissemble neurofilaments at the nerve terminals (Gilbert et al., 1975). Axon growth could be viewed according to this model simply as the extension of the cytoskeleton causing axonal elongation because the fibrillar proteins are not broken down. However, the response to axotomy is clearly more complex. The cell body responds to axotomy by altering the amounts of fibrillar proteins transported down the axon. This implies a feedback from the end of the axon, in the transmitting or growing state.

Increase in tubulin and actin relative to neurofilament proteins is appropriate for growth since actin is required for the motility of the growth cone and for axonal elongation (Sanger, 1975). Increase in fibre diameter of developing neurones is associated with a disproportionate increase in neurofilament content. Since cross-sectional area of mature axons is proportional to numbers of neurofilaments, increased supply of neurofilaments may be responsible for the increase in axonal diameter during maturation of developing or regenerating neurones. There is recent evidence to suggest that the decrease in neurofilamental protein transported by axotomized neurones may account for the decrease in their axonal diameter. Diameter decreased in a proximo-distal direction after axotomy at the same rate as the rate of transport of SCa proteins (Hoffman et al., 1980). This implied that axonal diameter was directly proportional to the amounts of neurofilaments supplied by SCa transport. Therefore it appears that the axotomized cell can shift its metabolic direction from stability to plasticity with regard to its fibrillar proteins. Physico-chemical properties of the fibrillar proteins which are consistent with this hypothesis are discussed in depth by Lasek (1981).

2.3. Physiological parameters of axotomy

2.3.1. Synaptic transmission

Evidence for shift in metabolic direction of axotomized neurones from neurosecretion has come from studies of reduction in transport of transmitter related proteins in the axons (cf., Section 2.2.3). This shift is also evident in the dendrites and soma where synaptic efficacy between presynaptic terminals and axotomized neurones declines as a result of depressed postsynaptic chemosensitivity, loss of postjunctional thickening and loss of functional synaptic contacts.

Synaptic depression is a very generalized finding for transmission between presynaptic neurones and axotomized neurones. It has been described as a reduction in amplitude of evoked monosynaptic responses in motoneurones in cats (Eccles et al., 1958; Downman et al., 1953; McIntyre et al., 1959; Kuno and Llinas, 1970b; Mendell et al., 1976) and frogs (Shapovalov and Grantyn, 1968; Farel, 1980) as well as sympathetic and ciliary ganglia in mammals and amphibia (e.g., Acheson and Remolina, 1955; Hunt and Riker, 1966; Purves, 1975; Brenner and Johnson, 1976). Functionally, synaptic depression results in a marked depression of firing of axotomized neurones within the first month after axotomy (Acheson et al., 1942; Gordon et al., 1980a). At the ultrastructural level, the number of synaptic profiles onto axotomized neurones declines sharply in the first few days following axon injury concurrent with the typical changes of chromatolysis (Blinzinger and Kreutzberg, 1968; Hamberger et al., 1970; Mathews and Nelson, 1975; Purves, 1975). This so-called synaptic disjunction occurs with roughly the same time course as electrophysiological depression in axotomized (Purves, 1975) or colchicine treated postganglionic neurones (Purves, 1976; Pilar and Landmesser, 1972) so that loss of synaptic contacts appears to be the major factor responsible for synaptic depression.

Following axotomy of cat motoneurones, three stages of synaptic dislocation or disjunction could be discerned electrophysiologically. There was no change in unitary EPSP amplitude or duration initially. EPSP amplitude was reduced and duration prolonged in a second intermediate stage but 1a connectivity, assessed by spike triggered averaging of the 1a afferent fibre, was unchanged. Finally, 1a connectivity was reduced in the third stage (Mendell et al., 1976). The gradual loss of synaptic contact which leads to loss of connectivity may be related to changes taking place in the axotomized cell. Synaptic thickenings below the synaptic contacts are lost from the postsynaptic membrane before the boutons become detached (Sumner, 1975; Mathews and Nelson, 1975). This suggests that changes in the axotomized cell precede and may be responsible, at least in part, for the gradual loss of boutons. Many of the detached boutons which are unapposed by postsynaptic elements become enveloped by glial cells which may also play some role in the separation of the bouton from the postjunctional membrane (Mathews and Nelson, 1975; Torvik and Skjorten, 1971; Sumner and Sutherland, 1973).

Loss of the postsynaptic thickening may occur in conjunction with or may possibly be related to the reduction of chemosensitivity of the axotomized neurones (Brown and Pascoe, 1954; Brenner and Martin, 1976). This reduction in chemosensitivity is consistent with a shift of the metabolism of the axotomized cell from chemical transmission to growth. The dendritic trees also emit growth cones like the damaged axons, and extensive sprouting of the trees has been recently demonstrated in isolated invertebrate ganglia with intracellular injection of lucifer yellow (Pitman and Rand, 1981). This sprouting is accompanied by a dendritic retraction or decline in width of the dendrites (Sumner and Sutherland, 1973), which is reversed when the axons make peripheral connections (Sumner and Watson, 1971).

Thus, decreased synaptic efficacy in axotomized cells can be accounted for by bouton displacement as a result of loss of postsynaptic substance and/or due to displacement by glial cells and by hyposensitivity of the postjunctional membrane. In experiments in which motoneurones have been axotomized by peripheral nerve section, at least some of the reduction in efficacy is also due to atrophy of afferent terminals with decline in release of transmitter (Knyihar and Csillik, 1976). Section of dorsal roots produced profound depression of monosynaptic reflexes even in intact motoneurones (Eccles and McIntyre, 1953; Gallego et al., 1979b). Reduced synaptic efficacy after cutting of the sensory fibres could not be mimicked by silencing intact sensory axons with localized tetrodotoxin block (Gallego et al., 1979b) or by tenotomy (Hník et al., 1963; Beránek and Hník, 1959). Efficacy of monosynaptic synapses onto motoneurones was increased and reflex output enhanced after silencing sensory fibre activity in contrast to reduction in the efficacy and reflex output after section of sensory fibres.

Some of the depression in synaptic transmission is partially compensated by an apparent hyperexcitability of axotomized motoneurones. Monosynaptic reflexes were either lost or very depressed but polysynaptic reflex discharges were even larger than normal (Campbell, 1944). This excitability can be partly accounted for by decrease in rheobasic current for axotomized motoneurones and the emergence of dendritic spikes (Kuno and Llinas, 1970a). Loss of monosynaptic reflexes and enhancement of polysynaptic reflexes may indicate selective loss of some terminals as suggested by ultrastructural studies (Sumner and Sutherland, 1973; Sumner, 1975) and by electrophysiological studies (Kuno and Llinás, 1970b).

2.3.2. Conduction velocity and nerve fibre diameter
Change in fibre diameter after peripheral nerve injury was first described histologically by Greenman (1913) who noted that fibre diameter decreased in the central nerve stump and that atrophy was greatest when reinnervation of peripheral endorgans was delayed or prevented. Most peripheral nerves contain myelinated fibres of different sizes which correspond to 2 to 3 classes of sensory fibres and 2 classes of motor fibres. The distribution of total fibre diameter (axon and myelin thickness) is normally bimodal (Boyd and Davey, 1968). For the 2 to 20 μm range of fibre sizes in cat and rabbit nerves (the range is smaller for smaller animals) there is about a 10-fold range of conduction velocity (Hursh, 1939). The normal bimodality is typically lost after axotomy. The fibre diameter distribution becomes unimodal as the largest fibre group merges into the smaller peak (Gutmann and Sanders, 1943; Weiss et al., 1945). Latencies of compound action potentials recorded on the proximal stumps of injured peripheral nerves increased. This showed that axon conduction velocities decreased consistent with reduction in fibre diameter (e.g., Acheson et al., 1942; Gutmann and Holubář, 1951; Király and Krnjević, 1959; Cragg and Thomas, 1961; Davis et al., 1978). Electrophysiological measurements were made in a number of different animals in each study and results were somewhat variable. Normal and even increased conduction velocities were reported by Sanders and Whitteridge (1946), 6 and 16 weeks after crush injury of rabbit nerves. Yet when Cragg and Thomas (1961) repeated the experiments, they found that similarly injured nerves conducted more slowly than normal until at least 150 days after crush when conduction velocities recovered to normal levels, at which time recovery of fibre diameters proceeds (Gutmann and Sanders, 1943).

Variability of the effects of nerve injury may have been due to differences in location and nature of the injury to the nerves, and in the extent and time course of reinnervation of target endorgans for each nerve. In addition, many of the measurements of axon diameter or conduction velocities were made when reinnervation would have been expected from regeneration rates of 3 to 4 mm/day for mammalian nerves (Gutmann et al., 1942). Reinnervation was not followed even qualitatively in most of these studies, and it is not possible to distinguish the effects of isolation of the nerves from the periphery from the effects of reformation of peripheral connections.

Recently, chronic extracellular electrodes have been used in studies of nerve injury in which each animal has served as its own control. Compound action potentials were recorded before and after nerve injury on the same peripheral nerves (Davis et al., 1978; Stein et al., 1980; Gordon and Stein, 1980; Gordon et al., 1980a). The time course of changes in the potentials were followed in nerves that were either crushed or sectioned. EMG activity and muscle contractions evoked by stimulation of the nerve proximal to the site of injury were recorded in all experiments to determine reinnervation, quantitatively. These experiments showed conclusively that compound potential amplitude, which is approximately equal to the fourth power of the latency of the potential, declined after axotomy and did not recover unless peripheral connections were remade (Davis et al., 1978). Amplitude declined and latency increased most rapidly immediately after axotomy. Decline in amplitude with time followed a simple exponential with a time constant of 1 to 2 months, and if the axons were prevented from remaking peripheral connections, the amplitude declined to an asymptotic level of approximately 25% of preoperative levels (Fig. 2). These findings confirmed earlier electrophysiological and histological findings that severed axons in the central stump undergo atrophy, but also showed for the first time that atrophy is most severe immediately following nerve injury but eventually levels off at a lower stable level. Atrophy is therefore not a continuous process and axotomized neurones

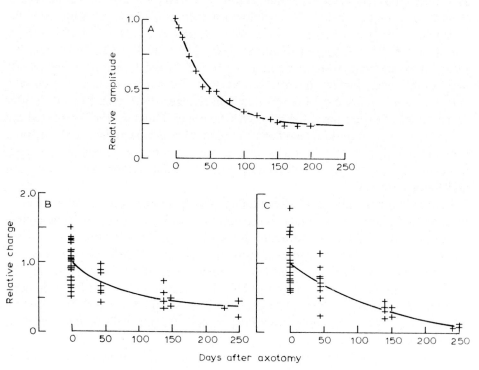

Fig. 2. Time course of changes in (A) amplitude of compound action potentials recorded chronically in cat hindlimb nerves, proximal to the site of section and ligation (Davis et al., 1978); and charge delivered to the ventral roots (B) and to the dorsal roots (C), in response to stimulation of ligated peripheral nerves. The area under the monophasic action potentials was divided by the root impendance to give units of charge. Decline in charge after axotomy gave a measure of the effect of axotomy on the diameter of sensory and motor nerves in each nerve (Hoffer et al., 1979). Cut nerves atrophied but finally reached a stable level (A) which was lower for sensory (C) than the motor (B) fibres.

continue to support viable axons for long periods of time. Although smaller in diameter, the nerves are still able to conduct action potentials. The median diameter of the cell bodies decreases very little and in the adult, the numbers of neurones did not change after axotomy (Carlson et al., 1979). Adult neurones are therefore not critically dependent on the periphery for their survival as embryonic neurones.

Gutmann and Sanders (1943) had suggested that the decline in fibre diameter after injury might be the consequence of outflow of axoplasm into the elongating axons sprouting from the site of injury. This explanation cannot hold, however, since axonal atrophy was similar and followed the same course in axotomized neurones whose axons were ligated to prevent regeneration and elongation, and neurones which were encouraged to regenerate after nerve repair (Davis et al., 1978). Yet the view of atrophic changes reflecting the regenerative response of the injured neurones may still be valid. It is now clear that axoplasm is not lost from the parent axons to supply the outgrowing axons. Material at the injury site may support regenerative sprouts in the early stages of regeneration before material moves from the soma to the axons. Fibrillar proteins, appropriate for growth, are transported in greater quantities (Lasek, 1981). Atrophy of parent axons above the injury site may be due to the alterations in the proportions of fibrillar proteins which are transported in the injured axon (Hoffman and Lasek, 1980) rather than a simple outflow of axoplasm from parent to regenerating axons (Section 2.2.4). In the first week after peripheral nerve section, ventral root nerves were significantly smaller near the cell body than more distally suggesting that atrophy proceeds in a proximo-distal direction after axotomy (Hoffman et al., 1980). The progression of atrophy was remarkably similar to the rate of transport of SCa, which transports neurofilament triplet protein. This suggested that the decline in diameter after axotomy might be a direct consequence of the decreased transport of neurofilaments protein.

All neurones are not equally affected by axotomy. Hoffer et al. (1979) found that sensory fibres continued to atrophy after the motor fibres had already reached an asymptotic level. The sensory fibres finally levelled off at significantly lower levels than motor fibres (Fig. 2). The greater atrophy of sensory fibres became evident about 1 month after axotomy which coincided with the loss of sensory activity during stereotyped movements such as walking (Gordon et al., 1980a). Although motor activity also falls after axotomy, probably as a result of reduced synaptic efficacy onto the motoneurones, activity continues indefinitely in axons without peripheral connections. Loss of neuronal traffic in the sensory and not the motor fibres was suggested as a possible factor responsible for the significantly more severe and more rapid atrophy of sensory than motor fibres after ligation (Hoffer et al., 1979; Milner and Stein 1981).

It was particularly striking that differential atrophy of sensory fibres became evident only when sensory activity had ceased but these results can only suggest loss of impulse traffic as one of several factors responsible for denervation atrophy of axons. The differential atrophy of the fastest conducting axons could be determined by inherent properties of the neurones and their particular supply of fibrillar proteins which in turn could be influenced by their activity. The fastest conducting parent axons atrophy most rapidly (Milner and Stein, 1981) and their regenerating sprouts also mature most rapidly (Devor and Govrin-Lippman, 1979a). Decrease or increase in neurofilaments transported could underline the changes in axon diameter during degeneration and regeneration respectively. Loss of peripheral contact results in alterations in proportions of fibrillar proteins to be appropriate for axonal growth. The atrophy could therefore represent part of the dedifferentiation of axotomized neurones to a growing state. Since the cell body can maintain the atrophied axons for long periods of time, adult neurones appear to be able to maintain this 'growing state' almost indefinitely, even if axons are prevented from elongating. The inherent properties of the cell body and possibly the cellular activity can maintain axons at a fraction of their normal diameter. The normal diameter is attained only by making functional connections between neurones and peripheral targets.

Myelin thickness or numbers of turns of myelin appears to be insensitive to atrophic changes taking place in the axons following axotomy. Recent ultrastructural studies of axotomized axons have shown that myelin thickness remains constant as fibre diameter declines (Gillespie and Stein, 1982). This is consistent with the light microscopic observation of a greater reduction in axonal than total fibre diameter (axon plus myelin) (Sanders, 1948; Cragg and Thomas, 1961).

Total fibre diameter has usually been used to measure the size of degenerating and regenerating nerve fibres. Total diameter is also the most extensively used parameter of nerve size for determining conduction velocity (Hurst, 1939; Rushton, 1951; Jack, 1955). Conduction velocity distributions were recently determined physiologically from compound action potentials and for intact nerves, there was good agreement with measurements of fibre diameters from cross-sections of the same nerves (Milner et al., 1981). Conduction velocity of atrophic nerves on the other hand, varied with axon diameter and not total diameter (Gillespie and Stein, 1982). Consequently, early histological studies of injured nerves underestimated the atrophic changes in the axons, by measuring total fibre diameter (e.g., Figs. 3 and 4).

2.3.3. Electrophysiological changes in axotomized motoneurones
Axotomy reduces but does not abolish characteristic differences in the action potentials in motoneurones supplying fast and slow muscles. The duration of the afterhyperpolarization is normally 2 to 3 times longer in slow motoneurones and falls in slow but not fast motoneurones after axotomy (Kuno et al., 1974a). The duration was shortened by more than 35% in soleus motoneurones. This change was regarded as a process of dedifferentiation of the motoneurone properties

since normal differentiation takes place as a result of a change in AHP duration in the slow soleus motoneurones and not the fast motoneurones (Huizar et al., 1975; Gallego et al., 1978).

Other changes in electrophysiological properties of axotomized fast and slow motoneurones include an increase in the action potential overshoot and increased motoneurone excitability. Hyperexcitability is associated with a decrease in rheobase current and the development of dendritic spikes (Kuno and Llinás, 1970a). The resting membrane potential of fast but not slow motoneurones falls after axotomy and input resitance increases (Kuno et al., 1974a). These changes are all complete within 2 to 7 weeks after section of the peripheral nerves and they cannot be readily changed in intact neurones. The afterhypolarization on the other hand appears to be a particularly labile property which changes in soleus motoneurones within the first 2 weeks after axotomy and which can readily be altered in intact motor units by altering the condition of the innervated muscle fibres. The duration of the afterhyperpolarization decreased under a number of different experimental conditions which are known to induce atrophy in the soleus muscle. Cordotomy in adult cats and kittens (Czéh et al., 1978; Gallego et al., 1978), immobilization of the muscle in a shortened position (Gallego et al., 1979a), partial denervation (Huizar et al., 1977), and nerve conduction block (Czéh et al., 1978) induced muscle atrophy and reduced the duration. Other electrophysiological properties were unaffected. These findings are discussed in more detail in Section 4.

3. Regeneration and reinnervation of target organs

Changes in axotomized neurones begin to reverse toward normal when the regenerating axons make peripheral connections. In as much as axotomy promotes dedifferentiation of neurones to a growing state, reformation of peripheral connections halts growth and promotes maturation, functional synaptic transmission and elaboration of peripheral processes.

3.1. Two stage regeneration: elongation and increase in diameter

Numerous fine nerve fibres of 1 μm or less in diameter, emerge from the parent axons in the central nerve stump of injured nerves (Speidel, 1935). Their growth and elongation is very similar to the growth of the fine nerve processes of similar dimensions in the embryo. There are two stages of axonal growth. During the first stage, nerve fibres grow in length while in the second stage of maturation they grow in diameter. During the first stage in embryogenesis, the growth potential is an inherent property of neurones and elongation proceeds as a consequence of supply of material from the cell body. Weiss (1950) anticipated later conclusive demonstrations of transport of fibrillar proteins and stated that: 'It seems hardly questionable that the capacity to form a nerve process is predicated on the production of long filamentous protein chains'. The growing nerve does not in itself determine the direction of the advance of the growth cone which depends on interactions with the micro-environment. Nerves do not grow in a homogeneous unstructured medium but will grow only along interfaces (Harrison, 1935) which may be provided in the animal by pioneer fibres or locally derived cells (reviewed by Weiss, 1950; Landmesser, 1980). Interaction with peripheral endorgans begins the second stage of development with maturation and increase in diameter of the nerve fibres. Axons also continue to grow in length in this stage as the animal grows and appear to be 'towed' or dragged by the endorgans (Weiss, 1950). Neurones which do not make peripheral connections during development cannot be maintained by their cell bodies in the first stage and must proceed into the second stage to survive.

Regenerating axons in adult animals recapitulate both stages of growth described for the embryonic neurones. During the first stage of outgrowth, the growth and dimensions of the regenerating axons are determined by the parent fibres and direction of growth is provided by vacated Schwann tubes in the distal nerve stump. The adult cell body can maintain viable axons for long periods of time without functional connections (Davis et al., 1978) in contrast to the embryonic neurone. But like embryonic neurones, maturation of the regenerating axons cannot proceed to completion in this stage without formation of peripheral connections. Interaction with the periphery promotes, in the second stage, complete maturation of the regenerating axons and recovery of parent axons from denervation atrophy.

The two stages of regeneration were finely distinguished in the definitive experiments of Weiss and Young and their colleagues in Chicago and London. In the experiments of Weiss et al. (1945), one of two branches of the crushed sciatic or buccal nerves in the rat hindlimb or neck, was capped to prevent regenerating fibres from making peripheral connections. The other branch remained connected to the periphery so that the distal nerve stump guided regenerating axons back to the periphery. In the experiments of Aitken et al. (1947), the medial gastrocnemius nerve was crushed bilaterally and nerve fibres were permitted to grow back to the muscle in one leg. Regenerating fibres in the other leg were prevented from making peripheral connections by cutting the nerve more distally and suturing it to muscle fascia. Crushed axons sent out numerous sprouts into the distal nerve stump. In the obstructed nerves, they grew as far as the imposed obstruction to their growth by the cap or ligature. They were maintained despite ligation. Regenerating axons were 3 to 5 times more numerous than parent axons (c.f., also Greenman, 1913; Gutmann and Sanders, 1943) showing that each parent axon supported more than one sprout, two regenerating axons being the average for sensory fibres (Horch and Lisney, 1981). The regenerating axons grew in diameter but did not approach control levels. One year after crush injury the regenerating axons in the capped nerve had exceeded the original one micrometer size and approached the size of the atrophic parent axons (Fig. 3A). Regenerating axons can therefore

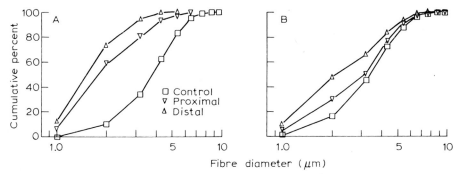

Fig. 3. Cumulative probability distributions of nerve fibre diameters (axon + myelin) in (A) the capped mandibular nerve branch and (B) the connected buccal branch of the facial nerve, 1 year after crush injury to the facial nerve. Fibres in the capped mandibular nerve regenerated and ended blindly in a neuroma while fibres in the intact buccal branch regenerated and made peripheral connections. Axons 5 mm proximal to the original crush (∇) atrophied without peripheral connections (A). Their regenerating axons, 10 mm distal to the crush (△), did not mature to diameters greater than their atrophied parent axons when they ended blindly in a neuroma. Regenerating axons which made peripheral connections (B) approached the size of the control nerves (□) but were still slightly smaller than their parent axons 1 year after crush injury. (The data is replotted from Weiss et al., 1945.)

mature in the first stage without peripheral connections but their size is ultimately limited by the size of the parent axons. The conduction velocity and therefore diameter of regenerating axons is always proportional to the conduction velocity of the parent axons from the earliest time when sprouts first grow out from the proximal stump (Devor and Govrin-Lipmann, 1979a). This suggests that the parent axons may determine the size of the regenerating axons. Less specific experiments of Simpson and Young (1945) and Evans (1947) also pointed to this conclusion. They found that somatic nerves which regenerated in the narrow tubes of degenerating sympathetic fibres in the mesenteric nerve or adrenal gland attained larger diameters than the non-myelinated fibres that previously occupied the tubes. The nerve and not the tube appeared to determine the diameter of regenerating fibres.

Maturation of regenerating axons did not proceed to completion in the first stage unless peripheral connections were made (cf., also Nageotte and Guyon, 1818; Sanders and Young, 1944, 1946). Regenerating axons which made peripheral connections proceeded to the second stage of maturation with almost complete restoration of the normal distribution for fibres proximal and distal to the crush injury (Fig. 3B) (Weiss et al., 1945; Gutmann and Sanders, 1943; Berry and Hinsey, 1946). Within 100 days of regeneration fibres began to show the bimodal fibre diameter distribution characteristic of normal nerve. The axons which ended in neuroma remained small and the fibre distribution was unimodal (Aitken et al., 1947). The numbers of regenerating axons declined to normal after formation of peripheral contacts and finally each parent axon supported only one axon distal to the injury site (Gutmann and Sanders, 1943; Devor and Govrin-Lipmann, 1979b).

3.2. Outgrowth and mechanical barriers to regenerating fibres

After crush injury, numbers and diameter of myelinated fibres are often fully reconstituted in the second maturation stage some time after peripheral connections are made (Gutmann and Sanders, 1943). Regeneration is, however, not always complete in adults particularly after nerve severance. Much of the failure of regeneration is due to the failure of the regenerating axons to overcome mechanical barriers to their growth. Mechanical barriers prevent them from reaching the periphery and making functional connections and thereby prevents complete maturation in the second stage.

To reach peripheral end-organs, growing nerve fibres must find empty Schwann tubes in peripheral stumps or have some form of mechanical guidance towards the peripheral endorgans they will reinnervate. The changes in the distal nerve stump are initially degenerative and associated with the removal of degenerative axonal materials (see Gutmann, 1958). Schwann cells proliferate during this degenerative phase and take part in the active phagocytosis of waste materials (Aguayo et al., 1976). These Schwann cells also play a vital role in the mechanical guidance of fibres in the distal nerve stump by formation of long, multinucleated bands, the so-called 'bands of Bungner' (Bungner, 1891). They, together with the fibroblasts accumulating at the injury site, extend towards the central stump and help bridge the gap between the central and peripheral stumps as well as provide conducting protoplasmic bands inside the neurilemmal sheaths (Cajal, 1928; Abercrombie and Johnson, 1946). Schwann cells therefore serve to guide the regenerating axons from the central stump, across the injury site and through the distal stump to peripheral target organs. They finally myelinate the regenerating axons (Simpson and Young, 1945; Aguayo et al., 1976). Schwann cells also play an essential supportive role for embryonic nerves because the nerves do not survive or grow in culture in their absence. A similar trophic role in regeneration has been suggested but has not yet been established (Bunge, 1980).

Axons are guided to the periphery by their former Schwann tubes if the continuity of the tubes is left uninterrupted by the nerve injury. This is the case after crush injury but axons seldom find their original Schwann tubes when the peripheral nerve is severed and the proximal and distal stumps surgically apposed. The nerves recover completely only in the former case. Even 1 year after nerve end-to-end suture, the distribution of fibre diameters, both proximal and distal to the suture, remained unimodal. There were always fewer large myelinated fibres which had formerly constituted the second peak of the bimodal fibre spectrum of normal hindlimb nerves (Gutmann and Sanders, 1943). Compound action potentials recorded proximal to the suture did not recover fully and projections based on simple exponential recovery curves predicted final values 60 to 70% of control (Davis et al., 1978). Their latency recovered completely which, together with the poorer recovery of amplitude, was consistent with recovery of some large fibres in the spectrum but in insufficient numbers to restore the bimodal distribution of fibre diameter. Incomplete recovery of sutured nerves is probably due almost entirely to mechanical barriers which prevent appropriate and functional connections between regenerating axons and target cells. These include scar tissue at the suture line, shrinkage of Schwann tubes, and connective tissue proliferation in denervated target organs. Recovery is also reduced by misdirection of axons to inappropriate target organs, or by severe atrophy of the target cells.

3.2.1. Suture line
The number of regenerating fibres in the peripheral stump varies considerably but can roughly be correlated with the closeness of apposition of the proximal and distal nerve stumps after suture (Gutmann and Sanders, 1943; Sunderland, 1978). The largest number of regenerating fibres was found in the distal stump across very closely apposed junctions where sprouts should be less likely to be lost in scar tissue before finding an empty Schwann sheath in the distal stump.

3.2.2. Conditions in peripheral stumps
Longitudinal sections of the suture line show extensive criss-crossing of fibres (Gutmann and Sanders, 1943). As a result the chances of axons finding their original Schwann sheaths is remote. Most axons will most likely enter tubes which were formerly occupied by others and they can be readily directed to different and often inappropriate target organs. The loss of fibres in inappropriate target organs is particularly high in large mixed nerves and in musculocutaneous branches such as the tibial and common peroneal nerves in the hindlimbs. Nerve recovery is consistently lower than for muscle nerves which supply single or small groups of muscles (Davis et al., 1978). Sensory fibres which are misdirected to muscle for example, will grow but do not make functional connections and do not influence the muscle properties (Gutmann, 1945).

Connective tissue proliferates in the peripheral stumps and empty Schwann tubes shrink in diameter with time (cf., Sunderland, 1978). Regenerating fibres mature less after delayed suture of cut nerves than after immediate suture. Poor maturation was correlated with the greater narrowing of the Schwann tubes in the distal stump after longer delays (Holmes and Young, 1942). The time course and success of reinnervation was not measured in these experiments. There is now evidence that the fibres which make peripheral contacts recover completely with a time constant of 1 to 2 months (Gordon and Stein, 1982). Fibres which would be delayed in making peripheral contacts by later suturing would be expected to mature later. This could account for the differences found in maturation of fibres in Holmes and Young's (1942) experiments.

3.2.3. Conditions of the denervated end-organs
Success of reinnervation and of recovery of muscle declines if reinnervation is delayed. Progressive shrinkage of Schwann tubes, atrophy of muscle fibres and loss of postsynaptic specialization in the muscle may account for progressive failure of reinnervation and of muscle recovery (Gutmann and Young, 1944; Gutmann, 1948, 1973, 1976). Axons normally follow existing tubes in the denervated muscle to the old endplate region where postjunctional membrane folding, high density of acetylcholine receptors and concentrations of cholinesterase are maintained for some time. The endplate region is the preferred site of reinnervation but the postjunctional specializations deteriorate with time after denervation. As a result, more regenerating axons attempt to make ectopic synapses. Reinnervating axon can make extrajunctional synapses under a number of conditions and these synapses can become as complex as the former endplate (Ip and Vrbová, 1973; Lomo and Slater, 1978; Gordon et al., 1980b). However, in atrophic muscle fibres, ectopic synapses are often unsuccessful (Gutmann and Young, 1944). Denervation atrophy of muscles is quite severe if reinnervation is delayed. Schwann tubes shrink and many regenerating nerve fibres do not follow the tubes. They run between and across the muscle fibres and often end blindly. Reinnervation and recovery of muscle was poor under these conditions but there have been no detailed studies on the reinnervating nerves (Gutmann, 1948, 1964).

Aitken et al. (1947) studied the influence of the muscle condition on maturation of regenerating axons. Muscles were denervated by peripheral nerve section and also tenotomized to promote severe muscle atrophy. Maturation of regenerating nerve fibres was exceedingly poor when the fibres regenerated across a suture line into these atrophic muscles (Aitken et al., 1947). The regenerating nerve fibres under these conditions were as small as fibres that ended blindly in a neuroma (Fig. 4). Nerve fibres which regenerated into tenotomized muscles after crush injury also matured less well than when they grew into denervated muscles. The effect was less dramatic than after nerve suture. The detrimental effect of muscle atrophy on regeneration is amplified after section of the nerve. The suture line provides further mechanical barriers and delays regenerating nerve fibres. Axons are also misdirected by inappropriate Schwann tubes (cf., also Evans, 1953).

3.3. Functional connections with end-organs and maturation

Fibres that succesfully remake connections, recover completely (Gordon and Stein, 1982a). Mechanical obstructions are therefore responsible for reducing the

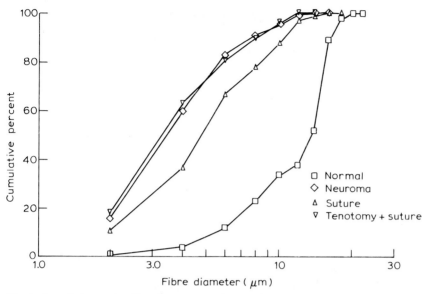

Fig. 4. Cumulative probability distributions of nerve fibre diameters (axon plus myelin) distal to a crush injury of the medial gastrocnemius nerve 100 days previously (Aitken et al., 1947). Regenerating axons which grew across a suture line into a tenotomised and denervated medial gastrocnemius muscle (▽) did not mature in the 100 day period. They remained as atrophic as regenerating axons which ended blindly in neuroma (◇). Axons which regenerated across a suture line into denervated muscles (△) were significantly larger than axons which had regenerated into the atrophic tenotomised muscles, and moved to the right along the X-axis toward the control nerves (□).

number of axons which reach the periphery and make functional contacts. They do not appear to influence the recovery of those axons which reinnervate target organs, as originally suggested. All α-motoneurones were similarly affected by axotomy. The axons proximal to the suture line were considerably smaller than normal when functional contacts with muscle were first made. Two to three months after end-to-end suture of severed nerves in the cat hindlimb, these axons had atrophied almost to the asymptotic level of the axons which never made connections (Fig. 5). Reinnervated axons recovered completely by 6 months and their distribution was equal to the distribution obtained for the same population of fibres, preoperatively. These experiments distinguished for the first time regenerating axons which had and had not made functional connections. They showed that functional nerve–muscle contacts completely reversed the atrophic effects of axotomy and reconstituted the normal distribution of α-motoneurones (Gordon and Stein, 1982). Thus, the distribution of all motor axons in resutured nerves, proximal to the suture line, is bimodal with clearly separated large and small peaks for axons that have not remade peripheral connections, respectively.

Recovery of single motor nerves with muscle connections followed a simple exponential time course with a time constant of 1 to 2 months (Fig. 6). This was

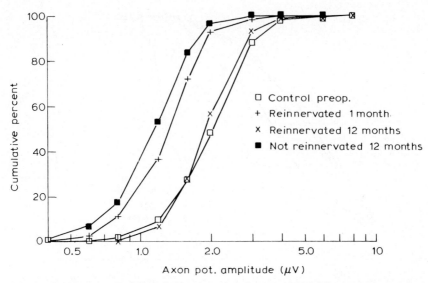

Fig. 5. Cumulative probability distribution of axon potential amplitudes recorded from the same axonal population, before (□), 2 to 3 months (+) and 1 year after nerve section and suture (× and ■) (from Gordon and Stein, 1982a). Axon potential amplitude recorded proximal to the injury site were smaller than normal when regenerating axons first made functional nerve–muscle connections (+) but recovered fully (×). Axons which did not remake nerve–muscle contacts (■) remained significantly smaller than normal. Note that all motor axons were similarly affected by denervation and reinnervation.

similar to the recovery of compound action potentials after crush injury where the entire fibre distribution is often fully reconstituted (Davis et al., 1978; Stein et al., 1980). The recovery of the compound potential therefore accurately reflects recovery of all the fibres. Recovery curves of compound potentials in resutured nerves, were severely distorted by the ongoing decay of unconnected nerves (Davis et al., 1978). As a result, the time course of recovery of the reinnervated nerves was overestimated. On the other hand, recovery of emg and muscle tension, even at the single unit level, predicted quite accurately the recovery of reinnervated motor nerves. This suggested that the nerve and muscle recover in parallel after mutual interaction.

Motor units always recovered preoperative tension and never became significantly larger than control even when whole muscle tension remained depressed. Residual muscle atrophy therefore indicated, to the first approximation, the fraction of motor axons that had been unsuccessful in reinnervating muscle (Gordon and Stein, 1982a). Complete muscle recovery indicated that almost all motor axons had grown back to muscle successfully. Incomplete recovery of the compound nerve potential could in that case be due to incomplete recovery of the sensory fibres. Sensory fibres can recover completely (Devor and Govrin-Lippmann, 1979a,b; Horsh and Lisney, 1981) but the time course of their recovery

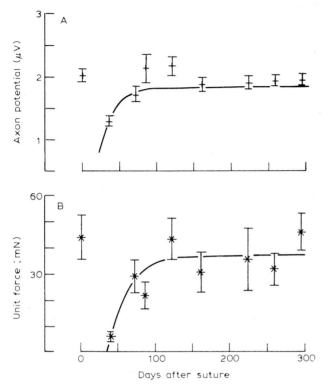

Fig. 6. Recovery of reinnervated motor units. (A) Axon potential amplitude of lateral gastrocnemius soleus axons proximal to site of section and suture, and (B) motor unit force. All recordings were made in the same motor unit population in one animal. Note that the recovery of nerve and muscle follow a similar time course. The time constants of the fitted exponentials were (A) 23 and (B) 27 days.

might be longer than for motor fibres (Brown and Butler, 1976). Muscle reinnervation can totally reverse the atrophy of nerve and muscle and promotes full maturation in the second stage of growth. Recovery of peripheral nerves and their endorgans is therefore related to the fraction of axons which successfully make functional nerve–muscle contacts.

3.3.1. Rematching of nerve and muscle properties after reinnervation
Cross-innervation studies of Buller et al. (1960) showed for the first time that the motor nerve determines muscles properties because the contractile properties were altered according to the nerve that supplied the muscle. Motor nerves determine many muscle characteristics including the structural, biochemical, histochemical, contractile and membrane properties (reviewed by Vrbová et al., 1978; Jolesz and Sreter, 1981). It follows that the properties of muscle fibres supplied by any one axon in single motor units will be homogeneous (Kugelberg

et al., 1970; Nemeth et al., 1970; Nemeth et al., 1981). Most muscles contain three motor units which have been distinguished histochemically and physiologically (Burke et al., 1973; Burke and Edgerton, 1975; Edington and Edgerton, 1976; Burke, 1981). The three motor unit types are distinct in their force output; fast-fatiguable units exerting significantly more tension than fast fatigue resistant units which in turn are larger than the slow units. The fast motoneurones are larger than the slow motoneurones. These differences may account for the finding that muscle force and speed are directly correlated with motoneurone or axon size in many muscles (e.g., Appelberg and Emonet-Denand, 1960; Jami and Petit, 1975; Stephens and Stuart, 1975; Henneman, 1980; Gordon and Stein, 1980, 1982b; Fleshman et al., 1981).

The characteristic electrophysiological properties of motoneurones to fast and slow muscles dedifferentiate after axotomy, but return to normal after reinnervation (Kuno et al., 1974a,b). This recovery is independent of the type of muscle which is reinnervated. The chances of nerves findings their original muscle fibres after nerve section is remote and each reinnervating axon supplies muscle fibres of different sizes and properties which formerly belonged to several different motor units (Kugelberg et al., 1970; Gordon and Stein, 1980). There was no correlation between unit size and their motor axons in these heterogeneous motor units. As nerves and muscles recovered from denervation, force and contractile speed of reinnervated motor units again correlated with axon size (Gordon and Stein, 1980; 1982b). These size relationships returned as the motoneurone properties recovered and when the histochemical properties were respecified (Kuno et al., 1974b; Warzawski et al., 1975; Gordon and Stein, 1982b). Rematching of nerve and motor unit size and contractile properties, was striking. Each motor unit type recovered its distinct properties and nerve size and force output returned to preoperative levels (Gordon and Stein, 1982a,b).

3.3.2. Influence of muscle on nerve
Many of the changes of reinnervation are the converse of denervation changes, and the maturation of the regenerating sprouts resembles the peripheral nerve maturation in developing animals. Once new motor terminals become functional, axotomized neurones begin to recover their former properties. Rate of recovery of axon size after reinnervation is very similar to the rate of atrophic changes in the axons after axotomy (Compare Figs. 2A and 6). If the diameter of the axons is determined by their neurofilament content, as suggested by Hoffman and Lasek (1980), axonal diameter might increase first near the cell body as increased amounts of neurofilament protein are supplied. Growth would proceed proximodistally at the rate of movement of SCa transport of approximately 1 to 5 mm/day. Recovery of axons at a distance of approximately 10 cm from the cell body was complete within 3 months (Gordon and Stein, 1982a). This recovery could be due to SCa transport but more direct measurements of regeneration are required to

establish whether there is a gradient from the cell body distally and whether regeneration is accompanied by increased transport of neurofilament triplet protein.

The size of the regenerating axons is proportional to their parent axons, but the axons are an order of magnitude smaller. Therefore, they need to grow considerably more in diameter to regain normal fibre size and equal their parent axon diameter.

Young had concluded in 1950 that 'the periphery exercises an important influence on the newly formed nerve fibres during regeneration and probably also during normal development, but the means by which the effect is produced remains completely obscure'. At that time, the evidence showed that regenerating fibres matured more completely if reinnervation was encouraged. We know now that peripheral connections can completely reverse the effects of axotomy but the nature of the influence of the periphery is still obscure and Young's conclusion is still valid. Activity of new transmitting nerve terminals may signal to the neurone cell body to express its inherent properties or the reinnervated muscle could determine the properties. Motoneurones recover their properties and appropriate axon size, irrespective of the muscles they reinnervate. This would argue against the idea that the muscle determines nerve properties. Yet, elaboration of transmitting terminals cannot promote maturation of regenerating axons without intact muscle. Sanes et al. (1978, 1980) have recently shown that regenerating axons acquire synaptic vesicles and thickened presynaptic membrane patches when they make contact with the extracellular basal lamina muscle sheath at the original end-plate sites. These terminals develop even if the muscle is irradiated and killed, and only the basal lamina remains. But the axons do not mature. Therefore, transmitting terminals and intact muscle are required for axonal maturation.

The effects of axotomy on the duration of the afterhyperpolarization can be mimicked by blocking nerve conduction with tetrodotoxin. Normal duration was restored by chronic stimulation of the nerves distal but not proximal to the conduction block (Czéh et al., 1978). This suggested that the activity of the neuromuscular junction or muscle contraction is important in determining motoneurone properties. Consistent with this suggestion, soleus motor nerves showed a significant reduction in fibre diameter when nerve and muscle activity was reduced by immobilization of the hindlimb. Conversely, soleus nerve hypertrophied in the overloaded contralateral limbs (Eisen et al., 1973). Functional overload of soleus motor units by denervation of its synergists also increased nerve fibre size (Wedeles, 1949; Edds, 1950b), but daily exercise had no effect on medial gastrocnemius nerves (Tomanek and Tipton, 1967). Tomanek and Tipton (1967) suggested that nerve fibre diameter was correlated with muscle mass rather than muscle activity since nerve and muscle hypertrophied after elimination of synergists but not in their exercised rats. Kuno and his colleagues have made similar conclusions for the duration of hyperpolarization (see Fig. 4

of Gallego et al., 1979b). Duration of the afterhyperpolarization in soleus motoneurones declined after spinal cord transection or peripheral nerve conduction block (Czéh et al., 1978). These changes were reduced by immobilizing the disused muscle in a lengthened position but not in a shortened position. Changes in duration could also be induced by immobilization of the soleus muscle in a shortened position and cutting the dorsal roots (Gallego et al., 1979b). Disuse atrophy in muscles is reduced by immobilization in a lengthened position and increased in a shortened position (Tabary et al., 1972). Properties of soleus motoneurones were therefore correlated with muscle mass under conditions where the nerve and muscle were relatively inactive.

Motoneurone properties and muscle mass are not necessarily casually related, however. Muscle wasting after cord transection could be prevented by chronic electrical stimulation at high and low frequencies of stimulation but soleus motoneurone properties recovered only after low frequency stimulation and not high frequencies of stimulation. Any pattern of stimulation can reduce muscle atrophy but low frequency stimulation maintains the characteristic properties of the slow soleus muscle (Vrbová et al., 1978). This suggests that the pattern of neuromuscular activity might also be important in maintaining the motoneurone properties.

Edds (1949) suggested that axon size varied with the metabolic load of the motoneurones. He found that nerve fibres, which sprouted to supply denervated muscle fibres increased in size. Motor nerves can sprout and increase their peripheral fields up to five times their normal size (Edds, 1950a; Edds, 1953; Brown and Ironton, 1978). Edds considered that hypertrophy of soleus motor nerves after synergist muscle denervation could result from their excessive use and increased metabolic requirements. Young (1950) had noted that motor nerves which supplied muscles outside of the limbs, which usually carried little or no weight, were normally significantly smaller than motor nerves to the limb musculature. Muscles such as the infrahyoid, facial muscles and diaphragm in the rabbit contain few if any spindles and Young suggested that the axon size might be significantly influenced by its reflex activity. The experiments of Czéh et al. (1978) seemed to indicate that it was neuromuscular activity and not the activity of the cell body that was important. Much of the cellular metabolism is concerned with neurosecretion (Section 2.2.3) so that neuromuscular secretion may represent the most significant metabolic load of motoneurones. Nerve fibres in the tractus hypophyseus, which transport oxytocin and vasopressin to the posterior pituitary, also showed significant hypertrophy under physiological and pathological conditions of high secretory activity. This hypertrophy was accompanied by an increase in microtubular content (Grainger and Sloper, 1974, 1976). The experiments on hypo- and hyperactivity of motoneurones therefore suggest that secretory activity of axon terminals and/or muscle activity may regulate motor nerve size and their properties. Recovery of reinnervated nerves follows functional neuromuscular transmission. Silenced motor nerves can grow, elaborate transmitting terminals, and make functional connections with muscle (Williams and Gilliatt, 1977) but available evidence suggests that further maturation of the regenerating axons depends on the activity of the motor unit, as it does during embryogenesis (Drachman, 1967, 1968; Giacobini et al., 1973). Neuromuscular activity appears to influence nerve as well as muscle properties.

Acknowledgements

Support from the Alberta Heritage Foundation is gratefully acknowledged. I would also like to thank Drs. Wayne Aldridge, Doug Eaton and Linda Bambrick for providing valuable suggestions for revision of the text; and Mrs. Linda Chambers, Rosalind Miller, Mrs. Sharon Kilback and Susan Buller for patiently organizing references and typing the manuscript.

References

Abercrombie, M. and Johnson, L.L. (1946) J. Anat. *80*, 37–50.
Abrams, J. and Gerard, R.W. (1933) Am. J. Physiol. *104*, 590–593.
Acheson, G.H., Lee, E.S. and Morison, R.S. (1942) J. Neurophysiol. *5*, 269–273.
Acheson, G. and Remolina, J. (1955) J. Physiol. (London) *127*, 603–616.
Aguayo, A.J., Epps, J. and Charron, L. (1976) Brain Res. *104*, 1–20.
Aitken, J.T., Sharman, M. and Young, J.Z. (1947) J. Anat. *81*, 1–22.
Anderson, H.K. (1902) J. Physiol. (London) *28*, 499–513.
Anderson, L.L. and McClure, W.O. (1973) Proc. Natl. Acad. Sci. USA *70*, 1521–1525.
Appelberg, B. and Émonet-Dénand, F. (1960) J. Neurophysiol. *30*, 154–160.
Austin, A. and Langford, C.J. (1980) Trends Neurosci. *3*, 130–132.
Benowitz, L.I., Shashoua, V.E. and Yoon, M.G. (1981) J. Neurosci. *1*, 300–307.
Beránek, R. and Hník, P. (1959) Science *130*, 981–982.
Berry, C.M. and Hinsey, J.C. (1946) Ann. N.Y. Acad. Sci. *47*, 559–573.
Birks, R., Katz, B. and Miledi, R. (1960) Physiol. (London) *150*, 145–168.
Bisby, M.A. (1978) Exp. Neurol. *61*, 281–300.
Bisby, M.A. (1980) Neurosci. Let. *21*, 7–11.
Bisby, M.A. (1981) J. Neurochem. *36*, 741–745.
Bittner, G.D. (1973) Am. Zool. *13*, 379–409.
Black, M.M. and Lasek, R.J. (1979) Exp. Neurol. *63*, 108–119.
Black, M.M. and Lasek, R.J. (1980) J. Cell. Biol. *86*, 616–623.
Blinzinger, K. and Kreutzburg, G. (1968) Z. Zellforsch. Mikrosk. Anat. *85*, 145–147.
Boeke, J. (1950) in Genetic Neurology, International Conference on the Development, Growth, and Regenerating of the Nervous System (Weiss, P., ed.), pp. 78–91. University of Chicago Press, Chicago.
Boyd, I.A. and Davey, M.R. (1968) Composition of Peripheral Nerves, Livingstone, London.
Boyle, F.C. and Gillespie, J.S. (1970) Eur. J. Pharmacol. *12*, 77–84.
Brattgard, S.O., Edström, J.E. and Hydén, H. (1957) J. Neurochem. *1*, 316–325.

Brattgard, S.O., Edström, J.E. and Hydén, H. (1958) Expl. Cell. Res. Suppl. 5, 185–200.
Bray, D. and Gilbert, D. (1981) Ann. Rev. Neurosci. 4, 505–523.
Brechter, M.S. and Raff, M.C. (1975) Nature (London) 258, 43–49.
Brenner, H.R. and Johnson, E.W. (1976) J. Physiol. (London) 260, 143–158.
Brenner, H.R. and Martin, A.R. (1976) J. Physiol. (London) 260, 159–175.
Brown, M.C. and Butler, R.G. (1976) J. Physiol. (London) 260, 253–266.
Brown, M.C. and Ironton, R. (1978) J. Physiol. (London) 261, 387–422.
Brown, G.L. and Pascoe, J.E. (1954) J. Physiol. (London) 123, 565–573.
Buller, A.J., Eccles, J.C. and Eccles, R.M. (1960) J. Physiol. (London) 150, 417–439.
Bunge, R.P. (1980) in Nerve Repair and Regeneration: Its Clinical and Experimental Basis (Jewett, D.L. and McCaroll, H.R., eds.), pp. 58–75, Mosby, St. Louis.
von Bungner, O. (1891) Zieglers Beitr. z. Path. u. path. Anat. 10, 321.
Burke, R.E. (1981) in Handbook of Physiology. Vol. IV, pp. 345–422. American Physiology Society, Bethesda.
Burke, R.E. and Edgerton, V.R. (1975) in Exercise and Sports Sciences Reviews (Wilmore, J.H., ed.), Vol.3, pp. 69–94, Academic Press, New York.
Burke, R.E., Levine, D.M., Salcman, M. and Tsairis, P. (1974) J. Physiol. (London) 238, 503–514.
Burke, R.E., Levine, D.N., Tsairis, P. Zajac, F.E. (1973) J. Physiol. (London) 234, 723–748.
Burrel, H.R., Dokas, L.A. and Agronoff, B.W. (1978) J. Neurochem. 31, 289–298.
Cajal, S.R. (1928) Degeneration and Regeneration of the Nervous System. II. (Translated and edited by May, R.M.) Oxford University Press.
Campbell, B. (1944) Anat. Rec. 88, 25–37.
Carlson, J., Lais, A.C. and Dyck, P.J. (1979) J. Neuropathol. Exp. Neurol. 38, 579–585.
Carmel, P.W. and Stein, B.M. (1969) J. Comp. Neurol. 135, 145–166.
Cavanaugh, M.W. (1951) J. Comp. Neurol. 94, 181–218.
Cheah, T.B. and Geffen, L.B. (1973) J. Neurobiol. 4, 443–452.
Cook, D.D. and Gerard, R.W. (1931) Am. J. Physiol. 77, 412–425.
Cragg, B.G. (1970) Brain Res. 23, 1–21.
Cragg, B.G. and Thomas, P.K. (1961) J. Physiol. (London) 157, 315–327.
Cuénod, M., Boesch, J., Marko, P., Perisic, M., Sandri, C. and Schonbach, J. (1972) Int. J. Neurosci. 4, 77–87.
Czéh, G., Gallego, R., Kudo, N. and Kuno, M. (1978) J. Physiol. (London) 281, 239–252.
Dahlström, A. (1965) J. Anat. 99, 677–689.
Dahlström, A. (1967) Acta. Physiol. Scand. 69, 158–166.
Dahlström, A. (1971) Phil. Trans. R. Soc. London, Ser B: 261, 325–358.
Davis, L.A., Gordon, T., Hoffer, J.A., Jhamadas, J. and Stein, R.B. (1978) J. Physiol. (London) 285, 543–559.
Del Cerro, M. (1974) J. Comp. Neurol. 157, 245–280.
Devor, M. and Govrin-Lippman, R. (1979a) Exp. Neurol. 64, 260–270.
Devor, M. and Govrin-Lippman, R. (1979b) Exp. Neurol. 65, 243–254.
Dowman, C.B.B., Eccles, J.C. and McIntyre, A.K. (1953) J. Comp. Neurol. 98, 9–36.
Drachman, D.B. (1967) Arch. Neurol. 17, 206–218.
Drachman, D.B. (1968) in CIBA Foundation Symposium on Growth of the Nervous System (Wolstenholme, G.E.W. and O'Connor, M., eds.), pp. 253–273, Churchill, London.
Droz, B. (1975) in The Basic Neurosciences (Tower, D.B., ed.), Vol. 1, pp. 111–127, Raven Press, New York.
Droz, B. and Koenig, H.L. (1970) in Protein Metabolism of the Nervous System (Lajtha, A., ed.), pp. 93–108, Plenum Press, New York.
Droz, B., Koenig, H.C. and Di Giamberardino, L. (1973) Brain Res. 60, 93–127.
Droz, B., Rambourg, A. and Koenig, H.L. (1975) Brain Res. 93, 1–13.

Duchen, L.W. and Strich, S.J. (1968) Q. J. Exp. Physiol. 53, 84–89.
Dziegielewska, K.M., Evans, C.A.N. and Saunders, N.R. (1980) J. Physiol. (London) 304, 83–98.
Eccles, J.C. and McIntyre, A.K. (1953) J. Physiol. (London) 121, 492–516.
Eccles, J.C., Libet, B. and Young, R.R. (1958) J. Physiol. (London) 143, 11–40.
Edds, M.V. (1949) J. Exp. Zool. 112, 29–47.
Edds, M.V. (1950a) J. Exp. Zool. 113, 517–552.
Edds, M.V. (1950b) J. Comp. Neurol. 93, 259–275.
Edds, M.V. (1953) Q. Rev. Biol. 28, 260–276.
Edgerton, V.R. (1978) Am. Zool. 18, 113–125.
Edington, D.W. and Edgerton, V.R. (1976) Biology of Physical Activity, Houghton Mifflin, Boston.
Eisen, A.A., Carpenter, S., Karpati, G. and Bellavance, A. (1973) J. Neurol. Sci. 20, 457–469.
Elam, J.S. and Maxwell, J.K. (1978) Trans. Am. Soc. Neurochem. 9, 132.
Engh, C.A. and Schofield, B.H. (1972) J. Neurosurg. 37, 195–203.
Evans, D.H.L. (1947) J. Anat. 81, 225–232.
Evans, D.H.L. (1953) J. Comp. Neurol. 99, 561–594.
Farel, P.B. (1980) Brain Res. 189, 67–77.
Feldberg, W. (1943) J. Physiol. (London) 101, 432–435.
Fernand, V.S.V. and Young, J.Z. (1951) Proc. R. Soc. London Ser. B: 139, 38–58.
Fischer, H.A. and Schmatolla, E. (1972) Science 176, 1327–1329.
Fleshman, J.W., Munson, J.B., Sypert, G.W. and Friedman, W.A. (1981) J. Neurophysiol. 46, 1326–1338.
Forman, D.S. and Borenberg, R.A. (1978) Brain Res. 156, 213–225.
Friede, R.L. and Samorajski, T. (1970) Anat. Rec. 167, 379–388.
Frizell, M. and Sjöstrand, J. (1974a) Brain Res. 78, 109–123.
Frizell, M. and Sjöstrand, J. (1974b) J. Neurochem. 22, 845–850.
Gallego, R., Huizar, P., Kudo, N. and Kuno, M. (1978) J. Physiol. (London) 281, 253–265.
Gallego, R., Kuno, M., Núñez, R. and Snider, W.D. (1979a) J. Physiol. (London) 291, 179–189.
Gallego, R., Kuno, M., Núñez, R. and Snider, W.D. (1979b) J. Physiol. (London) 291, 191–205.
Garner, J.A. and Lasek, R.J. (1981) J. Cell Biol. 88, 172–178.
Gerard, R.W. (1932) Physiol. Rev. 12, 469–592.
Giacobini, G., Filogamo, G., Weber, M., Boquet, P. and Changeux, J.P. (1973) Proc. Natl. Acad. Sci. USA 70, 1708–1712.
Gilbert, D.S., Newby, B.J. and Anderton, B.H. (1975) Nature (London) 256, 586–589.
Gillespie, M.J. and Stein, R.B. (1982) Brain Res. (in press).
Gordon, T., Hoffer, J.A., Jhamandas, J. and Stein, R.B. (1980a) J. Physiol. (London) 303, 243–263.
Gordon, T. and Stein, R.B. (1980) in Plasticity of Muscle (Pette, D., ed.), pp. 283–296, Walter de Gruyter, Berlin.
Gordon, T., Niven-Jenkens, N. and Vrbová, G. (1980b) Neuroscience 5, 597–610.
Gordon, T. and Stein, R.B. (1982a) J. Physiol. (London) 323, pp. 307–323.
Gordon, T. and Stein, R.B. (1982b) J. Neurophysiol. (in press).
Grafstein, B. (1975) Exp. Neurol. 48, 32–51.
Grafstein, B. (1977) in The Handbook of Physiology, Vol. 1 (Brookhart, J.M. and Mountcastle, V.B., eds.), pp. 697–717, American Physiological Society, Washington, D.C.
Grafstein, B. and Forman, D.S. (1980) Physiol. Rev. 60, 1167–1283.
Grafstein, B. and Murray, M. (1969) Exp. Neurol. 25, 494–508.
Grainger, F. and Sloper, J.C. (1974) Cell Tiss. Res. 153, 101–113.
Grainger, F. and Sloper, J.C. (1976) Cell Tiss. Res. 169, 405–414.
Greenman, M.J. (1913) J. Comp. Neurol. 23, 479–513.
Griffin, J.W., Price, D.C. and Drachman, D.B. (1976) J. Neurobiol. 7, 355–370.
Gunning, P.W., Kaye, P.L. and Austin, L. (1977) J. Neurochem. 28, 1245–1248.

Gutmann, E. (1942) J. Neurol. Psychiat. 5, 81–95.
Gutmann, E. (1945) J. Anat. 79, 1–8.
Gutmann, E. (1948) J. Neurophysiol. 11, 279–294.
Gutmann, E. (1958) Die Functionelle Regeneration der Peripheren Neuren. Akademie-Verlag, Berlin.
Gutmann, E. (1964) in Progress in Brain Research (Singer, M. and Schade, J.P., eds.), Vol. 13, pp. 72–114, Elsevier, Amsterdam.
Gutmann, E. (1973) in Methods of Neurochemistry (Rainer Fried, ed.), pp. 189–254, Marcel Dekker, New York.
Gutmann, E. (1976) Ann. Rev. Physiol. 38, 177–216.
Gutmann, E., Guttmann, L., Medawar, P.B. and Young, J.Z. (1942) J. Exp. Biol. 19, 14–44.
Gutmann, E. and Holubar, J. (1951) Arch. Int. Stud. Neurol. 1, 1–11.
Gutmann, E., Jakoubek, B., Rohlicek, V. and Akaloud, V. (1962) Physiol. Bohemoslov. 11, 437–442.
Gutmann, E. and Sanders, F.K. (1943) J. Physiol. (London) 101, 489–518.
Gutmann, E. and Young, J. (1944) J. Anat. 78, 15–43.
Hamberger, A., Hansson, H.A. and Sjöstrand, J. (1970) J. Cell. Biol. 47, 319–331.
Härkönen, M. (1964) Acta Physiol. Scand. 63, Suppl. 237, 1–94.
Härkönen, M.H.A. and Kauffman, F.C. (1974) Brain Res. 65, 141–157.
Harrison, R.G. (1935) Proc. R. Soc. London, Ser. B: 118, 155–196.
Heiwell, P.O., Dahlström, A., Larsson, P.A. and Booj, S. (1979) J. Neurobiol. 10, 119–136.
Henneman, E. (1980) in Medical Physiology (Mountcastle, V.B., ed.), Vol. 1, pp. 718–741, Mosby, St. Louis.
Heslop, J.P. (1975) Adv. Comp. Physiol. Biochem. 6, 75–163.
Hník, P., Beranek, R., Vyklický, L. and Zelená, J. (1963) Physiol. Bohemoslov. 12, 23–29.
Hoffer, J.A., Stein, R.B. and Gordon, T. (1979) Brain Res. 178, 347–361.
Hoffman, P.N., Griffen, J.W. and Price, D.C. (1980) Neuroscience 6, 33.7.
Hoffman, P.N. and Lasek, R.J. (1975) J. Cell Biol. 66, 351–366.
Hoffman, P.L. and Lasek, R.J. (1980) Brain Res. 202, 317–333.
Holmes, W. and Young, J.Z. (1942) J. Anat. 77, 63–96.
Holtzman, E. (1977) Neuroscience 2, 327–355.
Horch, K.W. and Lisney, S.J.W. (1981) J. Physiol. (London) 313, 275–280.
Huizar, P., Kudo, N., Kuno, M. and Miyata, Y. (1977) J. Physiol. (London) 265, 175–191.
Huizar, P., Kuno, M. and Miyata, Y. (1975) J. Physiol. (London) 252, 465–479.
Hunt, C.C. and Riker, W.K. (1966) J. Neurophysiol. 29, 1096–1114.
Hursh, J.B. (1939) Am. J. Physiol. 27, 131–139.
Hydén, H. (1960) in The Cell (Brachet, J. and Mirsky, A.E., eds.), pp. 215–323, Academic Press, New York.
Ingoglia, N.A., Stutman, J.A. and Eisner, R.A. (1977) Brain Res. 130, 433–445.
Ip, M.C. and Vrbová, G. (1973) Z. Zellforsch. Mikrosk. Anat. 146, 261–279.
Jack, J.J.B. (1955) Br. J. Anaesth. 47, 173–182.
Jami, L. and Petit, J. (1975) Brain Res. 96, 114–118.
Jockusch, H. and Jockusch, B.M. (1981) Exp. Cell. Res. 131, 345–352.
Jolesz, F. and Sreter, F.A. (1981) Ann. Rev. Physiol. 43, 531–552.
Karlstrom, L. and Dahlström, A. (1973) J. Neurobiol. 4, 191–200.
Kaye, P.L., Gunning, P.W. and Austin, L. (1977) J. Neurochem. 28, 1241–1243.
Kiraly, J.K. and Krnjević, K. (1959) Q. J. Exp. Physiol. 64, 244–257.
Knyihar, E. and Csillik, B. (1976) Exp. Brain Res. 26, 73–87.
Kuczmarski, E.R. and Rosenbaum, J.L. (1974) J. Cell Biol. 80, 356–371.
Kugelberg, E., Edström, L. and Abbruzzese, M. (1970) J. Neurol. Neurosurg. Psychiat. 33, 319–329.
Kuno, M. and Llinás, R. (1970a) J. Physiol. 210, 807–821.

Kuno, M. and Llinás, R. (1970b) J. Physiol. (London) 210, 823–838.
Kuno, M., Miyata, Y. and Muñoz-Martinez, E.J. (1974a) J. Physiol. (London) 240, 725–739.
Kuno, M., Miyata, Y. and Muñoz-Martinez, E.J. (1974b) J. Physiol. (London) 240, 725–739.
Landmesser, L.T. (1980) Ann. Rev. Neurosci. 3, 279–302.
Lasek, R.J. (1968) Exp. Neurol. 21, 41–51.
Lasek, R.J. (1970) Int. Rev. Neurobiol. 13, 289–324.
Lasek, R.J. (1981) Neurosci. Res. Prog. Bull. 19, 7–32.
Lasek, R.J. and Black, M.M. (1977) in Mechanisms, Regulation and Special Functions of Protein Synthesis in the Brain (Roberts, S., Lujtha, A. and Gispen, W.H., eds.), pp. 161–169, Elsevier Biomedical Press, Amsterdam.
Lasek, R.J., Gainer, H. and Barker, J.L. (1977) J. Cell Biol. 74, 501–523.
Lasek, R.J. and Hoffman, P.N. (1976) in Cell Motility. Book C, Microtubules and Related Proteins (Goldman, R., Pollard, T. and Rosenbaum, J., eds.), pp. 1021–1049, Cold Spring Harbour Laboratory, New York.
Lieberman, A.R. (1971) Int. Rev. Neurobiol. 14, 49–124.
Lieberman, A.R. (1974) in Essays on the Nervous System. A Postscript for Professor J.Z. Young (Bellairs, R. and Gray, E.G., eds.), pp. 71–105, Oxford University Press, Oxford.
Lømo, T. and Slater, C.R. (1978) J. Physiol. (London) 275, 391–402.
Lubínska, L. (1964) in Progress in Brain Research (Singer, M. and Schade, J.R., eds.), Vol. 13, pp. 1–71, Elsevier Amsterdam.
Lubínska, L. (1975) Int. Rev. Neurobiol. 17, 241–296.
McEwen, B. and Grafstein, B. (1968) J. Cell Biol. 38, 494–508.
MacIntosch, F.C. (1938) Arch. Int. Physiol. 47, 321–324.
McIntyre, Bradley, K. and Brock, L.G. (1959) J. Gen. Physiol. 42, 931–958.
McQuarrie, I.G. and Grafstein, B. (1973) Arch. Neurol. 29, 43–55.
McQuarrie, I.G., Grafstein, B. and Gershon, M.D. (1977) Brain Res. 32, 433–435.
Marinesco, G. (1896) C. R. Soc. Biol. III, 930.
Mathews, M.R. and Nelson, V.H. (1975) J. Physiol. (London) 245, 91–135.
Mathews, M.R. and Raisman, G. (1972) Proc. R. Soc. London, Ser. B: 181, 43–79.
Mendell, L.M., Munson, J.B. and Scott, J.G. (1976) J. Physiol. (London) 255, 67–79.
Miledi, R. and Slater, C.R. (1968) Proc. R. Soc. London, Ser. B: 169, 289–306.
Miledi, R. and Slater, C.R. (1970) J. Physiol. (London) 207, 507–528.
Milner, T.E. and Stein, R.B. (1981) J. Neurol. Neurosurg. Psychiat. 44, 485–596.
Milner, T.E., Stein, R.B., Gillespie, J. and Hanley, B. (1981) J. Neurol. Neurosurg. Psychiat. 44, 476–484.
Murray, M. (1973) Exp. Neurol. 39, 489–497.
Nageotte, J. and Guyon, L. (1818) C.R. Soc. Biol. Paris. 81, 571–574.
Nemeth, P., Pette, D. and Vrbová, G. (1981) J. Physiol. (London) 311, 489–495.
Nissl, L.F. (1982) All. Z. Psychiat. Ihre Grenzg. 48, 197–198.
O'Brien, R.A.D. (1978) J. Physiol. (London) 282, 91–103.
Ochs, S. (1972) J. Physiol. (London) 227, 627–645.
Palade, G.E. (1975) Science 189, 347–357.
Peters, A., Palay, S.L. and Webster, H. de F. (1976) in The Fine Structure of the Nervous System: The Neurons and Supporting Cells. W.B. Saunders, Philadelphia.
Peters, A. and Vaughn, J.E. (1967) J. Cell Biol. 32, 113–119.
Pilar, G. and Landmesser, L. (1972) Science 177, 1116–1118.
Pitman, R.M. and Rand, K.A. (1981) Neuroscience 15, 3.
Purves, D. (1975) J. Physiol. (London) 252, 429–463.
Purves, D. (1976 J. Physiol. (London) 259, 27–49.
Ranson, S.W (1912) J. Comp. Neurol. 22, 487–546.

Reis, D.J. and Ross, R.A. (1973) Brain Res. 57, 307–326.
Rushton, W.A.H. (1951) J. Physiol. (London) 115, 101–122.
Sanders, F.K. (1948) Proc. R. Soc. London, Ser. B: 135, 323–357.
Sanders, F.K. and Young, J.Z. (1944) J. Physiol. (London) 103, 119–136.
Sanders, F.D. and Young, J.Z. (1946) J. Exp. Biol. 22, 203–212.
Sanders, F.K. and Whitteridge, D. (1946) J. Physiol. (London) 105, 152–174.
Sanes, J.R., Marshall, L.M. and McMahan, U.J. (1978) J. Cell Biol. 78, 176–198.
Sanger, J.W. (1975) Cell. Tiss. Res. 161, 431–444.
Saunders, N.R. (1975) in Cholinergic Mechanisms (Waser, P.G., ed.), pp. 177–185, Raven Press, New York.
Schubert, P., Kreutzberg, G.W. and Lux, H.D. (1972) Brain Res. 47, 331–343.
Schwartz, J.H. (1979) Ann. Rev. Neurosci. 2, 467–504.
Shapovalov, A.I. and Grantyn, A.A. (1968) Biophysics 13, 308–319.
Sieler, N. (1973) in Polyamines in Normal and Neoplastic Growth (Russell, D.H., ed.), pp. 137–156, Raven Press, New York.
Simpson, S.A. and Young, J.Z. (1945) J. Anat. 79, 48–65.
Sinicropi, D.V. and Kauffman, F.C. (1979) J. Biol. Chem. 254, 3011–3017.
Speidel, C.C. (1935) J. Comp. Neurol. 61, 1–82.
Stein, R.B., Gordon, T., Hoffer, J.A., Davis, C.A. and Charles, D. (1980) in Nerve Repair and Regeneration: Its Clinical and Experimental Basis, (Jewett, D.L. and McCaroll, H.R., eds.), pp. 166–176, Mosby, St. Louis.
Stephens, J.A. and Stuart, D.G. (1975) Brain Res. 91, 117–195.
Sumner, B.E.H. (1975) Exp. Neurol. 46, 605–615.
Sumner, B.E.H. and Sutherland, F.I. (1973) J. Neurocytol. 2, 315–328.
Sumner, B.E.H. and Watson, W.E. (1971) Nature (London) 233, 273–275.
Sunderland, S. (1978) Nerves and Nerve Injuries, Livingstone, Edinburgh and London.
Tabary, J.C., Tabary, C., Tardieu, C., Tardieu, G. and Goldspink, G. (1972) J. Physiol. (London) 224, 321–324.
Tomenek, R.J. and Tipton, C.M. (1967) Anat. Rec. 159, 105–112.
Tonge, D.A. (1974a) J. Physiol. (London) 241, 127–139.
Tonge, D.A. (1974b) J. Physiol. (London) 241, 141–153.
Torvik, A. and Skjorten, F. (1971) Neuropath. 17, 265–282.
Vrbová, G., Gordon, T. and Jones, R. (1978) Nerve–Muscle Interaction, Chapman and Hall, London.
Waller, A.V. (1850) Phil. Trans. R. Soc. London, Ser. B: 140, 423.
Warzawski, M., Telerman-Toppet, N., Durdu, J., Graff, G.L.A. and Coërs, C. (1975) J. Neurol. Sci. 24, 21–32.
Watson, W.E. (1965) J. Physiol. (London) 180, 741–753.
Watson, W.E. (1968) J. Physiol. (London) 196, 655–676.
Watson, W.E. (1969) J. Physiol. (London) 202, 611–630.
Watson, W.E. (1970) J. Physiol. (London) 210, 321–343.
Watson, W.E. (1972) J. Physiol. (London) 225, 415–435.
Watson, W.E. (1974) Brain Res. 65, 317–322.
Watson, W.E. (1976) Cell Biology of Brain, Chapman and Hall, New York.
Webster, H. de F. (1962) J. Cell Biol. 12, 361–383.
Wedeles, C.H.A. (1949) J. Anat. 83, 57.
Weiss, P. (1950) in Genetic Neurology, International Conference on the Development, Growth, and Regeneration of the Nervous System (Weiss, P., ed.), pp. 1–39, University of Chicago Press, Chicago.
Weiss, P., Edds, M.V. and Cavanaugh, M. (1945) Anat. Rec. 92, 215–233.

Weiss, P.A. and Mayr, R. (1971) Proc. Natl. Acad. Soc. USA 68, 846–850.
Willard, M. (1977) J. Cell Biol. 75, 1–11.
Willard, M., Cowan, W.M. and Vagelos, P.R. (1974) Proc. Natl. Acad. Sci. USA 71, 2183–2187.
Willard, M. Wiseman, M., Levine, J. and Skene, P. (1979) J. Cell Biol. 81, 581–591.
Williams, I.R. and Gilliatt, R.W. (1977) J. Neurol. Sci. 33, 267–273.
Wujek, J.K. and Lasek, R.J. (1980) Neuroscience 6, 6.
Yamada, K.M., Spooner, B.S. and Wessels, N.K. (1971) J. Cell Biol. 49, 614–635.
Young, J.Z. (1946) Lancet 2, 109–113.
Young, J.Z. (1950) in Genetic Neurology, International Conference on the Development, Growth and Regeneration of the Nervous System (Weiss, P., ed.), pp. 92–104, University of Chicago Press, Chicago.
Zelená, J. (1972) Z. Zellforsch. Mikcrosk. Anat. 124, 217–229.

CHAPTER 11

Pathology of autonomic nerve–smooth muscle mechanisms in the gut of man

E.R. HOWARD and J.R. GARRETT*

Department of Surgery and Department of Oral Pathology and Oral Medicine*, King's College Hospital, London, SE5 9RS, England

The autonomic innervation of the gastrointestinal tract in man may be affected by a large number of pathological processes which include congenital, degenerative, inflammatory and parasitic diseases. Further, whereas the neurones of the central nervous system are partially protected by the 'blood-brain' barrier, the autonomic nerves of the gut are more exposed to a wide variety of drugs and metabolic abnormalities.

Some well defined diseases of the gut however may be caused by functional disorders of the intrinsic nerves without gross histological damage; examples of these disorders include diffuse spasm of the oesophagus and some types of pseudo-obstruction of the small bowel.

Whatever the cause of the pathological process which affects the nerves of the gut the end result is commonly hypertrophy and dilatation of any affected segment which has lost normal peristalsis. A partial or complete loss of relaxation reflexes in sphincters adjacent to affected bowel may also occur and the consequent obstruction leads to further dilatation of proximal bowel. This sequence of events is clearly seen in achalasia of the cardia and congenital aganglionosis of the distal bowel (Hirschsprung's disease).

This chapter opens with a brief description of the normal histological appearances and regional variations seen in the normal nerve plexuses of mammalian gut and a number of abnormalities affecting each region of the gastrointestinal tract are then described. It is clear from a survey of the literature that the sophisticated techniques currently available for the study of autonomic nerves have yet to be applied to many gastrointestinal diseases and at present the reported observations on human disease are relatively few in number.

1. Anatomy of autonomic innervation

1.1. Intrinsic nerves

Meissner (1857) was the first to describe neurones and a nerve plexus in the submucosa of the bowel wall. This discovery was acknowledged by Lister (1858) who believed from experiments on the gut of rabbits that muscle contractions were regulated through this plexus. Later Auerbach (1864) described ganglion cells connected by unmyelinated fibres which formed a larger plexus between the longitudinal and the circular layers of the muscularis externa – the myenteric plexus. Two nerve plexuses are now recognised in the submucosa – Meissner's close to the muscularis mucosa and Henle's which lies adjacent to the inner aspect of the circular muscle (Gunn, 1968).

Detailed study of the neurones of the bowel wall by Dogiel (1899), using methylene blue staining, suggested a classification of the cells into 3 types: Type I with a single axon and short dendrites; Type II with long processes which ended in the submucosa; and Type III possessing short dendrites in communication with other neurones. Dogiel believed that Type I neurones were motor and Type II sensory in function. The neurones were further classified after silver staining by Hill (1927) into 2 types, the cells either staining darkly and possessing short dendrites or staining lightly with long dendrites. The latter were unipolar, bipolar or multipolar. Honjin et al. (1959), from experiments with mice also based their classification on the affinity of neurones for silver and divided them into argyrophil or argyrophobe cells. Many similar types of classification have now been published (see reviews of Schofield, 1968; Gabella, 1976).

The numbers of neurones per unit area of bowel vary from region to region in all mammals (Gabella, 1979). For example, in the myenteric plexus of guinea pig the duodenum contains more neurones than the ileum, and the colon more than the duodenum. In monkey, however, the myenteric plexus of the ileum contains more than either the duodenum or the colon. Smith (1972) has reviewed the variation in numbers and types of neurones in different levels of the gut in guinea pig, dog, mouse and rabbit and the variation in neuronal counts within single organs such as the stomach. Further, she described the human myenteric plexus in detail from studies with silver staining. In summary, it comprises a mesh of nerve fibres and the neurones are aggregated at the corners of the mesh. Only a small proportion of the nerve cells are argyrophil, the dendrites of which end around non-argyrophil cells. The fine nerve fibres of the plexus appear to be a mixture of intrinsic and extrinsic axons but thick nerve fibres were considered to be the axons of the intrinsic neurones. The proportion of argyrophil cells in the ganglia varies from region to region; for example, it is small in the jejuno ileum and larger in the colon and rectum (Smith, 1972).

Histochemical studies of the bowel with catecholamine fluorescence techniques have revealed a complete absence of adrenergic neurones from the gut of mammals, except the proximal colon of the guinea pig (Furness and Costa, 1974). Cholinesterase studies of human gut revealed intermyenteric ganglia to be irregular in size and shape and to show variable staining. The individual neurones also varied in size and staining and many appeared to be unstained. In the circular muscle moderate numbers of cholinesterase positive nerves were present but there were fewer in the longitudinal muscle (Garrett et al., 1969).

It has been suggested that the cholinesterase positive cells which are the majority of intramural neurones (Koelle et al., 1950) are the argyrophobe cells identified by silver staining techniques (Smith, 1972) and that these are responsible for the motor innervation of the muscle. Argyrophil cells, on the other hand, may coordinate peristalsis through their intraneuronal axons and Leaming and Cauna (1961) suggested that in cat argyrophil cells are cholinesterase negative.

Electron microscopical studies of human myenteric ganglia of the large bowel (Howard and Garrett, 1970) were similar to those described by Richardson (1958) in the small intestine of the rabbit, and the complexity of the neurones, axons and supporting cells suggested a resemblance to primitive brain tissue. The neurones were situated at the periphery of the ganglia with much of their outer surface exposed, unsupported, and close to adjacent muscle cells. The density of cytoplasm varied and occasional cells were seen with a particularly dense cytoplasm which suggested there might be two types of cell perhaps relating to the argyrophobe and argyrophil cells described previously (Smith, 1972). The majority of axons within the ganglia were between 0.4 and 1 μm diameter but many were outside this range. The ganglia contained no collagen and a paucity of definitive synaptic structures were detected (Howard, 1970; Howard and Garrett, 1970).

1.2. Extrinsic nerves

1.2.1. Sympathetic nerves

Although adrenergic neurones are absent from the bowel wall of all mammals except the guinea pig it has a rich supply of adrenergic nerves with the majority of terminal fibres arranged as 'capsules' around the ganglia (Norberg, 1964; Jacobowitz, 1965; Garrett et al., 1969). Adrenergic nerves are comparatively sparse in the muscle layers except in sphincter regions of certain mammals. For example, there is a dense adrenergic innervation of the lower oesophageal sphincter of the guinea pig (Costa and Gabella, 1971), the pyloric region of the rat (Gillespie and Maxwell, 1971) and the anal canal and internal anal sphincter of the cat (Howard and Garrett, 1973). There is little published information on man but the internal anal sphincter is less heavily innervated (Baumgarten, 1967).

The postganglionic sympathetic nerves to the gut arise from adrenergic neurones in the prevertebral ganglia and the preganglionic fibres to these ganglia originate in the thoraco-lumbar spinal cord. Segments T_2–T_7 supply the oesophagus via the stellate ganglia and thoracic sympathetic chain, and segments T_6-T_9 innervate the stomach through the greater splanchnic nerves and coeliac ganglia. The small bowel and large bowel are supplied by segments T_6-T_{11} and $T_{12}-L_2$ respectively.

1.2.2. Parasympathetic nerves

The majority of vagal fibres are preganglionic cholinergic unmyelinated axons which arise in the medulla, the jugular and the nodose ganglia. There are a few noncholinergic inhibitory fibres (Langley, 1898) and a few adrenergic fibres (Gabella, 1979). The vagal innervation extends from oesophagus to transverse colon, and the remainder of the bowel receives a parasympathetic supply from the second and third sacral segments via the pelvic nerves. The parasympathetic supply, acting via enteric ganglia, is motor to most of the gut but seems to have little effect on the smooth muscle of the internal anal sphincter (Garrett et al., 1974). The large area supplied by the vagus means that relatively few vagal fibres control a very large number of intrinsic neurones. Experiments on the mouse colon (Smith, 1972) showed that at least 48 neurones were supplied by one parasympathetic fibre.

1.3. Other types of innervation

Immunofluorescence histochemistry, used originally for studying enterochromaffin cells in the gut, has identified several putative peptide transmitters, although their roles remain to be proven. Some of the peptides, such as substance P, are present throughout the gastrointestinal tract in the neurones of Auerbach's plexus (Pearse and Polak, 1975). The reduced amount of substance P in segments of aganglionic bowel (Hirschsprung's disease) also suggests a close relationship with ganglion cells (Ehrenpreis and Pernow, 1952). Another polypeptide, Vasoactive Intestinal Peptide (VIP) has been found in neurones and postganglionic nerve fibres, especially in Meissner's plexus, and may have a role in smooth muscle relaxation (Polak and Bloom, 1979; Furness and Costa, 1980). Enkephalin, Somatostatin, Neurotensin, Bombesin and Gastrin/CCK have all been identified in the gut wall (Schultzberg et al., 1980) and tissue culture of mouse embryo intestine has shown production of these substances from intrinsic neurones (Schultzberg et al., 1978). There is now evidence that nerves containing these peptides are abnormal in disease states, for example, Crohn's disease and Chagas' disease (Polak and Bloom, 1979).

Pharmacological studies have demonstrated neurones in the gut wall which are associated with relaxation reflexes in intestinal muscle but which are unaffected by antagonists of cholinergic or adrenergic transmission. Adenosine-5-triphosphate has been considered as the transmitter released from these nerves (Burnstock, 1972, 1975a,b) and it has been suggested that they control the non-cholinergic non-adrenergic relaxation of, for example, the lower oesophageal sphincter (Castell, 1975; El Sharkawy and Diamant, 1975). The disturbance of oesophageal motility, known as Achalasia of the cardia may be in part due to interruption of these inhibitory pathways.

The remainder of this chapter is devoted to a survey of the observations which have been reported from studies of autonomic nerves in various disorders of the gut and their relationship to the structure of normal bowel which has been briefly outlined above. Many techniques have been used to elucidate the complex structure of the autonomic innervation of the bowel but as yet few of these have been applied to the study of nerves in gastrointestinal disease.

2. Oesophagus

2.1. Anatomy

The oesophagus begins at the level of the sixth cervical vertebra (Crico-pharyngeus), where it is continuous with the pharynx. It pierces the diaphragm opposite the tenth thoracic vertebra and joins the stomach 2 cm below the diaphragm. The average length of the human adult oesophagus is 24 cm. The upper 25% is composed of striated muscle supplied by fibres from the glossopharyngeal and vagus nerves. Whilst the smooth muscle of the lower half of the oesophagus receives an extrinsic parasympathetic supply through the vagus nerves, the striated and smooth muscle portions are separated by a mixed muscle zone of variable length (Smith, 1972).

Horseradish peroxidase injected into the lower oesophageal sphincter (L.O.S.) of cat has been used to study the origin of the extrinsic nerve supply (Niel et al., 1980). Retrograde axonal transport of the enzyme has shown the neurones of the preganglionic parasympathetic fibres to be situated in the vagal dorsal motor nucleus and the nucleus ambiguus of the medulla. Sympathetic postganglionic neurones were identified in the ventro-medial area of the stellate ganglion and in thoracic ganglia from T_3 to T_{11}. The parasympathetic and stellate ganglion sympathetic axons reached the L.O.S. through the thoracic vagi, but the remainder of the sympathetic axons arrived via the splanchnic nerves.

The muscle coats of the oesophagus are arranged into outer longitudinal and inner circular layers and the myenteric plexus lies between. According to Smith (1972) argyrophil cells account for 20 to 40% of the plexus neurones, which is a greater percentage than elsewhere in the gut (e.g., 15 to 20% in colon). The nerve fibres are also thick, contain many axons, and communicate freely with extrinsic nerves.

2.2. Normal function

Primary peristalsis ('stripping' wave) in the body of the oesophagus is initiated by neural stimulation (Diamant and El Sharkawy, 1977) and maintained by a mixture of neural and myogenic activity (Sarna et al., 1977). Atkinson (1980) has concluded that intact innervation controlled through the medullary swallowing centre is necessary for normal peristalsis. Not infrequently oesophageal peristalsis weakens with age probably due to a weakening of muscle fibres (Hollis and Castell, 1974) and to disappearance of neurones from the myenteric plexus (Eckhardt and Lecompte, 1977).

The striated upper oesophageal sphincter is a barrier to regurgitation of food into pharynx and larynx and oesophageal dilatation causes a reflex rise in sphincter pressure. Relaxation during swallowing is a result of interruption of nervous impulses which originate in the nucleus ambiguus. Damage to the neurones of the nucleus ambiguus in poliomyelitis or motor neurone disease causes the sphincter to remain relaxed with the danger of aspiration pneumonia.

The lower oesophageal sphincter (L.O.S.) is morphologically insignificant but clearly demonstrable with intraluminal recording devices. The sphincter is a specialised area of circular muscle and, as well as the intrinsic properties of muscle fibres, it is influenced by both neural and hormonal mechanisms. In the resting, closed state the sphincter maintains an intraluminal pressure of 15 to 35 mmHg (Cohen, 1979) which falls to zero during coordinated relaxation.

L.O.S. has been investigated in the cat (Gonella et al., 1977) and shown to be under the control of intramural excitatory neurones and non-adrenergic inhibitory neurones both activated by preganglionic cholinergic fibres from the vagi. Active inhibition of the sphincter is initiated by swallowing. This dual control of the L.O.S. by the vagus was first recognised by Langley (1898) in the rabbit.

Interruption of the vagi at cervical or hilar level in dog reduces L.O.S. pressure although the sphincter remains closed (Hwang et al., 1947; Higgs and Ellis, 1965). However, division of these nerves in opossum produces a rise in L.O.S. pressure (Rattan and Goyal, 1974) and the reason for this species difference is not clear. In man, truncal vagotomy at the hiatus has little effect although the rise in sphincter pressure which normally accompanies a rise in abdominal pressure (Lind et al., 1966) is reduced. However, the sphincter is still closed (Angorn et al., 1977).

The L.O.S. has been shown by several workers to be partially controlled by non-adrenergic inhibitory nerves and these were thought by some to be of a 'purinergic' type (Castell, 1975; El Sharkawy and Diamant, 1975). However, recent studies in opossum have shown neither ATP nor adenosine to be the inhibitory transmitter in the vagal pathways (Rattan and Goyal, 1980).

A number of gut hormones may affect L.O.S. pressure. Pentagastrin increases the intersphincteric pressure in man (Giles, 1969) and in dog this effect has been shown to be a direct action on smooth muscle (Zwick et al., 1976). However, observations in patients with reduced sphincter pressures and reflux showed no reduction in fasting serum gastrin levels (Farrell et al., 1974). The role of gastrin in the physiological control of the L.O.S. in man is not yet proven. Secretin, Cholecystokinin, Glucagon and Motilin have also been shown to affect L.O.S. pressure (Atkinson, 1980).

2.3. Oesophageal pathology

2.3.1. Lower oesophageal sphincter incompetence (gastro-oesophageal reflux)

Gastro-oesophageal reflux causes pain (heartburn), oesophagitis, and occasionally oesophageal stricture. The major abnormality is loss of tone in the L.O.S. and abnormalities of muscle, gut hormones and nerves have been considered in the aetiology. A primary abnormality of muscle seems unlikely as the muscle of the sphincter responds well to cholinergic drugs (Lipshutz et al., 1973). Further, patient studies have also failed to reveal evidence of hypogastrinaemia in cases of reflux (Dent and Hansky, 1976). The evidence for a defect in innervation is stronger in that the normal rise in sphincter pressure during increases of abdominal pressure is absent (Lind et al., 1966); the pressure-rise reflex is diminished after truncal vagotomy (Angorn et al., 1977); gastric emptying is delayed in cases of hiatus hernia (Donovan et al., 1977) and reduced L.O.S. pressures with reflux may accompany diabetic autonomic neuropathy, (Stewart et al., 1976).

Heatley et al. (1980), studied vagal function in 28 patients with gastro-oesophageal reflux using the effects of insulin hypoglycaemia on gastric secretion, and pulse rate variation with respiration. Twenty-five percent of the patients with reflux showed evidence of impaired vagal efferent function in the upper alimentary tract whilst 12 of the patients showed some evidence of dysfunction of cardiac vagal fibres.

Reduced L.O.S. pressure may also be a feature of infancy and in its extreme form presents as severe reflux in the presence of a relaxed L.O.S. and known as Chalasia. The sphincter muscle shows normal responses to cholinergic drugs (Moroz et al., 1974) and spontaneous resolution during the first few months of life is usual.

2.3.2. Achalasia of the cardia (cardiospasm)

Achalasia is caused by a neurogenic lesion which results in a loss of normal oesophageal peristalsis and an impairment of L.O.S. relaxation. It presents most commonly in middle life, but all ages may be affected, including the newborn (Thompson, 1950). Dysphagia with food sticking in the chest is the usual presenting symptom, although in older age groups aspiration pneumonia may be a problem. The severity of the dysphagia may vary with acute exacerbations. Barrett (1953) described 3 phases in the natural history of achalasia: (1) the body of the oesophagus shows more contractions than normal and there is some hold-up at the cardia; (2) the muscle of the body is hypotonic, normal peristalsis is absent and ineffectual 'tertiary' contractions occur; and (3) the oesophagus is tortuous enough to produce mechanical obstruction at the L.O.S. region.

Rake (1927) was the first to describe a lesion of the myenteric plexus in achalasia. He reported a loss of neurones and round cell infiltration between the longitudinal and circular muscle layers. The loss of neurones from the body of the oesophagus has been confirmed by several observers (Trounce et al., 1957; Adams et al., 1960). The fate of neurones in the L.O.S. seems more variable. Adams et al. (1960) and Cassella et al. (1964) reported some retention of neurones in the lower oesophagus, whilst Misiewicz et al. (1969) described a loss throughout the oesophagus. Smith (1972) confirmed the variability of the neuronal loss when she described the complete loss of neurones in one oesophagus and the retention of argyrophobe cells in a biopsy from the sphincter area of a second patient.

It is not clear whether a similar variability in pathology affects the extrinsic vagus nerves. Degenerative changes in vagus nerves and the dorsal vagal nucleus in the medulla have been reported in achalasic patients (Casella et al., 1964), but not confirmed in other reports. Lendrum (1937) examined the vagus nerves in 10 cases and found no abnormality, and Smith (1972) examined the vagus in her case using osmium and silver staining but could find no definite lesion.

Pharmacological evidence for an abnormality of the vagus nerves has been provided through the use of the Hollander test in which acid secretion is stimulated during insulin hypoglycaemia. The reflex depends on intact vagi and is therefore abolished by the operation of truncal vagotomy. Abnormal Hollander tests were reported in 6 out of 13 (Iordanskaya, 1962) and 14 out of 32 (Woolam et al., 1967) patients with achalasia, suggesting that there is indeed a vagal lesion in some cases.

Pharmacological studies reflect the histological observations. For example, 5 out of 6 strip-biopsy specimens taken from the L.O.S. region of patients with achalasia responded by contracting to the application of nicotine after eserine, suggesting that neurones were present (Trounce et al., 1957). Similar experiments on the dilated body of the oesophagus, however, showed no response to nicotine, and electrical stimulation experiments also showed an absence of nerve responses (Ellis et al., 1960).

The results of the degeneration of intrinsic nerves of the oesophagus, revealed by histological and pharmacological techniques, are seen on barium and manometric studies as a loss of coordinated peristalsis in the body of the oesophagus although non-peristalsis low-amplitude contractions continue. The L.O.S. has a raised resting pressure which may be more than twice that recorded in normals, and relaxation, which is complete in normals is frequently less than 40% in patients with achalasia (Cohen, 1978, 1979). Hormonal influences may be responsible for some of the L.O.S. dysfunction and from observation of patients Cohen et al. (1971) have provided evidence for supersensitivity to endogenous gastrin. Pressure reduction can be shown by gastric acidification, which reduces gastrin release, and an increased sensitivity to gastrin I has also been demonstrated.

The dysfunction of the L.O.S. in achalasia is probably due to a combination of neural and hormonal effects. The complex control that the vagus normally exerts through communications with myenteric cholinergic neurones and non-adrenergic inhibitory neurones (Gonella et al., 1977) is destroyed by the degenerative process (Tuch and Cohen, 1973) and a denervation supersensitivity to gastrin may well be an additional factor.

The aetiology of achalasia remains unknown, although most cases seem to be acquired rather than congenital. Cassella et al. (1964) suggested that a brain stem lesion is the primary defect and that the myenteric damage represents a transsynaptic degeneration. A neurotropic viral infection, such as herpes zoster attacking neurones in the vagal nucleus and then travelling along vagal axons has also been suggested (Smith, 1970). The presence of lymphocytes in ganglion sites (Cassella et al., 1964; Misiewicz et al., 1969) has been cited as evidence for an infective or auto-immune process.

2.3.3. Tumour induced achalasia
Neoplastic invasion of the myenteric plexus can produce severe dysfunction of the L.O.S. giving the appearances of achalasia (Tucker et al., 1978). Treatment of the neoplasm may return the oesophageal motor activity to normal (Davis et al., 1975).

2.3.4. American trypanosomiasis (Chagas' disease)
This type of megaoesophagus is caused by the protozoon parasite Trypanosoma cruzi endemic in parts of South America and Southern Mexico. The vector is Triatoma infestans ('Kissing Bug'), a blood sucking insect which spreads the parasite by faecal contamination. The acute phase of infestation usually occurs within the first 5 years of life with fever splenomegaly and lymphadenopathy, and the parasite lodges in the nervous plexuses throughout the gastro-intestinal tract. The heart, biliary tract and urinary tract may also be affected (Raia, 1955; Ferreira-Santos, 1961; Köberle, 1963). Köberle (1956) suggested that neurotoxin released from dead parasites destroyed the autonomic nerves but more recently the nerve damage has been thought to be caused by immunological injury (Santos-Buch, 1979). Lymphocytes sensitized to T. cruzi may be reactive to components of the nerve plexuses and anti-parasympathetic-neurone IgG has been described in 83% of patients. Vagal ganglion cells are also destroyed and replaced with lymphocytic infiltrates.

The physiological abnormalities have features similar to those found in Achalasia of the Cardia. The muscle of the body responds to cholinergic drugs (e.g.,

Mecholyl) (Santos-Buch, 1979) and the L.O.S. pressure has been reported as either normal or raised (Bettarello and Pinotti, 1976); or reduced, when compared with normal controls (Padovan et al., 1980). Using dose–response studies to intravenous pentagastrin the latter authors also showed that chagasic patients possess a lower sensitivity to the stimulant than did the controls, and this was considered as evidence that the full stimulatory effect of pentagastrin on the L.O.S. depends on an interaction between the hormone and cholinergic nervous excitation, similar to the synergism between gastrin and cholinergic nervous excitation on the gastric oxyntic glands (Davison, 1974).

2.3.5. Diffuse oesophageal spasm
The diagnosis of diffuse spasm depends on clinical and manometric studies (Cohen, 1979). The condition is characterized by chest pain, which may be mistaken for angina pectoris, dysphagia and high amplitude, non-peristaltic contractions occurring with at least 30% of swallows (Dimarino and Cohen, 1974).

Diffuse spasm may change into primary achalasia in 3 to 5% of cases (Vantrappen et al., 1979), which may help to explain the pathology. Abnormal relaxation or raised L.O.S. pressure are seen in 30% of patients (Dimarino and Cohen, 1974) and swallowing may cause repetitive contractions that are prolonged in duration. Myenteric neurones are present in oesophageal spasm, but are infiltrated with inflammatory cells. The oesophageal branches of the vagi show degenerative changes, and there is increased sensitivity to cholinergic drugs (Mellow, 1977) and to intravenous gastrin (Eckhardt and Weigand, 1974; Cohen, 1975). As with achalasia, the aetiology is unknown.

2.3.6. Systemic sclerosis (scleroderma)
The oesophagus may be involved in a variety of collagen diseases including systemic sclerosis, Raynaud's disease, dermatomyositis etc. The lesion is a patchy atrophy of smooth muscle with progressive fibrosis and some loss of ganglion cells (Goetz, 1945). Physiological abnormalities can be explained from the observations. In early stages of the disease manometric studies demonstrate reduced peristaltic force which may progress to complete loss of oesophageal function. Similarly, reduced pressure in the L.O.S. eventually leads to total incompetence and consequent reflux produces oesophagitis and stricture in 40% of patients (Cohen, 1979). In early cases L.O.S. muscle reacts to cholinergic drugs but in late cases this response is lost (Cohen et al., 1972). Gastrin responses are similarly impaired.

Hurwitz et al. (1976), noting the close relationship with Raynaud's phenomenon, suggested that the muscle damage and fibrosis are a result of ischaemia on the oesophagus (Earlam, 1972).

3. Stomach and pylorus

3.1. Stomach

Smith (1972) has given a complete description of the anatomy of the stomach. The muscle of the stomach is composed of 3 interconnected layers. The outer longitudinal muscle splits into 2 bands which run along the lesser and greater curves and is continuous with the longitudinal muscle of the oesophagus. The longitudinal bands converge at the pylorus to form a complete coat of muscle. The middle circular layer is a continuation of the outer fibres of the circular muscle of the oesphagus and at the distal end of the stomach it forms the pyloric sphincter. The inner oblique muscle coat is continuous with fibres on the luminal side of the oesophageal circular muscle, and it spreads out over the anterior and posterior gastric surfaces, gradually disappearing. The myenteric plexus has a very irregular distribution particularly amongst muscle bundles of the circular coat, both in the stomach and pyloric sphincter and the ganglia tend to be widely spaced. This plexus is continuous with the myenteric plexus of the duodenum. Most of the neurones in the fundus of the stomach are described as argyrophobic, but the numbers of argyrophil cells increase towards the pylorus.

The intrinsic nerves do not seem to be essential for gastric motility and multiple divisions of the plexus or ganglion blockade with nicotine failed to abolish peristalsis (Cannon, 1911; Thomas and Kunz, 1926). Duodenal peristalsis is particularly important for efficient gastric emptying and the prevention of reflux, and if damaged then gastric emptying is severely disturbed (Edwards, 1961).

Stimulation of vagal nerves increases gastric peristalsis, increases secretion, and inhibits pyloric sphincter activity and vagotomy alters gastric motility by impairing reservoir function, increasing basal tone and speeding the emptying rate for fluids. One year after vagotomy the basal pressure is still raised and the strength of gastric contractions is still diminished (Wilbur and Kelly, 1973; Stadaas et al., 1974; Stadaas, 1980).

Sympathetic nerve activity reduces peristalsis, causes pyloric closure and probably reduces secretion. Splanchnic (sympathetic) nerve section, however, probably has little effect on gastric motility (Cannon, 1906).

3.1.1. Peptic ulceration
Whilst the aetiology of most peptic ulcers remains unknown gastro-duodenal ulceration may occur in association with lesions in the central nervous system. Peptic ulceration after intracranial operation in man was first reported by Cushing (1932) and has been produced in animals after damage to the hypothalamus (Long et al., 1962). Sympathectomy gave protection from ulceration when the lesions were confined to the anterior hypothalamus whilst vagotomy was effective for lesions in the posterior hypothalamus (Keller, 1936a,b; Maire and Patton, 1968).

3.1.2. American trypanosomiasis (Chagas' disease)

Degeneration of myenteric neurones in the stomach usually occurs in association with megaoesophagus in Chagas' disease. In a study of 22 patients, 4 showed delay in gastric emptying with barium retained for more than 6 h. Three of these patients showed a hyperactive muscle response to subcutaneous methacholine. Decreased secretion of both acid and pepsin accompany these changes (Padovan et al., 1977).

3.2. Pyloric innervation

Experiments in animals have shown the pylorus to have greater spontaneous muscle activity than either the antrum of the stomach or the duodenum (McCrea and McSwiney, 1926; Mir et al., 1977, 1979). Vagal stimulation caused an increase in pyloric activity but the response was converted to inhibition-relaxation after atropine. The relaxation was unaffected by adrenergic blocking agents. Isolated muscle strips from the pylorus of opossum, cat, dog and man showed similar inhibitory responses (Anuras et al., 1974).

The experiments have been repeated in dogs (Telford et al., 1979) and again vagal stimulation increased motor activity in the pylorus and produced inhibition after atropine. Division of the nerve of Latarjet abolished the effect proving that the inhibitory neurone pathway lies within the vagus nerves. The transmitter for this non-adrenergic inhibition is unknown, but the experiments of Telford et al. (1979) ruled out histamine.

In summary, the pyloric sphincter exhibits spontaneous, rhythmic, contractile activity which is modified by both motor and inhibitory vagal impulses.

3.3. Pyloric pathology

3.3.1. Hypertrophic pyloric stenosis of infancy

Hypertrophy of the muscle of the pyloric sphincter causing gastric outlet obstruction (Hirschsprung, 1888b), occurs within 3 months of age in 3 out of every 1000 live births. There is a male preponderance of approximately 4:1 and a genetic predisposition is well documented. The genetic basis appears to involve multiple genes and does not follow Mendelian principles (Carter and Evans, 1969). The clinical features include forceful projectile vomiting, gastric distension with constipation secondary to dehydration. Delayed gastric emptying and an elongated pyloric canal showing no contractile activity are the radiological features (Runström, 1939).

Hypertrophic pyloric stenosis does not seem to be congenital in origin. Wallgren (1946) examined 1000 new born male infants with barium meals, all of which showed no abnormality. Five of the children developed pyloric stenosis within the following few weeks.

Operative procedures which bypass the obstructed pylorus lead to persistence of the muscle hypertrophy even into adult life (Armitage and Rhind, 1951), but division of the muscle (myotomy) allows resolution of the condition within 3 months.

The smooth muscle hypertrophy affects mainly the circular coat of the pyloric sphincter with secondary thickening of the gastric muscle (Belding and Kernohan, 1953; Friesen and Pearse, 1963).

Many varying observations have been made on the myenteric plexus. Herbst (1934), in 3 cases, found evidence of disintegrating nucleoli and a loss of normal chromatin patterns in pyloric neurones and an infiltration of inflammatory cells in the plexus. An extensive study by Belding and Kernohan (1953) of material from infants and adults, with pyloric stenosis revealed normal ganglia in the stomach but the majority of neurones in the pylorus showed indistinct nuclear membranes, altered cytoplasm and often appeared as 'ghost' cells. The appearances were likened to the changes seen after excessive vagal stimulation. Increased numbers of dead neurones and disorganisation of ganglia were reported by Alarotu (1956) but Friesen et al. (1956) believed that the appearances were caused by many histologically immature cells and a relative absence of mature neurones, suggesting the plexus seen in premature infants. Whilst agreeing that neurones were sparser and smaller than normal in hypertrophic pyloric stenosis Roberts (1959) disputed that 'maturation' interpretation and pointed out that the condition is not more common in premature infants.

Silver staining techniques showed a striking absence of argyrophil cells in the abnormal pylorus whilst argyrophobic cells appeared normal (Rintoul and Kirkman, 1961); Smith (1972) suggested that failure of argyrophil maturation in infancy leads to failure of pyloric contractility which is so important for gastric emptying. She explained the muscle hypertrophy as an effect of denervation.

Histochemical studies of pyloric muscle from 15 affected infants revealed abundant cholinesterase activity in ganglion cells which therefore appeared to be living and not degenerate or dead cells. However, only a minority of neurones had the appearance of large mature cells lending support to the idea that there is an arrest of normal development in the myenteric plexus of the pylorus (Friesen and Pearse, 1963).

Ultrastructural observations on ganglia in hypertrophic pyloric stenosis did not reveal morphological abnormalities in the neurones, and were in basic agreement with the data of Friesen and Pearse (Challa et al., 1977). Numerous, swollen, degenerating axons were seen although the authors thought that these changes might be caused by the compression of the hypertrophied muscle, individual fibres of which appeared to be normal.

Many authors have suggested a neurogenic basis for hypertrophic pyloric stenosis (McCrea, 1924; Bendix and Necheles, 1947; Belding and Kernohan, 1953), whilst Dodge (1970, 1976) has shown that the administration of pentagastrin to foetuses of bitches induced hypertrophic pyloric stenosis in some of the puppies. Some of the puppies also developed peptic ulcers. Further studies showed that ganglion cell changes were present in the abnormal pyloruses, and that pyloric motility was disorganised (Dodge, 1976).

Fasting plasma gastrin levels in 14 affected infants were found by Rogers et al. (1974) to be similar to those of controls although greater volumes of gastric juice were measured in the patients. They speculated that pyloric hypertrophy might be the result of increased cholecystokinin and secretin action on the pyloric muscle.

The measurement of immunoreactive gastrin levels in infants with pyloric stenosis has given conflicting results to date. Spitz and Zail (1976) reported raised levels compared with controls, but Grochowski et al. (1980), found no significant difference in the two groups.

3.3.2. Hypertrophic pyloric stenosis in adults

Hypertrophic stenosis may occur at any time during adult life and the appearances of the pylorus at operation are very similar to the infantile disease. However, a male preponderance is not seen in adults (Du Plessis, 1966) and the condition is often associated with gastric ulcer, which may be a secondary phenomenon (Bateson et al., 1969). Examination of 2 adult cases revealed an absence of argyrophil cells and some increase in Schwann cells, reminiscent of the changes seen in achalasia of the cardia (Smith, 1972).

4. The small intestine

In a review of the morphology of the normal human small bowel Smith (1972) describes the myenteric plexus as lying directly on the longitudinal muscle. The neurones are described as predominantly argyrophobic with argyrophil cells increasing in numbers towards the distal ileum. The mesh of the small bowel plexus is wider than in the colon and there is a paucity of thick axons which are usually associated with argyrophil cells. The parasympathetic nerve supply does not seem essential to small bowel activity as this is rarely disturbed by the operation of truncal vagotomy. Increased activity of adrenergic nerves, however, may lead to prolonged cessation of effective peristalsis and intestinal obstruction (paralytic ileus). Furness and Costa (1974) reviewed the nervous pathways involved in ileus and concluded that they involve peripheral, prevertebral ganglionic, and spinal routes. Many causes of paralytic ileus are recorded, but the common reasons include handling of the bowel during abdominal procedures and peritoneal infection. The local treatment of prolonged ileus by blocking noradrenaline release or blocking the action of sympathetic nerves on myenteric ganglia has been used successfully in post-operative patients (Neely and Catchpole, 1971; Petri et al., 1971). The stimulation of intestinal motility by the anticholinesterase neostigmine is also effective in less severe cases (Heimbach and Crout, 1971).

4.1. Duodenum

4.1.1. Megaduodenum

Reported lesions of the myenteric plexus of the small bowel are rare but dilatation of the duodenum down to the area crossed by the superior mesenteric vessels is well described. The cause of the hypertrophy and dilatation of the duodenum has been ascribed to the mechanical obstruction by the vessels but 2 cases have occurred in patients with non-rotation of the colon where the superior mesenteric vessels did not cross the bowel (Nell, 1933). Further, bypass operations such as duodeno-jejunal anastomosis may cause no diminution in duodenal size (Barnett and Wall, 1955). In the latter case histological examination revealed an absence of ganglia in the duodenal wall although a normal myenteric plexus was present in the jejunum.

4.1.2. Congenital aganglionosis (Hirschsprung's disease)

Congenital aganglionosis of distal bowel may rarely be associated with a lack of ganglia in gut as far proximal as the duodenum. Bodian et al. (1951) described one case in a large series in which the aganglionosis extended from duodenum to rectum. In a review by Talwalker (1976) of long segment cases with small bowel involvement, 6 of 11 showed aganglionosis extending up to the distal duodenum whilst in 1 case the entire duodenum was aganglionic and associated with ganglion cell loss in the oesophagus.

4.1.3. Systemic sclerosis (scleroderma)

This condition which has been described in the oesophagus may affect the duodenum causing dilatation and obstructive symptoms. An investigation by Dimarino et al. (1973) showed normal smooth muscle contractions and abnormal cholinergic reflexes during distension. Although histology is not available ganglion cell damage is a likely explanation (Goetz, 1945).

4.1.4. American trypanosomiasis (Chagas' disease)

Raia et al. have described 12 cases of megaduodenum in Chagas' disease with muscle hypertrophy and dilatation, and histology confirmed degeneration of the myenteric plexus. Three of these cases were associated with megaoesophagus and megacolon.

4.1.5. Autonomic neuropathy of unknown origin

Autonomic neuropathy may occur as an acute phenomenon at any age. The aetiology is unknown. The lesions are widespread throughout the autonomic nervous system and may include ileus, retention of urine, failure of lacrimation and sweating, and postural hypotension (Hopkins et al., 1974). Achalasia of the cardia may develop causing dysphagia and at a later stage massive dilatation of the duodenum may be associated with poor gastric emptying (Howard, unpublished observations). Histology is not available.

4.2. Jejuno-ileum

Motor activity of the small bowel may be disordered in a wide range of conditions. Sepsis, electrolyte imbalance and brain injuries may all be associated with ileus (Gellis, 1963), and obstruction may occur in the newborn from the transplacental transmission of maternal drugs such as ganglion blocking agents and heroin (Hallum et al., 1954).

Small bowel dysfunction has also been described in older patients in association with diabetes mellitus (Malins and French, 1957; Katz and Spiro, 1966) and in systemic amyloidosis (Legge et al., 1970).

A large group of chronic obstructive conditions in the small bowel have now been reported the clinical presentations of which suggest a mechanical cause. Investigations, however, have revealed motility disorders and the aetiologies of these rare lesions have included smooth muscle degeneration (Schuffler et al., 1977), abnormality of neural control mechanisms (Sullivan et al., 1977), abnormal non-adrenergic inhibition (Lewis et al., 1978) and neuronal abnormalities (Tanner et al., 1976).

Four cases of chronic adynamic bowel which remains thin and dilated, have been reported in infants and study of the intrinsic nerves has shown no obvious abnormality with conventional histological techniques (Kapila et al., 1975).

Many of these conditions have previously been grouped under the title 'chronic idiopathic intestinal pseudo-obstruction', but with the aid of histochemical, ultrastructural and physiological investigations it may be possible to pinpoint abnormalities in the bowel wall in most cases.

4.2.1. Chronic idiopathic intestinal obstruction
This diagnosis should now be reserved for cases in which ganglion cells are present with normal morphology. Shaw et al. (1979) reviewed 75 cases gathered from the literature. The condition starts in adolescence or young adulthood with attacks of pain distension and vomiting and there is frequently a family history of similar symptoms.

Plain X-rays suggest mechanical obstruction but barium studies show disordered motility only. Occasionally a similar disorder is seen in infancy. The adolescent type is typically intermittent and sometimes accompanied by a motility disorder in the urinary tract (Faulk et al., 1978). Detailed barium studies have shown normal gastric emptying, hypotonic duodenum and a slow transit through the small bowel which may take 24 h.

Although ganglion cells are present in the bowel wall studies by Sullivan et al. (1977) of myoelectrical activity suggested a neural dysfunction. As yet there are no reported quantitative studies on numbers of neurones in the myenteric plexuses of these patients. Intestinal resection has helped to relieve symptoms in some cases.

4.2.2. Segmental dilatation
Nine cases of intestinal obstruction in which there was a single well defined segment of dilated bowel without mechanical obstruction or obvious neurological lesion were reviewed by Irving and Lister (1977). A common finding was heteroplastic tissue (striated muscle, lung, etc.) in the walls of the dilated segments and it was suggested that interruption of the myenteric plexus produced the functional disorder. X-rays may be deceptive in suggesting a diagnosis of congenital aganglionosis.

4.2.3. Familial neuronal disease
Small bowel 'pseudo-obstruction' may occur in association with generalised neurological disease (ataxia, dysarthria, etc.) (Schuffler et al., 1978). Post mortem studies of 2 adult siblings showed degenerative changes in the myenteric plexus throughout the gastro-intestinal tract. The sections were compared with normal controls and the muscle was noted to be normal.

One third of the neurones contained eosinophilic intra-nuclear inclusions which appeared at the ultra-structural level to be non-viral, non membrane-bounded filaments. Neurones of the central nervous system contained similar intra-nuclear inclusions. The numbers of neurones and axons in the myenteric plexus were reduced whilst the dendrites were swollen.

4.2.4. Crohn's disease
Crohn's disease is a transmural inflammatory lesion causing symptoms related to inflammation and malabsorption. Ulceration may progress to stricture formation, fistulae and abscesses. The bowel is frequently affected in a patchy manner ('skip' lesions) and fissuring is characteristic with clefts passing deeply into the bowel wall and lined with granulation tissue. Non-caseating granulomata are common.

Nerve fibres in the myenteric plexus of affected areas of bowel show thickening. Study of the peptidergic substance V.I.P. has revealed severe abnormalities in Crohn's disease. The V.I.P. nerves are thickened and disorganised and are excessively immunoreactive. There seems to be as much as a 4-fold increase in V.I.P. containing nerves and the myenteric plexus appears larger and more spread out than normal (Polak et al., 1978; Polak and Bloom, 1979). It is not yet clear whether the changes in the autonomic nerves are primary or secondary effects and very little has been published on the morphology of the myenteric plexus in Crohn's disease studied by conventional histological or histochemical techniques.

5. Large bowel and anal sphincter

The human colon and rectum are approximately 165 cm in length. From the caecum to the recto-sigmoid junction the outer longitudinal muscle coat is arranged mainly in 3 thick bands, the taeniae coli. The longitudinal muscle between the taeniae is very thin and sacculations or haustrations, are produced by the tonus and contractions of the taeniae. The circular muscle of the distal rectum is modified to form the internal anal sphincter which, unlike the lower oesophageal sphincter, is easy to identify. This sphincter is intimately related, both physiologically and anatomically to the voluntary muscle of the pelvic floor (levator ani) and external anal sphincter, so that neurological or muscular lesions in either smooth or skeletal muscle structure can have severe effects on the control of faecal continence. The internal sphincter contributes 85% of the pressure in the anal canal at rest but only 40% after rectal distension when continence is maintained by the striated muscle (Frenckner and von Euler, 1975).

The motor innervation of the voluntary muscles arises in the 4th sacral segment of the spinal cord and reaches the muscle via the pudendal nerves. The parasympathetic supply of proximal colon is vagal whilst that for the distal bowel emerges from the anterior roots of the sacral nerves, joins the pelvic nerves (nervi erigentes) and intermingles with post ganglionic sympathetic nerves in the hypogastric plexuses. The sympathetic system of nerves includes the splanchnics, the preaortic ganglia, and the superior and inferior hypogastric plexuses. Both sympathetic and parasympathetic nerves reach the distal bowel wall with the rectal and colonic arteries (see review by Garry, 1957).

Intrinsic neurones in the large bowel are present in normal numbers almost to the internal anal sphincter. In a study of 20 infants and children Aldridge and Campbell (1968) showed that a hypoganglionic zone extended cranially from the pectinate line of the anal canal for a distance of 4 mm in the myenteric plexus, 7 mm in the deep submucous plexus and 10 mm in the superficial submucous plexus. The ganglia in premature infants were not dissimilar from those of mature children.

The specialised neural control of the internal anal sphincter has been investigated in man and animals. From observations with spinal anaesthesia in man it was concluded that resting tone was maintained by a tonic excitatory sympathetic discharge. The results also indicated that there was no parasympathetic discharge affecting tone (Frenckner and Ihre, 1976). Cat experiments have also shown an excitatory effect of noradrenaline on sphincter muscle and a lack of effect with acetyl choline (Garrett et al., 1974; Bouvier and Gonella, 1981). The internal anal sphincter of cat and vervet monkey also showed prominent non-adrenergic non-cholinergic inhibitory responses (Garrett et al., 1974; Garrett and Howard, 1981; Rayner, 1979). The conclusion from these experiments is that at any instant numerous mechanisms are affecting the state of contraction or relaxation in the internal anal sphincter. These include direct effects on the muscle and indirect effects via the ganglia supplying the nerves to the muscle (Garrett and Howard, 1981).

The influence of extrinsic autonomic nerves on the morphology and functions of intrinsic plexuses in the large bowel is well illustrated by the results of a resection of the nervi erigentes in a patient with multiple sclerosis performed in an attempt to cure detrusor muscle spasm in the bladder. There was inability to defaecate, transit time in the colon was prolonged and the left colon was eventually resected. Histology showed changes in Auerbach's plexus with loss of neurones, diminution in neuronal size, and Schwann cell hyperplasia. Meissner's plexus was unaffected and the changes were thought to be caused by 'trans synaptic' degeneration (Devroede and Lamarche, 1974). Similar changes have been recorded in patients who sustained damage to extrinsic nerves in the lumbo-sacral area from trauma (Devroede et al., 1979).

As in other areas of the gut the function of muscle and nerves of the large bowel may be affected by disorders of other organs in the body. Constipation for example is a feature of cretinism or myxoedema, excess vitamin D and hypercalcaemia (Hendersen, 1968) and porphyria (Goodall, 1967).

5.1. Developmental anomalies (neuronal dysplasias)

5.1.1. Congenital aganglionosis (Hirschsprung's disease)

After the classical description of congenital megacolon associated with chronic constipation was made by Hirschsprung (1888a) attention was concentrated for a long time on the proximal dilated megacolon. Three early reports of histological abnormalities in Auerbach's plexus of bowel distal to the megacolon (Tittel, 1901; Dalla Valle, 1920; Cameron, 1928) were ignored although all 3 suggested that there might be a distal segment of bowel incapable of normal peristalsis. Three later papers defined a constant histological abnormality in a variable segment of undilated rectum and colon which consisted of an absence of ganglion cells and a presence of large nerve trunks between the muscle layers of the bowel (Whitehouse and Kernohan, 1948; Zuelzer and Wilson, 1948; Bodian et al., 1949).

Early histochemical studies of aganglionic bowel showed the large nerve trunks to be strongly positive for acetylcholinesterase (AChE) and they were believed to be cholinergic and parasympathetic (Kamijo et al., 1953; Adams et al., 1960). Further work demonstrated AChE-positive nerves within the circular muscle (Niemi et al., 1961; Meier-Ruge, 1968) and this suggested that the muscle might be innervated by cholinergic nerves.

Later investigations (Garrett et al., 1969; Howard, 1970, 1972) revealed that bowel resected from patients with Hirschsprung's disease contained a pattern of variation in the distribution of AChE-positive nerves at different levels within each specimen. There were also wide variations in the numbers of nerves between different cases. The circular muscle of distal rectum usually contained more AChE-positive nerves than normal rectum and the severity of the presentation had a direct relationship with the numbers of such nerves. It is likely that the nerves are exerting a motor activity for spinal anaesthesia causes relaxation (Ehrenpreis, 1946). Ascending the bowel these nerves gradually diminished in number until the most proximal aganglionic tissue contained fewer nerves than normal. In addition, the zone above the aganglionic bowel, where ganglia first appear, also contained fewer muscular nerves than normal, but it has been noted that this area of bowel may show maximal contraction on barium enema examination (Ehrenpreis, 1970; Garrett and Howard, 1981). The factors responsible for this zone of maximal contraction are unknown and it is of interest to speculate on whether they are neural, hormonal, or a mixture of the two.

The normal arrangement of adrenergic nerves around ganglia is, of course, absent in the aganglionic segment of Hirschsprung's disease. Instead, the nerves show a variable distribution throughout both muscle layers. As with AChE-positive nerves the largest number of adrenergic fibres was found in distal rectum and these gradually decrease cranially (Bennett et al., 1968; Gannon et al., 1969).

Ultrastructural examination of the large nerve trunks of the aganglionic intermuscular zone revealed the presence of a few myelinated axons (Howard and Garrett, 1970). The structure of the non-myelinated axons within the muscle layers showed no major difference from normal except a tendency for the numbers of axons making up each Schwann axon bundle to be increased. Sites similar to the neuro-effector junctions of normal bowel were easily identifiable, suggesting that the nerves are functional.

The majority of aganglionic segments start at the internal anal sphincter and extend cranially for a variable distance. The aganglionosis may extend throughout the colon (Zuelzer–Wilson syndrome) (Meier-Ruge et al., 1972) and in these cases the colon is innervated as far as the splenic flexure by sacral parasympathetic nerve fibres.

In some cases the aganglionosis may involve the whole of the gastro-intestinal tract (Talwalker, 1976).

Occasional reports have appeared suggesting that aganglionosis may occur as a 'skip' lesion (MacIver and Whitehead, 1972; Martin et al., 1979). In the latter case both the ascending and descending colon were aganglionic and the ganglionic transverse colon was used for surgical reconstruction.

Recent work by Bishop et al. (1980) has shown that peptide-containing nerves are less abundant but not completely absent in aganglionic segments but there is no available data yet on any possible variation in these nerves along the length of the abnormal gut.

A rare association of total colonic aganglionosis and a failure of automatic control of ventilation in infants has been reported in 4 cases (Stern et al., 1981).

A common defect of stem serotonergic nerve cells has been postulated to be the cause of these combined brain and gut abnormalities (Haddad et al., 1978).

Physiological studies of aganglionic bowel show contractions which bear no relation to coordinated peristalsis in proximal ganglionic regions (Swenson et al., 1949; Hiatt, 1951). Further, pressure studies of the ano-rectal region in Hirschsprung's disease demonstrate an absence of the relaxation reflex in the smooth muscle of the internal sphincter which normally occurs during rectal distension (Howard and Nixon, 1968). In summary there are at least 3 physiological abnormalities in aganglionic bowel that contribute to the intestinal obstruction which characterises the condition and which causes a proximal megacolon. There is an absence of coordinated peristalsis caused by the absence of ganglion cells, there are uncoordinated contractions of variable strength and duration which may be related to the degree of cholinergic innervation to the muscle layers, and inhibitory reflexes between the rectum and internal anal sphincter are absent.

The histochemical studies suggest that surgical resections of affected bowel should always include adequate lengths of the dilated ganglionic as well as the undilated aganglionic bowel to include the 'transitional' zone with its reduced muscle innervation. It is possible that inadequate removal of this tissue may lead to the incomplete relief of symptoms. The clear identification of ganglia using a rapid non-specific esterase staining technique has been suggested for this purpose (Garrett and Howard, 1969).

This histochemical assessment of innervation in the submucosa, which is frequently increased in aganglionic bowel, has been used for the diagnosis of Hirschsprung's disease on mucosal biopsy specimens (Meier-Ruge, 1968). An increase in AChE-positive nerves in the mucosa was considered diagnostic of aganglionosis. Gannon et al. (1969) depended on an increased adrenergic innervation of the muscularis mucosae, but the technique is probably too sophisticated for general use. In our hands neither method on mucosa alone has given unequivocal results in all cases.

5.1.2. Hypoganglionosis

The occurrence of reduced numbers of ganglion cells in the 'transitional' zone of Hirschsprung's disease is well documented (Garrett et al., 1969; Meier-Ruge, 1974), but reduced numbers of ganglia in the absence of an aganglionic segment have now been identified in some patients with severe constipation (Bentley, 1964, 1971; Ehrenpreis, 1970; Meier-Ruge, 1974; Munakata et al., 1978; Garrett and Howard, 1981). Far fewer and smaller ganglionic masses than normal are present in the distal bowel and there is a sparse innervation in the muscle layers. The condition could be missed on routine histology with inadequate biopsy specimens and could easily be considered as short segment Hirschsprung's disease. Anorectal myotomy (Bentley, 1964) and enzyme histochemical staining techniques are recommended for exact diagnosis (Garrett and Howard, 1981). Meier-Ruge

(1974) stated that 'hypoganglionosis cannot be confirmed with the desired reliability by examination of mucosal biopsies'. We have found that acid-phosphatase staining is particularly revealing for ganglion cells in the intermyenteric zone (previously unpublished observation).

5.1.3. Hyperganglionosis
Reports of increased numbers of ganglion cells in a few children with Hirschsprung-like symptoms have been recorded by Meier-Ruge (1974) and Garrett and Howard (1981). Histochemistry showed an excessive amount of ganglionic tissue in the myenteric plexus and in the submucosa from all regions of resected bowel, and from biopsies of the small bowel in one case. Neurones were also seen in the lamina propria of the mucosa. Although Meier-Ruge (1974) described an associated aplasia or hypoplasia of the sympathetic innervation Garrett and Howard (1981) found a relatively normal distribution around ganglia. Electronmicroscopy has revealed vast numbers of axons in all parts of the bowel wall and many contained peptide-like vesicles. The bowel tends to be incoordinate in its activity and, despite the excesses of nerves, the 'wiring' is presumably incorrect.

Puri et al. (1977) have reported a case of aganglionosis in the distal bowel and hyperganglionosis in the proximal colon, and thus it would seem that a wide range of developmental anomalies of intrinsic nerves in the gut may eventually be described.

5.1.4. Neurofibromatosis of the colon
Neurofibromata of the gut wall may occur in Von Recklinghausen's disease (Chalkley and Bruce, 1942; Dahl et al., 1957; Lukash et al., 1966) and are sometimes associated with constipation, intestinal obstruction or intussusception. Congenital megacolon may be associated with a more diffuse change in the myenteric plexus described as plexiform neurofibromatosis (Ternberg et al., 1958; Staple et al., 1964). The latter case showed diffuse hyperplasia of axons, but normal neurones. Phat et al. (1980) studied a further case with silver staining and this revealed areas of hyper- and hypoganglionosis, and hyperplasia of Schwann cells. The ganglion cells were irregular in shape and dendrites were described as hypertrophic. The patient who had presented with intestinal obstruction responded to colonic resection.

5.2. Acquired disorders

5.2.1. American trypanosomiasis (Chagas' disease)
Destruction of the myenteric plexus of the colon and rectum is a common lesion in Chagas' disease (Köberle, 1956, 1963) leading to a great reduction in nerve cell counts. The severity of the resulting megacolon may need surgical intervention and it is of interest that colonic dilation regresses if colostomy is performed

(Köberle, 1963). This suggests that the smooth muscle cells are not severely affected and that dysfunction of the ano-rectum is most important in the genesis of intestinal obstruction. Smith (1972) studied 3 cases with severe colonic dilation and loss of haustration. Many neurones and Schwann cells had disappeared completely leaving a line of residual connective tissue. A residual inflammatory reaction was present around the plexus. When argyrophil cells were recognised they were very abnormal. Hypertrophy of smooth muscle was severe.

5.2.2. Ulcerative colitis

As the inflammatory lesion of ulcerative colitis, which starts in the mucosa, progresses there is increasing damage to the myenteric plexus, leading eventually to an irreversible impairment of colonic motility (Meier-Ruge, 1974). The morphological changes in severe cases are widespread (Robertson and Kernohan, 1938; Storsteen et al., 1953; Okamoto et al., 1964; Orf, 1965). A 2-fold increase in numbers of neurones may be accompanied by Schwann cell proliferation and axon proliferation. At a later stage the neurones may be decreased in number.

Increased numbers of neurones and axons in the myenteric plexus have been noted in obstructive bowel lesions such as stenosis (Filgamo and Vigliana) and conditions of chronic irritation (Smith, 1972) around colostomy sites. It therefore seems to be a non specific response and caution should be exercised in relating some of these observations to the aetiology of diseases in the gastro-intestinal tract.

5.2.3. Irritable colon

Abdominal pain and exacerbations of constipation or diarrhoea are the symptoms of irritable colon which are widely accepted as being caused by abnormal colonic motility. Abnormal myoelectrical activity has been recorded consistently in affected bowel and precipitating factors include emotional stress (Taylor et al., 1980). In a recent ultrastructural study mucosal biopses were taken from 10 patients with irritable bowel syndrome and compared with normal controls (Riemann et al., 1980). Calculation of numbers of neurosecretory granules revealed an increase in cholinergic vesicles in affected subjects but no other abnormalities.

5.2.4. Pathology of central nervous system

Changes in the morphology of neurones and axons in the central nervous system in severe neurological disease such as the neuronal storage diseases (e.g., amaurotic idiocy) and the leucodystrophies (e.g., metachromatic leuco-encephalopathy) can also be seen within nerve plexuses of the gut (Bodian and Lake, 1963; Smith et al., 1976). Stored glycolipid has been demonstrated within the neurones of Auerbach's plexus more easily than in Meissner's plexus and metachromatic material may be seen within the nerve trunks of the bowel wall. Biopsies from the rectum are obviously safer than brain biopsies and are of importance for diagnostic reasons in children with these progressive disorders.

5.2.5. Autonomic neuropathy in diabetes mellitus

Insulin-dependent diabetics may develop a wide range of neuropathic symptoms caused by damage to peripheral sensorimotor nerves and visceral autonomic nerves. Symptoms ascribed to autonomic neuropathy include postural hypotension, diarrhoea, gastric atony, gustatory sweating, bladder atony and impotence. Duchen et al. (1980) studied autonomic nerves in 5 diabetic patients who died from a variety of causes. Sympathetic ganglia were reported to contain enlarged and vacuolated neurones, the vacuoles consisting of distended endoplasmic reticulum. Inflammatory changes were present within the ganglia and club-shaped argyrophilic masses of axons or dendrites were found adjacent to the neurones.

Severe losses of myelinated axons were found in the vagus nerves of all 5 patients although the vagal nuclei did not show any obvious abnormalities.

Detailed studies have yet to be made of the intrinsic nerves in the gut of diabetics but it seems likely that structural changes may occur which would account for the diverse symptoms seen in these patients.

6. Drugs and the myenteric plexus

Many drugs can affect the autonomic innervation of the gut and the effects can range from a temporary episode of constipation which may be caused for example by the anticholinergic side-effects of tricyclic or tetracyclic antidepressants (Milner and Buckler, 1964; Milner and Hills, 1966) to destruction of neurones in the myenteric plexus by chronic laxative abuse (Smith, 1972). The nervous tissues of the gut are not protected by the specialised blood-brain barrier found in the central nervous system.

Most clinical observations have been made on the effects of long term laxative agents such as anthraquinones and bisacodyl, which are stimulant cathartics. Long term medication may give rise to a 'cathartic' colon producing symptoms of increasing abdominal discomfort and even increasing constipation. The bowel wall may be thinned and pigment deposited in macrophages in the submucosa, (melanosis coli). Ultrastructural examination of mucosal biopsies from affected colons showed ballooning of axons, reduction of Schwann cell organelles and a significant increase in axonal area. Nerve endings showed a decrease in neurosecretory granules (Riemann et al., 1980). Axons that escape complete destruction apparently have the power to regenerate when laxatives are discontinued. The clinical symptoms appear to correlate with the morphological abnormalities.

Experimental administration of anthraquinones to animals via the intraperitoneal route (Smith, 1967, 1968) produces similar damage to Auerbach's plexus and intestinal dilatation. The transit time in the gut is increased in rats (Keeler et al., 1966).

The mode of action of the stimulant cathartics in man seems to depend on their absorption by the small bowel and reexcretion into the large bowel (Straub and Triendl, 1937).

Chronic atropine therapy was used at one time for the treatment of Parkinson's disease. Long term administration caused severe constipation with dilatation and hypertrophy of the colon (Siegmund, 1935). Chronic administration to mice again caused constipation and the most striking change in the myenteric plexus was the increased numbers of swollen argyrophil cells (Smith, 1972).

Intraperitoneal chlorpromazine in rats also produces megacolon (Zimmerman, 1962) and histology shows increased numbers of argyrophil cells and fragmentation of axons (Smith, 1972).

Mepacrine and Daunorubicin (Keeler, 1966; Smith, 1972) produce obvious myenteric damage after intraperitoneal administration in rodents and plexus disintegration is caused by the Vinca alkaloid-group of drugs. Vincristine and Vinblastine are widely used in the management of neoplastic disease and abdominal distension with constipation are common side-effects. Neurological damage can be widespread, for example, tendon reflexes are frequently lost (Tobin and Sandler, 1966) whilst ultrastructural studies showed damage to neurotubules in man and animals (Shezawski and Wisniewski, 1969). Post mortem observations in man revealed an increase in numbers of argyrophil cells but no evidence of plexus disintegration (Smith, 1972)

Conclusion

It is now appreciated that disordered gastro-intestinal motor activity can be associated with a wide range of neuro-morphological abnormalities. We believe that with the increasing application of new sophisticated cytochemical, immunohistochemical, and ultrastructural techniques an even greater number of variations and abnormalities will be detected in the autonomic innervation of the human gastro-intestinal tract, which will help to explain the functional disorders that result from neuro-muscular imbalance.

References

Adams, C.W.M., Brain, R.H.F., Ellis, F.G., Kauntze, R. and Trounce, J.R. (1961) Guy's Hosp. Rep. *110*, 191–236.
Adams, C.W.M., Marples, E.A. and Trounce, J.R. (1960) Clin. Sci. *19*, 473–481.
Alarotu, H. (1956) Acta. Paediat. Scand. 45, Suppl. *107*, 1–131.

Aldridge, R.T. and Campbell, P.E. (1968) J. Pediat. Surg. *3*, 475–490.
Angorn, I.B., Dimopoulos, G., Hegarty, M.M. and Moshal, M.G. (1977) Br. J. Surg. *64*, 466–469.
Anuras, S., Cooke, A.R. and Christensen, J. (1974) J. Clin. Invest. *54*, 529–535.
Arimura, A., Sato, H., Dupont, A., Nishi, N. and Schally, A.W. (1975) Science *189*, 1007–1009.
Armitage, G. and Rhind, J.A. (1951) Br. J. Surg. *39*, 39–43.
Atkinson, M. (1980) in Recent Advances in Gastroenterology (Bouchier, I.A.D., ed.), pp. 1–22, Churchill Livingstone, London.
Auerbach, L. (1864) Virchows Arch. Path. Anat. *30*, 457–460.
Barnett, W.O. and Wall, L. (1955) Ann. Surg. *141*, 527–535.
Barrett, N.R. (1953) Ann. Roy. Coll. Surg. Engl. *12*, 391–402.
Bateson, E.M., Talerman, A. and Walrond, E.R. (1969) Br. J. Radiol. *42*, 1–8.
Baumgarten, H.G. (1967) Z. Zellforsch. *83*, 133–146.
Baumgarten, H.G., Holstein, A.-F. and Owman, C. (1970) Z. Zellforsch. *106*, 376–397.
Belding, H.H. and Kernohan, J.W. (1953) Surg. Gynec. Obstet. *97*, 322–334.
Bendix, R.M. and Necheles, H. (1947) J. Am. Med. Ass. *135*, 331–333.
Bennett, A., Garrett, J.R. and Howard, E.R. (1968) Br. Med. J. *1*, 487–489.
Bentley, J.F.R. (1964) Dis. Colon Rect. *7*, 462–470.
Bentley, J.R.F. (1971) Gut *12*, 85–90.
Bettarello, A. and Pinotti, H.W. (1976) Clin. Gastroenterol. *5*, 103–117.
Bishop, A., Polak, J., Lake, B., Bryant, M.G. and Bloom, S.R. (1980) Regulatory Peptides, Suppl. *1*, S. 11.
Bodian, M. and Lake, B.D. (1963) Br. J. Surg. *50*, 702–714.
Bodian, M., Stephens, F.D. and Ward, B.C.H. (1949) Lancet *1*, 6–11.
Bodian, M., Carter, C.O. and Ward, B.C.H. (1951) Lancet *1*, 302–309.
Bouvier, M. and Gonella, J. (1981) J. Physiol. (London) *310*, 445–456.
Bouvier, M. and Gonella, J. (1981) J. Physiol. (London) *310*, 457–469.
Burnstock, G. (1972) Pharmacol. Rev. *24*, 501–581.
Burnstock, G. (1975a) J. Exp. Zool. *194*, 103–134.
Burnstock, G. (1975b) in Handbook of Psychopharmacology (Iversen, L., Iversen, S. and Snyder, S., eds.), pp. 131–194, Plenum Press, New York.
Cameron, J.A.M. (1928) Arch. Dis. Childn. *3*, 210–211.
Cannon, W.B. (1906) Am. J. Physiol. *17*, 429–442.
Cannon, W.B. (1911) Am. J. Physiol. *29*, 250–266.
Carter, C.O. and Evans, K.A. (1969) J. Med. Genet. *6*, 233–254.
Cassella, R.R., Brown, A.L., Sayre, G.P. and Ellis, F.H. (1964) Ann. Surg. *160*, 474–486.
Cassella, R.R., Ellis, F.H. and Brown, A.L. (1965) J. Am. Med. Ass. *191*, 379–382.
Castell, D.O. (1975) Ann. Int. Med. *83*, 390–401.
Challa, V.R., Jona, J.Z. and Markesbery, W.R. (1977) Am. J. Pathol. *88*, 309–315.
Cohen, S. (1975) Ann. Intern. Med. *82*, 714–715.
Cohen, S. (1978) in Gastrointestinal Pathophysiology (Brooks, F.P., ed.), pp. 71–96, Oxford University Press, New York.
Cohen, S. (1979) N. Engl. J. Med. *301*, 184–191.
Cohen, S., Fisher, R., Lipshutz, W., Turner, R., Myers, A. and Schumacher, R. (1972) J. Clin Invest. *51*, 2663–2668.
Cohen, S., Fisher, R. and Tuch, A. (1972) Gut *13*, 556–558.
Costa, M. and Gabella, G. (1971) Z. Zellforsch. *122*, 357–377.
Cushing, H. (1932) Surg. Gynec. Obstet. *55*, 1–34.
Dalla Valle, A. (1920) Pediatria (Napoli) *28*, 740–752.
Davis, J.A., Kantrowitz, P.A., Chandler, H.L. and Schatzki, S.C. (1975) N. Engl. J. Med. *293*, 130–132.

Davison, J.S. (1974) Gastroenterology 67, 558–559.
Dent, J. and Hansky, J. (1976) Gut 17, 144–146.
Devroede, G. and Lamarche, J. (1974) Gastroenterology 66, 273–280.
Devroede, G., Arhan, P., Duguay, C., Tetreault, L., Akoury, H. and Perey, B. (1979) Gastroenterology 77, 1258–1267.
Diamant, N.E. and El-Sharkawy, T.Y. (1977) Gastroenterology 72, 546–556.
Dimarino, A.J., Carlson, G., Myers, A., Schumacher, H.R. and Cohen, S. (1973) N. Engl. J. Med. 289, 1220–1223.
Dimarino, A.J. and Cohen, S. (1974) Gastroenterology 66, 1–6.
Dodge, J.A. (1970) Nature (London) 225, 284–285.
Dodge, J.A. (1976) in Topics in Paediatric Gastroenterology (Dodge, J.A., ed), pp. 88–91, Pitman Medical, London.
Dogiel, A.S. (1899) Arch. Anat. Physiol. Anat. Abt. 130–158.
Donovan, I.A., Harding, L.K., Keighley, M.R.B., Griffin, D.W. and Collis, J.L. (1977) Br. J. Surg. 64, 847–848.
Dragstedt, L.R. (1945) Ann. Surg. 122, 973–989.
Duchen, L.W., Anjorin, A., Watkins, P.J. and Mackay, J.D. (1980) Ann. Intern. Med. 92, 301–303.
Du Plessis, D.J. (1966) Br. J. Surg. 53, 485–492.
Earlam, R.J. (1972) Am. J. Dig. Dis. 17, 255–261.
Eckhardt, V.F. and Lecompte, P.M. (1977) Gastroenterology 72, 1055.
Eckhardt, V.F. and Weigand, H. (1974) Gut 15, 706–709.
Edwards, D.W.A. (1961) Proc. Roy. Soc. Med. 54, 930–933.
Ehrenpreis, T. (1946) Acta. Chir. Scand. 94, Suppl. 112.
Ehrenpreis, T. (1970) Hirschsprung's Disease, pp. 41–78, Year Book Medical Publishers, Chicago.
Ehrenpreis, T. and Pernow, B. (1952) Acta. Physiol. Scand. 27, 380–388.
El-Sharkawy, T.Y. and Diamant, N.E. (1975) in Proceedings of Fifth International Symposium on Gastrointestinal Motility (Vantrappen, G., ed.), pp. 176–180, Typoff Press, Belgium.
Ellis, F.G., Kauntze, R., Nightingale, A. and Trounce, J.R. (1960) Quart. J. Med. 29, 305–312.
Farrell, R.L., Castell, D.O. and McGuigan, J.E. (1974) Gastroenterology 67, 415–422.
Faulk, D.L., Anuras, S., Gardner, G.D., Mitros, F.A., Summers, R.W. and Christensen, J. (1978) Ann. Intern. Med. 89, 600–606.
Ferreira-Santos, R. (1961) Proc. Roy. Soc. Med. 54, 1047–1053.
Filgamo, G. and Vigliana, F. (1954) Riv. Pat. Nerv. Ment. 75, 1–32.
Frenckner, B. and Ihre, T. (1976) Gut 17, 306–312.
Frenckner, B. and VonEuler, C. (1975) Gut 16, 482–489.
Friesen, S.R., Boley, J.O. and Miller, D.R. (1956) Surgery 39, 21–29.
Friesen, S.R. and Pearse, A.G.E. (1963) Surgery 53, 604–608.
Furness, J.B. and Costa, M. (1974) Rev. Physiol. 69, 1–51.
Furness, J.B. and Costa, M. (1980) Neuroscience 5, 1–20.
Gabella, G. (1976) Structure of the Autonomic Nervous System, Chapman and Hall, London.
Gabella, G. (1979) Int. Rev. Cytol. 59, 129–193.
Gannon, B.J., Noblett, H.R. and Burnstock, G. (1969) Br. Med. J. 3, 338–340.
Garrett, J.R. and Howard, E.R. (1969) Proc. R. Micr. Soc. 4, 76–78.
Garrett, J.R. and Howard, E.R. (1975) J. Physiol. (London) 247, 25–27.
Garrett, J.R. and Howard, E.R. (1981) in Development of the Autonomic Nervous System (CIBA Foundation Symposium 83) (Burnstock, G., ed.), pp. 326–344, Pitman Medical London.
Garrett, J.R., Howard, E.R. and Nixon, H.H. (1969) Arch. Dis. Child. 44, 406–417.
Garrett, J.R., Howard, E.R. and Jones, W. (1974) J. Physiol. (London) 243, 153–166.
Garry, R.C. (1957) Br. Med. Bull. 13, 202–206.
Gellis, S.S. (1962–63) in The Year Book of Pediatrics (Gellis, S.S., ed.) p. 232, Year Book Medical Publishers, Chicago.

Giles, G.R., Mason, M.C., Humphries, C. and Clark, C.G. (1969) Gut *10*, 730–734.
Gillespie, J.S. and Maxwell, J.D. (1971) J. Histochem. Cytochem. *19*, 676–681.
Goetz, R.H. (1945) Clin. Proc. *4*, 337–392.
Gonella, J., Niel, J.P. and Roman, C. (1977) J. Physiol. (London) *273*, 647–664.
Goodall, J. (1967) Proc. R. Soc. Med. *60*, 1001–1002.
Grochowski, J., Szafran, H., Sztefko, K., Janik, A. and Szafran, Z. (1980) J. Pediat. Surg. *15*, 279–282.
Gunn, M. (1968) J. Anat. *102*, 223–239.
Haddad, G.G., Mazza, N.M., Defendini, R., Blanc, W.A., Driscoll, J.M., Epstein, M.A., Epstein, R.A. and Mellins, R.B. (1978) Medicine *57*, 517–526.
Hallum, J.L. and Hatchuel, W.L.F. (1954) Arch. Dis. Child. *29*, 354–356.
Heatley, R.V., Collins, R.J., James, P.D. and Atkinson, M. (1980) Br. Med. J. *1*, 755–757.
Heimbach, D.H. and Crout, J.R. (1971) Surgery *69*, 582–587.
Henderson, W. (1968) Postgrad. Med. J. *44*, 724–727.
Herbst, C. (1934) Z. Kinderheilk. *56*, 122–135.
Hiatt, R.B. (1951) Ann. Surg. *133*, 313–320.
Higgs, B. and Ellis, F.M. (1965) Surgery *58*, 828–834.
Hill, C.J. (1927) Phil. Trans. R. Soc. London, Ser. B: *215*, 355–387.
Hirschsprung, H. (1888a) Jb. Kinkerheilk. *27*, 1–7.
Hirschsprung, H. (1888b) Jb. Kinkerheilk. *28*, 61–68.
Hollis, J.B. and Castell, D.O. (1974) Ann. Intern. Med. *80*, 371–374.
Holsti, O. (1931) Acta. Med. Scand. *76*, 316–342.
Honjin, R., Izumi, S. and Osugi, H. (1959) J. Comp. Neurol. *111*, 291–319.
Hopkins, A.K., Neville, B. and Bannister, R.L. (1974) Lancet *1*, 769–771.
Howard, E.R. (1970) M.S. Thesis, London University, pp. 7–183.
Howard, E.R. (1972) Postgrad. Med. J. *48*, 471–477.
Howard, E.R. and Garrett, J.R. (1970) Gut *11*, 1007–1014.
Howard, E.R. and Garrett, J.R. (1973) Z. Zellforsch *136*, 31–44.
Howard, E.R. and Nixon, H.H. (1968) Arch. Dis. Child. *43*, 569–578.
Hurwitz, A.L., Duranceau, A. and Postlethwait, R. (1976) Am. J. Dig. Dis. *21*, 601–606.
Hwang, K., Essex, H.E. and Mann, F.C. (1947) Am. J. Physiol. *149*, 429–448.
Iordanskaya, N.I. (1962) Vestn. Khir. Grekov *88*, 24–28 (Eng. abstract).
Irving, I.M. and Lister, J. (1977) J. Pediat. Surg. *12*, 103–112.
Jacobowitz, D. (1965) J. Pharmacol. Exp. Ther. *149*, 358–364.
Johnson, R.H. and Spalding, J.M.K. (1974) Disorders of the Autonomic Nervous System, pp. 248–265, Blackwell Scientific Publications, Oxford.
Kamijo, K., Hiatt, R.B. and Koelle, G.B. (1953) Gastroenterology *24*, 173–185.
Kapila, L., Haberkorn, S. and Nixon, H.H. (1975) J. Pediat. Surg. *10*, 885–892.
Katz, L.A. and Spiro, H.M. (1966) N. Engl. J. Med. *275*, 1350–1361.
Keeler, R., Richarson, H. and Watson, A.J. (1966) Lab. Invest. *15*, 1253–1262.
Keller, A.D. (1936a) Arch. Pathol. (Chicago) *21*, 127–164.
Keller, A.D. (1936b) Arch. Pathol. (Chicago) *21*, 165–184.
Kirkpatrick, J.B. (1978) Gastroenterology *75*, 918–919.
Köberle, F. (1956) Virchows Arch. Path. Anat. *329*, 337–362.
Köberle, F. (1958) Gastroenterology *34*, 460–466.
Köberle, F. (1959) Z. Tropenmed. Parasit. *10*, 236–268.
Köberle, F. (1963) Gut *4*, 399–405.
Koelle, G.B., Koelle, E.S. and Friedenwald, J.S. (1950) J. Pharmacol. Exp. Ther. *100*, 180–191.
Kravitz, J., Snape, W.J. and Cohen, S. (1978) Am. J. Physiol. *234*, E359–E364.
Lane-Roberts, P.A. (1959) Proc. R. Soc. Med. *52*, 1022–1023.

Langley, J.N. (1898) J. Physiol. (London) 23, 407–414.
Leaming, D.B. and Cauna, N. (1961) J. Anat. 95, 160–169.
Legge, D.A., Wollaeger, E.E. and Carlson, H.C. (1970) Gut 11, 764–767.
Lendrum, F.C. (1937) Arch. Intern. Med. 59, 474–511.
Lewis, T.D., Daniel, E.E., Sarna, S.K., Waterfall, W.E. and Marzio, L. (1978) Gastroenterology 74, 107–111.
Lind, J.F., Warrian, W.C. and Wankling, W.J. (1966) Can. J. Surg. 9, 32–38.
Lipshutz, W., Tuch, A.F. and Cohen, S. (1971) Gastroenterology 61, 454–460.
Lister, J. (1858) Proc. R. Soc. London, Ser. B: 9, 87–98.
Long, D.M., Leonard, A.S., Chou, S.N. and French, L.A. (1962) Arch. Neurol. Psychiat. (Chicago) 7, 167–175.
Long, D.M., Leonard, A.S., Story, J. and French, L.A. (1962) Arch. Neurol. Psychiat. (Chicago) 7, 176–183.
MacIver, A.G. and Whitehead, R. (1972) Arch. Dis. Child. 47, 233–237.
Maire, F.W. and Patton, H.D. (1968) Am. J. Physiol. 184, 345–350.
Malins, J. and French, J. (1957) Quart. J. Med. 26, 467–480.
McCrea, E.D. (1924) J. Anat. 59, 18–40.
McCrea, E.D. and McSwiney, B.A. (1926) J. Physiol. (London) 61, 28–34.
Martin, L.W., Buchino, J.J., Le Coultre, C., Ballard, E.T. and Neblett, W.W. (1979) J. Pediat. Surg. 14, 686–687.
Meier-Ruge, W. (1968) Virchows Arch. Path. Anat. 344, 67–85.
Meier-Ruge, W. (1974) Hirschsprung's Disease: its Etiology, Pathogenesis and Differential Diagnosis, Current Topics in Pathology, Vol. 59, pp. 131–179, Springer, New York.
Meier-Ruge, W., Hunziker, O., Tobler, H.J. and Walliser, C. (1972) Beitr. Path. Bd. 147, 228–236.
Meissner, G. (1857) Z. Ration. Med. 8, 364–366.
Mellow, M. (1977) Gastroenterology 73, 237–240.
Milner, G. and Buckler, E.G. (1964) Med. J. Aust. 1, 921–922.
Milner, G. and Hills, N.H. (1966) Br. Med. J. 1, 841–842.
Mir, S.S., Mason, G.R. and Ormsbee, H.S. (1977) Gastroenterology 73, 432–434.
Mir, S.S., Telford, G.L., Mason, G.R. and Ormsbee, H.S. (1979) Gastroenterology 76, 1443–1448.
Misiewicz, J.J., Waller, S.L., Anthony, P.P. and Gummer, J.W.P. (1969) Quart. J. Med. 38, 17–30.
Moroz, S.P., Espinoza, J., Cumming, W.A. and Diamant, N.E. (1976) Gastroenterology 71, 236–241.
Munakata, K., Okabe, I. and Morita, K. (1978) J. Pediat. Surg. 13, 67–75.
Neely, J. and Catchpole, B. (1971) Br. J. Surg. 58, 21–28.
Nell, W. (1933) Beitr. Klin. Chir. 157, 401–413.
Niel, J.P., Gonella, J. and Roman, C. (1980) J. Physiol. (Paris) 76, 591–599.
Niemi, M., Kouvalainen, K. and Hjelt, L. (1961) J. Pathol. Bact. 82, 363–366.
Norberg, K.A. (1964) Int. J. Neuropharmacol. 3, 379–382.
Okamoto, E., Kakutani, T., Iwasaki, T., Namba, M. and Veda, T. (1964) Med. J. Osaka Univ. 15, 85–106.
Orf, G. (1965) Dtsch. Z. Nervenheilk. 187, 837–860.
Padovan, W., Godoy, R.A., Dantas, R.O., Menghelli, U.G., Oliveira, R.B. and Troncon, L.E.A. (1980) Gut 21, 85–90.
Padovan, W., Meneghelli, U.G. and Alves de Godoy, R. (1977) Am. J. Dig. Dis. 22, 618–622.
Pearse, A.G.E. and Polak, J.M. (1975) Histochemistry 41, 373–375.
Petri, G., Szenohradszky, J. and Porszasz-Gibiszer, K. (1971) Surgery 70, 359–367.
Phat, V.N., Sezeur, A., Danne, M., Dupuis, D., Vaissiere, G. and Camilleri, J.P. (1980) Pathol. Biol. 28, 585–588.
Polak, J.M., Bishop, A.E. and Bloom, S.R. (1978) Scand. J. Gastroent. 13, Suppl. 49, 144.

Polak, J.M. and Bloom, S.R. (1979) in Gut Peptides (Miyoshi, A., ed.) pp. 258–267, Kodansha Ltd., Tokyo.
Puri, P., Lake, B.D., Nixon, H.H., Mishalany, H. and Claireaux, A.E. (1977) J. Pediat. Surg. 12, 681–685.
Raia, A. (1955) Surg. Gynec. Obstet. 101, 69–79.
Raia, A., Acquaroni, D. and Netto, A.C. (1961) Am. J. Dig. Dis. 6, 757–771.
Rake, G.W. (1927) Guy's Hosp. Rep. 77, 141–150.
Rattan, S. and Goyal, R.K. (1974) J. Clin. Invest. 54, 899–906.
Rattan, S. and Goyal, R.K. (1980) Gastroenterology 78, 898–904.
Rayner, V. (1979) J. Physiol. (London) 286, 383–399.
Richardson, K.C. (1958) Am. J. Anat. 103, 99–135.
Riemann, J.F., Schmidt, H. and Zimmermann, W. (1980) Scand. J. Gastroent. 15, 761–768.
Rintoul, J.R. and Kirkman, N.F. (1961) Arch. Dis. Child. 36, 474–480.
Robertson, H.E. and Kernohan, J.W. (1938) Proc. Mayo Clin. 13, 123–125.
Rogers, I.M., Drainer, I.K., Moore, M.R. and Buchanan, K.D. (1975) Arch. Dis. Child. 50, 467–471.
Runström, G. (1939) Acta. Paediat. 26, 383–433.
Said, S.I. (1978) in Gut Hormones (Bloom, S.R., ed.) pp. 465–469, Churchill Livingstone, Edinburgh.
Santos-Buch, C.A. (1979) in International Review of Experimental Pathology (Richter, G.W. and Epstein, M.A., eds.) pp. 63–100, Academic Press, New York.
Sarna, S.K., Daniel, E.E. and Waterfall, W.C. (1977) Gastroenterology 73, 1345–1352.
Schofield, G.C. (1968) Handb. Physiol. 4, 1579–1727.
Schuffler, M.D., Bird, T.D., Sumi, S.M. and Cook, A. (1978) Gastroenterology 75, 889–898.
Schuffler, M.D., Lowe, M.C. and Bill, A.H. (1977) Gastroenterology 73, 327–338.
Schultzberg, M., Dreyfus, C.F., Gershon, M.D., Hökfelt, T., Elde, R.P., Nilsson, G., Said, S. and Goldstein, M. (1978) Brain Res. 155, 239–248.
Schultzberg, K., Hökfelt, T., Nilsson, G., Terenius, L., Rehfeld, J.F., Brown, M., Elder, R., Goldstein, M. and Said, S. (1980) Neuroscience 5, 689–744.
Shaw, A., Shaffer, H., Teja, K., Kelly, T., Grogan, E. and Bruni, C. (1979) J. Pediat. Surg. 14, 719–726.
Shelawski, M.L. and Wisniewski, H. (1969) Arch. Neurol. (Chicago) 20, 199–206.
Siegmund, H. (1935) Münch. Med. Wschr. 82, 453–454.
Smith, B. (1967) J. Neurol. Neurosurg. Psychiat. 30, 506–510.
Smith, B. (1968) Gut 9, 139–143.
Smith, B. (1968) Gut 11, 388–391.
Smith, B. (1972) Neuropathology of the Alimentary Tract, pp. 3–98, Edward Arnold, London.
Smith, P., Dickson, J.A.S. and Lake, B.D. (1976) Br. J. Surg. 63, 313–316.
Spitz, L. and Zail, S. (1976) J. Pediat. Surg. 11, 33–35.
Stadaas, J.O. (1980) Scand. J. Gastroent. 15, 799–804.
Stadaas, J., Aune, S. and Haffner, J.F.W. (1974) Scand. J. Gastroent. 9, 479–485.
Staple, T.W., McAlister, W.H. and Anderson, M.S. (1964) Am. J. Roentgenol. 91, 840–845.
Stern, M., Hellwege, H.H., Grävinghoff, L. and Lambrecht, W. (1981) Acta. Paed. Scand. 70, 121–124.
Stevens, M.B., Hookman, P., Siegel, C.I., Esterly, J.R., Shulman, L.E. and Hendrix, T.R. (1964) N. Engl. J. Med. 270, 1218–1222.
Stewart, I.M., Hosking, D.J., Preston, B.J. and Atkinson, M. (1976) Thorax 31, 278–283.
Storsteen, K.A., Kernohan, K.W. and Bargen, J.A. (1953) Surg. Gynec. Obstet. 97, 335–343.
Straub, W. and Triendl, E. (1937) Arch. Exp. Pathol. Pharmakol. 185, 1–19.
Sullivan, M.A., Snape, W.J., Matarazzo, S.A., Petroukubi, R.J., Jeffries, G. and Cohen, S. (1977) N. Engl. J. Med. 297, 233–238.
Swenson, O., Rheinlander, H.F. and Diamond, I. (1949) N. Engl. J. Med. 241, 551–556.

Talwalker, V.C. (1976) J. Pediat. Surg. *11*, 213–216.
Tanner, M.S., Smith, B. and Lloyd, J.K. (1976) Arch. Dis. Child. *51*, 837–841.
Taylor, I., Darby, C., Hyland, J. and Hammond, P. (1980) Scand. J. Gastroent. *15*, 237–240.
Telford, G.L., Mir, S.S., Mason, G.R. and Ormsbee, H.S. (1979) Am. J. Surg. *137*, 92–97.
Ternberg, J.L. and Winters, K. (1965) Am. J. Surg. *109*, 663–665.
Thomas, J.E. and Kuntz, A. (1926) Am. J. Physiol. *76*, 606–626.
Thompson, J. (1950) Arch. Dis. Child. *25*, 52–60.
Tittel, K. (1901) Wien. Klin. Wschr. *14*, 903–907.
Tobin, W.E. and Sandler, G. (1966) Nature (London) *212*, 90–91.
Trounce, J.R., Deuchar, D.C., Krantze, R. and Thomas, G.A. (1957) Quart. J. Med. *26*, 433–443.
Tuch, A. and Cohen, S. (1973) J. Clin. Invest. *52*, 14–20.
Tucker, H.J., Snape, W.J. and Cohen, S. (1978) Ann. Intern. Med. *89*, 315–318.
Uddman, R., Alumets, J., Edvinsson, L., Hakanson, R. and Sandler, F. (1978) Gastroenterology *75*, 5–8.
Vantrappen, G., Janssens, H., Hellemans, J. and Coremans, G. (1979) Gastroenterology *76*, 450–457.
Von Euler, U.S. and Gaddum, J.H. (1931) J. Physiol. (London) *72*, 74–87.
Wallgren, A. (1946) Am. J. Dis. Child *72*, 371–376.
Whitehouse, F.R. and Kernohan, J.W. (1948) Arch. Inter. Med. *82*, 75–111.
Wilbur, B.G. and Kelly, K.A. (1973) Ann. Surg. *178*, 295–302.
Woolam, G.L., Maher, F.T. and Ellis, F.H. (1967) Surg. Forum *18*, 362–365.
Zinnerman, G.R. (1962) Arch. Pathol. (Chicago) *74*, 47–51.
Zuelzer, W.W. and Wilson, J.L. (1948) Am. J. Dis. Child. *75*, 40–64.
Zwick, R., Bowes, K.L., Daniel, E.E. and Sarna, S.K. (1976) J. Clin. Invest. *57*, 1644–1651.

CHAPTER 12

Neuromuscular diseases viewed as a disturbance of nerve–muscle interactions

GERTA VRBOVÁ

Department of Anatomy and Embryology, Centre for Neurosciences, University College London, Gower Street, London WC1E 6BT, England

1. Introduction

The conventional distinction of neuromuscular diseases as neurogenic or myogenic shows the prevailing attitude of classifying a disturbance either as that of the nervous system, or muscle.

Since Erb (1884) first classified these diseases, the neurogenic disorders are thought to be diseases of the nervous system, while the myopathies are said to be diseases of the muscle. This division has been challenged by several workers (see McComas, 1977). McComas suggested that most primary myopathies are in fact diseases of the nervous system, and that the involvement of the muscle is secondary. His evidence in favour of the idea that primary muscle diseases are 'neurogenic' was the finding that, in a number of diseases that are traditionally considered to be primary diseases of the muscle whole motor units are destroyed. In addition it was claimed that fast motor units are more affected than slow ones. His suggestion provoked much discussion, which nevertheless left the questions unresolved (see Section 3.4). The previous chapters in this book clearly illustrate the interdependance of the motoneurone and skeletal muscle and show that nerve and muscle are so closely linked to each other that it is often difficult, or near impossible, to decide which is the primary disturbance in a disease of the neuromuscular system, and as fashions change different proposals are favoured.

In this chapter an attempt will be made to consider neuromuscular disorders in the light of the information contained in this book. The possibility that some clinical conditions could at least partly be explained by a failure of the appropriate interactions between nerve and muscle is seldom considered, yet such consideration could prove fruitful.

The majority of neurological disorders are crippling, largely because of the disturbance of locomotion they produce. Since a change of locomotion affects the muscle it can be expected that in most neurological conditions the functional state of the muscle will be altered. It would be an impossible task to describe such changes in a great number of neurological diseases, instead only two diseases will be dicussed, both occuring in infancy or childhood.

The two conditions discussed will be (a) the severe form of spinal muscular atrophy and (b) Duchenne muscular dystrophy. These were chosen because they are diseases of the developing neuromuscular system, and it is during development that the interdependance of the motoneurone and muscle is greatest.

The severe form of spinal muscular atrophy will be discussed in the light of new information contained in this book that elucidates the mechanism influencing the survival of developing motoneurones.

Duchenne dystrophy is the most thoroughly studied, so-called primary myopathy. Results from studies of this condition indicate that the maturation of muscles in patients suffering from Duchenne dystrophy is impaired. Consideration will be given to the possibility that the interaction between a normally developing nervous system and a slowly maturing muscle leads to the rapid deterioration of muscles seen in this condition.

2. Spinal muscular atrophy of the Werdnig–Hoffmann type

This is a familial disease and is thought to be due to a single autosomal recessive gene (see Campbell and Liversedge, 1981). The characteristic feature of the disease is massive motoneuronal death with severe muscle involvement. In view of the destruction of lower motoneurones it is considered to be a lower motoneurone disease.

2.1. Clinical features

The most conspicuous symptom is muscle weakness caused by complete or partial paralysis. This paralysis is due to the death of a large number of motoneurones and the resulting denervation of a proportion of the musculature. The disease varies in severity according to the extent of clinical weakness. In the most severe form the symptoms usually become apparent during the first weeks or months of the infant's life, sometimes even in late gestation. The clinical picture is generalized weakness and hypotonia, sparing only to some extent the diaphragm and the muscles supplied by the cranial nerves. The baby displays a characteristic 'jelly-like' posture when it lies supine with its arms and legs abducted. It is unable to lift its head and the tendon reflexes are absent.

Byers and Banker (1961) reported that the speed at which the disease progresses depends on the time of onset. They divided their patients into several groups according to the time of onset of the disease and found a faster progress in those infants where the symptoms appear early (up to 2 months of age) than in infants where the first signs appear later, i.e., between 2 months and 1 year. Dubowitz (1978) argues that the rate of progress depends on the initial severity of the symptoms rather than on the time of onset of the disease. The two views need not be contradictory, since it is possible that the symptoms are more severe the earlier the onset of the disease.

Finally, there is a very mild form of spinal muscular atrophies (Kugelberg--Welander syndrome) that starts relatively late (Wolfhart et al., 1955; Kugelberg and Welander, 1956). The rate of onset can vary from early childhood until adolescence and the progress of the disease can be very slow. Whether this very slowly progressing type of spinal muscular atrophy is a separte clinical entity from the severe form is disputed and the arguments for and against this suggestion are summarized by Dubowitz (1978). In this article only the severe Werdnig–Hoffmann form of the disease will be considered so that the discussion of other forms of spinal muscular atrophies is not relevant.

2.2. Histological changes in muscles and CNS

Histological examination of the diseased muscles reveals a strikingly abnormal picture. In addition to great numbers of small muscle fibres that are thought to be denervated, fibres ranging from normal to 4× normal size are also seen (Adams, 1969). Figure 1 illustrates the extremely wide variation in fibre size in the diseased muscle. Histochemical studies show that the normal correlation between fibre size and fibre type is lost (see Dubowitz and Brooke, 1973). Generally in healthy muscles, large fibres react weakly for oxidative enzymes, while small fibres have a strong reaction for these enzymes; in muscles from patients with lower motoneurone disease, however, large as well as small fibres react strongly for oxidative enzymes. The histochemical enzyme profile of a muscle fibre reflects its function, and there is ample evidence to show that increased muscle activity leads to an increase of oxidative enzymes (see Vrbová et al., 1978). Such changes can occur quite rapidly. It may be that in the diseased muscles the large fibres became transformed into oxidative ones due to excessive use. Equally likely is the possibility that the size of small fibres, normally rich in oxidative enzymes, increased.

When muscles from patients with lower motoneurone disease are stained for glycolytic enzymes, the staining pattern is again abnormal and does not conform to the usual expectation that large fibres react most strongly for these enzymes (see Dubowitz and Brooke, 1973). The irregular histochemical appearance of the large muscle fibres, that are thought to be still working, indicates that they are receiving an abnormal activity pattern from the motoneurones that supply them (see later).

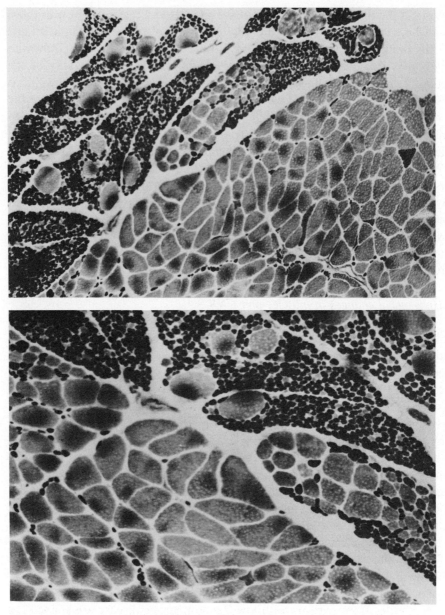

Fig. 1. Transverse section through muscle fibres from a biopsy from the quadriceps muscle of a patient suffering from the severe form of spinal muscular atrophy ATPase at 9.5. Magnification top × 104; bottom × 172.

The histochemical pattern of the small atrophic fibres is mixed, i.e., some fibres react strongly for oxidative enzymes while others do not, and their reaction for glycolytic enzymes and myosin ATPase is also varied (see Fig. 1). Interestingly, this is the case even in the most severe forms of Werdnig–Hoffmann disease when the infant is affected very early. Specialisation of muscle fibres into different types takes place during development (Nystrom, 1978) under the influence of innervation, but once established it persists for some time in the denervated muscle (see Dubowitz, 1969; O'Brien and Vrbová, 1980). Thus the finding that the small atrophic muscle fibres are differentiated into fibre types suggests that they have been innervated but have subsequently lost contact with their axons (see Dubowitz, 1981).

Perhaps the most interesting morphological feature of muscles from patients suffering from lower motoneurone disease is the structure of the motor nerve endings. The motor axons have a normal appearance right up to the last few millimeters of their course, but then they break up into a tangle of fine beaded fibres in the terminal nerve bundles and at the endplate (see Woolf, 1969). In view of the normal appearance of the axon while the terminal is 'sick' it was suggested that in these diseases the neurone is 'dying back'. Indeed, in no other malady is the 'dying back' of the neurone so conspicuous as in the most severe of spinal muscular atrophies, the Werdnig–Hoffmann disease (see Woolf, 1960, 1969), and it could be this 'dying back' phenomenon that will finally give us a clue to the aetiology of the disease (see later speculations).

The changes in the CNS are mostly confined to the anterior horn cells but sometimes cells of the motor nuclei of cranial nerves are also affected. A decrease in the number of motoneurones is particularly obvious in the cervical and lumbar region. Many remaining motoneurones show chromatolytic and degenerative changes (Werdnig, 1891; Byers and Banker, 1961) and these are similar to those seen after axon injury (Chou and Fahadej, 1971).

2.3. Abnormalities of EMG

There are not many reports of changes in EMG in patients suffering from Werdnig–Hoffmann disease. This is not surprising, since most of the patients are infants and it is difficult or near impossible to goad babies into producing graded EMG acitivity patterns. It may be for this reason that there is little information about EMG recruitment patterns in normal infants. Nevertheless there is good evidence that in patients suffering from lower motoneurone diseases the number of motor units in individual muscles is very much reduced and that this reduction is due to motoneurone death.

It is reported that the motor unit potential (MUP) is longer in duration and larger in size in patients suffering from spinal muscular atrophy. This prolonged duration and increased size is thought to be due to nerve sprouting, since it is also seen in patients with partially denervated muscles (for review see Desmedt, 1981). However in the severe Werdnig–Hoffmann disease large MUPs with long duration are rarely seen (McComas, 1977). This is particularly surprising since relatively large and long MUPs ought to be present in muscles of healthy infants where motor units of relatively large size were found (Sacco et al., 1962). The presence of large motor units during early stages of postnatal development is well documented in cats and rats (Bagust et al., 1973; Brown et al., 1975, 1976). This is because individual muscle fibres are supplied by branches from several axons and each axon innervates a large number of muscle fibres. Results reported by Sacco and co-workers (1962) showing that the size of the motor unit in infants is relatively large indicate that in infants too individual muscle fibres are supplied by more than one axon.

In addition to reports that MUP size and duration are increased in patients with lower motoneurone disease, Desmedt and Borenstein (1977) recently noticed that in the less severe form of spinal muscular atrophy of the Kugelberg–Welander type, the large MUP is often followed by a small potential at a constant latency, and they favour the interpretation that these 'linked' potentials are caused by the recruitment of additional muscle fibres innervated by long sprouts (Desmedt and Borenstein, 1977). However, the assumption that such extensive enlargement of the motor unit size due to sprouting takes place in spinal muscular atrophy of the Werdnig–Hoffmann type does not seem to be substantiated by histological findings. While it is well documented that in a partially denervated muscle the surviving neurones increase their peripheral field by sprouting (Edds, 1953; Van Harrefeld, 1952) it is questionable whether in motoneurone disease the neurones maintain their ability to enlarge their peripheral field to the same extent. The morphological evidence suggests that diseased motoneurones are indeed less able to enlarge their peripheral field than normal ones. When reviewing results from his large material on muscles from patients with spinal muscular atrophy Woolfe (1969) concludes: 'The tendency of the axon to put out sprouts in response to changes of its internal and external milieu is governed by the vitality of the anterior horn cell, of which it is a distal extremity. Where the cell is itself diseased the force behind the axonic flow is likely to be diminished so that the sprouts will be of slender caliber and attempts to form end-plates will be abortive. Thus far stouter collateral sprouts with much better formed end-plates are seen after poliomyelitis than in motor neurone disease, in which the surviving axons are extensions of the cells already diseased.' That sprouting may not be substantial in these diseased muscles is also suggested by results showing that motor units of patients suffering from lower motoneurone disease develop normal tensions in spite of their large MUPs (Milner–Brown et al., 1974). Although these results were obtained from patients with amyotrophic lateral sclerosis, the finding could also apply to genetic motor neurone diseases. Unfortunately, no data on the tensions developed by individual motor units in infants suffering from spinal muscular atrophy is available. Nevertheless, the morphological findings suggest that there is not a simple connection between the increased duration and size of the MUP and sprouting of the nerves.

The second characteristic feature of this group of diseases, and indeed of all motoneurone diseases, is the observation that there are spontaneously firing motor units in the muscle. This spontaneous firing correlates with the clinical observation of fasciculations, also present in all forms of motorneurone diseases (see Richardson and Borwick, 1969). Buchthal and Olsen (1970), in their study of EMG activities in infants suffering from Werdnig–Hoffmann disease, described units that fired spontaneously at rates of 10 to 15 Hz for long periods of time, even when the babies were asleep. These authors also found that the same units were used by the infants during spontaneous locomotion, and could therefore be voluntarily or reflexly activated.

The persistent spontaneous activity of motoneurones that are still used in voluntary movement is an interesting phenomenon; it could be that the normal inhibitory mechanisms are less efficient in these diseases. In motoneurone disease the Renshaw collateral, which is also an extension of the axon, could be retracting and thus reducing its inhibitory effect on the motoneurone pool. An additional factor may be a change in the biophysical properties of the dendrites and cell bodies of the neurones. It is known that chromatolytic motoneurones (after section of their axons) develop abnormal properties and some of these changes may be relevant to their spontaneous firing. Normally, dendrites are unable to conduct action potentials, but in motoneurones that are undergoing chromatolysis the dendrites acquire the ability to propagate action potentials (Kuno and Llinás, 1970a,b). This change may increase the likelihood of excitatory inputs causing an action potential to be fired by the motoneurone. Not only motoneurones that are themselves damaged, but also those that are intact but whose terminals are surrounded by denervated muscle fibres show changes of their membrane properties, and such changes could influence their excitability (Huizar et al., 1977). The results on the response of motoneurones to axon injury are reported and discussed in detail by Gordon in Chapter 10 of this volume.

Whichever the reason for the spontaneous and prolonged activity of motor units, it may explain the histochemical finding of large muscle fibres with high levels of oxidative enzymes since it is known that muscle activity leads to an increase of oxidative enzymes. Above all, continuous activity of the developing motoneurones may increase the likelihood of nerve terminals losing contact with muscle fibres, particularly in infants, for in developing animals neuromuscular activity causes a reduction of the number of nerve–muscle contacts (O'Brien et al., 1978).

2.4. Motoneurone disease viewed as a disturbance of nerve–muscle interaction

The considerations that follow are based on experimental results summarized in previous chapters and some of the features of the disease described above.

Motoneurone death during early stages of development takes place in the central nervous system of all vertebrates, and during early stages of development the motoneurones are critically dependant on their target organs, the skeletal muscle fibres; if they fail to establish contact with the muscle they die (Hamburger, 1975). This dependance of the neurone on muscles during embryonic and early postnatal life was much explored in amphibians and birds, and only to some extent on mammals. The results obtained for mammals are of particular importance in relation to lower motoneurone disease, for they show that even after birth the motoneurone continues to be dependant on its connections with the muscle. Romanes (1946) removed different parts of the hind leg at, or shortly after birth in rats and mice and found that motoneurones supplying the missing muscles disintegrated when the operation was performed on young animals, but that their chance of survival improved with age. Romanes and, later, others (see Lieberman, 1971, 1974) interpreted this as an indication of the great sensitivity of young motoneurones to injury to their axons rather than their critical dependance on the skeletal muscle. Nevertheless, these results on mammals taken together with those on birds and lower vertebrates clearly show that it is essential for the survival of the motoneurone to establish and maintain contact with the muscle for some time after birth. The precise nature of the influence that the muscle exerts on the motoneurone is as yet unknown, but some important facts are beginning to emerge. During early stages of normal vertebrate development a considerable number of motoneurones die, but it has become clear that even the motoneurones that are usually destined to die can survive if the nerve terminals are allowed to establish and maintain contact with the muscle (Pittman and Oppenheim, 1979; Laing and Prestige, 1978). On the other hand, motoneurones that are normally destined to survive will degenerate if their terminals lose contact with the muscle (Oppenheim and Nuńez, 1982).

It was found in chick embryos and newborn rats that muscle inactivity produced by paralysing drugs results in the maintenance of a greater than normal number of connections between nerves and muscle (Gordon et al., 1974; O'Brien and Vrbová, 1978; Srihari and Vrbová, 1978; Duxson, 1982) and these same procedures also increase the number of surviving motoneurones (Oppenheim, 1982).

Thus the mechanisms that regulate the maintenance and loss of connections between muscle and axon terminals during development are of great importance for the survival of the motoneurone. A mechanism that is involved in the maintenance and loss of connections between motor nerve endings and muscle was proposed by O'Brien et al. (1978, 1980), and its discussion in relation to lower motoneurone disease is pertinent. When nerve–muscle connections are first made the neurones are probably relatively inactive, and can synthesize and release only small amounts of transmitter (Diamond and Miledi, 1962; O'Brien and Vrbová, 1978). During this early developmental period a large number of nerve–muscle contacts is established, and those motoneurones that fail to make contact with the muscle are probably the ones destined to die. However, even after the early developmental period, the number of contacts between the surviving neurones and muscles is greater than in fully mature vertebrates, and most of the connections will become eliminated during further stages of development. Thus in newborn animals each muscle fibre is supplied by more than one axon terminal. In most mammalian muscles these axon terminals contact the muscle fibre at a single site, the end-plate. With increasing age all but one nerve ending lose contact with each muscle fibre, so that each muscle fibre is eventually supplied by only a single axon. This means that the peripheral field of individual motoneurones becomes much reduced during postnatal development. Although most of these observations were obtained on laboratory animals – rats, rabbits, mice and cats – it is likely that in infants too a similar process is occurring. A reduction of the peripheral field of individual motor units was seen to take place during the first 2 years of life in babies (Sacco et al., 1962). This observation was based mainly on the reduction of the size and duration of the MUPs with the age of the infant (Sacco et al., 1962).

Interestingly, it is during or just after the period when the motoneurones are so critically dependant on connections with their target organs that massive reorganisation of their nerve endings takes place. Some mechanisms that are involved in this reorganisation are now understood. For example, it is clear that the interaction of ACh with the cholinergic receptor is an essential step in the process of elimination of superfluous nerve endings. In experiments in which this interaction is reduced by using neuromuscular blocking drugs more nerve terminals stay in contact with the muscle fibre, while when the action of ACh is enhanced by increased nerve activity, exposing muscles to ACh, or by administering anticholinesterases, the reduction of nerve–muscle contacts is increased. A model to explain these results was proposed by O'Brien et al. (1978, 1980). According to this model proteolytic enzymes are released in response to ACh, and these enzymes digest the nerve terminal membrane. If the rate of replacement balances the rate of digestion, the connection between nerve ending and muscle will be maintained. If more proteolytic enzymes are released, either because of a greater number of ACh receptors or some other reason, then the rate of replacement has to increase proportionally. Thus more connections will be disrupted if either the responsiveness of the muscle to ACh is increased, or if the neurone is unable to replace its nerve endings. This regulation of nerve–muscle connections is of critical importance for the survival of the motoneurone. In situations where contacts are maintained, such as when the muscle is paralysed by neuromuscular blocking drugs, a greater number of motoneurones survive. On the other hand, procedures that are known to disrupt more than the usual number of nerve–muscle connections lead to increased motoneurone death. It is interesting that administration of anticholinesterase does not affect the number of surviving motoneurones during earlier stages of development, when little or no cholinesterase is present at the neuromuscular junction and little ACh is released from the nerve endings. Later, however, when the output of ACh from the nerve endings increases, the administration of anticholinesterases induces massive cell death even though at those later stages of development the normal process of cell death is complete (see Chapter 3).

From this brief account of the changes of synaptic inputs to skeletal muscle fibres during development of vertebrates and the critical dependance of the motoneurones on these inputs it is apparent that, during the first 2 years of an infant's life, a disturbance of the interaction between muscle and nerve may be disastrous for the motoneurone.

The study of the morphological characteristics of the nerve endings that are disconnected from their endplates during normal development indicates that they do not degenerate, but retract. Their appearance as seen from Fig. 2 is not unlike that found in infants suffering from Werdnig–Hoffmann disease, where the 'dying back' of the axon is one of the most conspicuous morphological features of the diseased muscle. The findings that the progress of the disease is more rapid the earlier it appears is consistent with the possibility that the normally occurring loss of connections between nerve and muscle proceeds too soon and too rapidly in these infants. A rapid loss of connections could be expected to lead to motoneurone degeneration. During later stages of development the dependance of the motoneurone on its target is less, so that it can survive a considerably greater loss of connections. Finally, once a considerable number of motoneurones has lost contact with its muscle fibres, the remaining neurones seem to be driven to greater activity (see Buchthal and Olsen, 1970), such an increase in activity may lead to an even more rapid disruption of contacts between muscle fibres and nerve endings, and add to the rapidly progressive nature of the disease.

If, as we suggest, lower motoneurone disease could be explained in terms of a disturbed interaction between muscle and motoneurone, it would be desirable to know a cause for the excessive loss of connections between nerve and muscle fibres. At present any proposal as to the mechanisms that may induce such a situation can only be based on speculation.

During normal muscle development the synthesis of muscle-specific proteins including ACh receptors increases so that, when the ingrowing neurone reaches the muscle, the muscle fibre membrane already contains a considerable number of ACh receptors that enable it to respond to the incoming branches of the motoneurone (Betz et al., 1980). The changes of distribution of these receptors during development are discussed in Chapter 4. It could be that for some reason the synthesis of ACh receptors and other muscle specific proteins is too rapid in

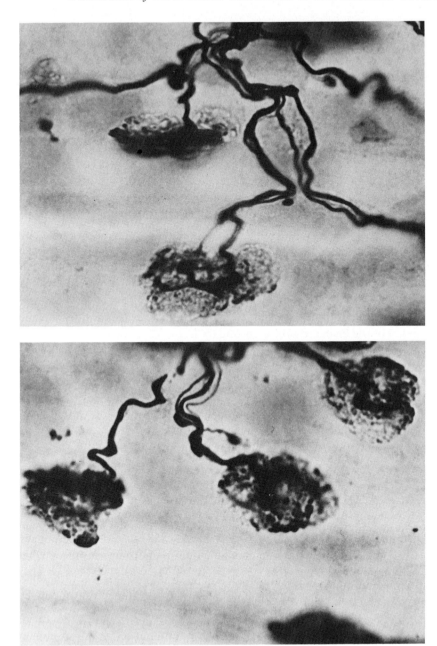

Fig. 2. Longitudinal sections through soleus muscles from a 12-day-old rat stained with a combined cholinesterase (siver method). Note the 'retracting' nerve terminals.

infants suffering from lower motoneurone disease so that their muscles have a very high concentration of ACh receptors, and thus a high sensitivity to ACh. If this was the case, even small amounts of ACh released from nerve endings could lead to an excess release of proteolytic enzymes and disruption of connections between nerve and muscle. If so, only nerve endings that are either very inactive or are replaced at a very high rate will be able to maintain contact with muscle fibres. The presence of extraordinarily large muscle fibres, up to $4 \times$ normal diameter, in the diseased muscles could be taken to support such a possibility, for their large size could indicate the ability of the muscle fibres to synthesize muscle specific proteins rapidly. Thus one could argue that the muscle develops too rapidly and is 'out of phase' with the development of the neurone. It is important to note that even in those infants where the disease starts early, the small muscle fibres that must have lost their innervation some time before nonetheless differentiated into fibre types. It suggests that these muscle fibres must have not only been innervated at some earlier time, but that they also underwent rapid development and differentiation. If this was the case then the classification of this disease as purely neurogenic would be inaccurate. However, it could also be that the motoneurone is unable to replace its nerve terminals and that this is due to some inherent inability of the diseased motoneurone to do so. This possibility, although just as likely, is less attractive, because, while the prospect of treatment of the disease would be encouraging if the disease was indeed caused by a disturbance of muscle maturation, it would be much less encouraging if it was the neurone that was unable to maintain its terminals. It is a challenge to distinguish between these two possibilities.

3. Duchenne dystrophy

3.1. Brief description of clinical features

In the last century the French physician Duchenne (1868) first described cases of a muscle wasting disease, which later was to carry Duchenne's name. The hereditary nature of the disease was recognised by Meryon (1852), who also noticed that it predominantly affected males. Duchenne (1868) and others after him described several forms of this hereditary wasting disease.

It was Erb (1884) who, while grouping these conditions together as primary diseases of the muscle, or, as he called them 'dystrophia muscularis progressiva', defined and classified the newly described syndromes. He postulated that they all were due to a 'complex nutritional disturbance of the muscle fibre' (Erb, 1891). The most up-to-date and comprehensive classification of the muscular dystrophies has been reviewed by Gardner-Medwin and Walton (1969).

The first symptoms of the Duchenne type dystrophy are usually difficulty in walking and difficulty in rising from the floor. When rising from the floor the patient does so in a characteristic manner, by climbing with his hands up his thighs and so levering his trunk to the upright position (Gowers' sign). The muscles appear enlarged due to the deposition of connective and fatty tissue in the muscle belly. With advancing age and relentless progress of the disease this enlargement is followed by wasting; the muscles harden and shorten and give rise to deformities. In most cases the patients are unable to walk by the age of 11. For detailed description of the clinical picture see Dubowitz (1978) and Campbell and Liversedge (1981).

3.2. Histological and physiological changes

The characteristic morphological features of the muscle change with the progress of the disease (Dubowitz and Brooke, 1973). Atrophy of some muscle fibres is seen together with hypertrophy of others, infiltration by phagocytic cells, degeneration and regeneration of muscle fibres are all 'typical features' of these diseased muscles (see Fig. 3). It is interesting, and may be significant, that some of these changes in the muscles are most marked in the 'preclinical' phase of the disease (Hudgson et al., 1967).

Histochemical examination of muscles from patients suffering from Duchenne dystrophy shows that they contain a higher then normal proportion of muscle fibres with high levels of oxidative enzymes and low levels of glycolytic ones. When reacted for myosin ATPase fibres containing ATPase typical of fast muscles are fewer and smaller in size (Dubowitz and Brooke, 1973; Buchthal et al., 1971a,b, 1974). These changes could be due either to a selective loss of fast fibres or to a transformation of the existing fibres into fibres with different histochemical characteristics.

The fibre types in muscles from patients appear to be grouped, and this has been taken to indicate that the remaining motor axons reinnervate some fibres that have lost their innervation (Dubowitz, 1969).

Physiological studies of the mechanical properties of the muscles showed that the time course of contraction and relaxation is slower in muscles from Duchenne children, a result consistent with the change of the histochemical properties of the muscle (Buchthal et al., 1971a,b, 1974). In view of these findings the suggestion was put forward by McComas et al. (1971) that Duchenne dystrophy is a primary disease of the nervous system and that motoneurones to fast muscles are preferentially involved. The arguments for and against this hypothesis are discussed in detail by McComas (1977).

Results obtained by studying the biochemical composition and physiological properties of muscles show that the characteristic features typical of fast mammalian muscles do not develop in patients with Duchenne dystrophy. Vertebrate fast muscles change and acquire a number of characteristic properties (physiologi-

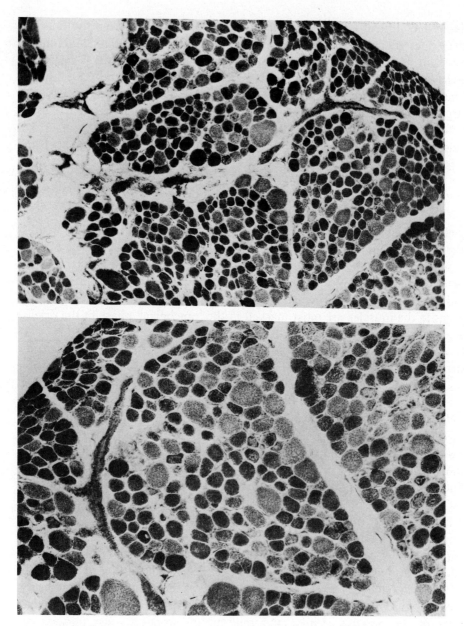

Fig. 3. Transverse section through muscle fibres from a biopsy taken from the quadriceps muscle of a patient suffering from Duchenne dystrophy NADH diaphorase. Top ×104; bottom ×170. Micrographs in this figure and in Fig. 1 were kindly provided by Professor V. Dubowitz.

cal and biochemical); the time course of contraction and relaxation becomes faster with age (Close, 1964; Buller et al., 1969a), and the activities of those enzymes that are concerned with rapid production of energy and transfer of phosphate to form ATP, increase (Kendrick-Jones and Perry, 1967; Perry, 1971). Other features of the muscle such as the ability of the sarcoplasmic reticulum to take up Ca^{2+} also change and become more efficient with age (Margreth et al., 1980; Zubrzycka-Gaarn and Sarzala, 1980). These changes are schematically illustrated in Fig. 4. In muscles from patients suffering from Duchenne dystrophy

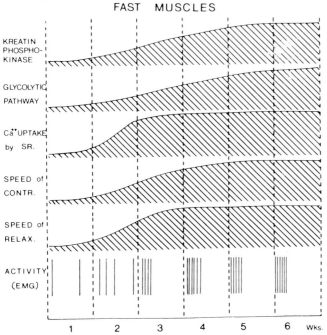

Fig. 4. Schematic representation of some changes taking place in fast skeletal muscles during postnatal development. The vertical lines in the lowest part of the graph indicate motor unit potentials. The shaded curves represent the approximate rate of change.

these developmental changes characteristic of maturation of fast muscles do not take place (see Perry, 1971). The enzyme activities characteristic of normal fast muscles remain low and resemble those of immature muscles (Dreyfus et al., 1962). The ability of the sarcoplasmic reticulum to accumulate Ca^{2+} remains low and does not attain the high levels typical of fast muscles (Samaha and Gergely, 1969; Peter et al., 1974); the time course of contraction and relaxation is also slower than normal (Buchthal et al., 1971a,b, 1974). Thus muscles from Duchenne patients resemble in many respects immature muscles, and it has therefore been suggested that they are unable to complete their maturation (Dubowitz, 1969; Perry, 1974).

Some, but not all these results can be attributed to the presence of a small number of regenerating muscle fibres (Lipton, 1978). These are formed from satellite cells released from the degenerating muscle fibres. However, in view of the small proportion of regenerating muscle fibres it is unlikely that they could account for the biochemical and physiological findings. In spite of these attempts of regenerative processes that are seen in the muscle the desintegration of the muscles is not halted but progresses relentlessly producing further loss of muscle fibres.

The arrangement of the nerves and nerve endings reflects the changes of the muscle fibres. Many nerve endings that had supplied the degenerating muscle fibres would have lost contact with them, and thus appear like the 'withdrawing' axons. Other nerve terminals may send sprouts to newly-formed muscle fibres. Thus the arrangements of axons consist of features typical of the withdrawal of nerve terminals, sprouting and the innervation of regenerating fibres (Jedrzejowska et al., 1965).

This finding is in apparent contrast with results obtained from ultrastructural studies which showed normal-looking axon terminals (Jeruzalem et al., 1974). However, light microscopical studies reveal different features of the innervation pattern from the ultrastructural studies, where mainly those axons that are in contact with the muscle fibre are examined and retracting axons would not be sampled.

3.3. Changes of EMG

The interference EMG activity pattern of children suffering from Duchenne dystrophy shows that their motor units are recruited during movement and that the number of motor units recruited does not appear to be reduced. However, the duration and size of the action potential of individual motor units are reduced. Recently, Desmedt and Borenstein (1976) noticed that, in muscles from children suffering from Duchenne dystrophy, fibrillation potentials as well as 'linked potentials' can be recorded. The 'linked potentials' fire at a constant latency after the originally activated motor unit, and are therefore considered to be caused by long sprouts from the same axon. While there are in dystrophic muscles regenerating muscle fibres that are acquiring new innervation, or fragments of muscle fibres that have lost connection with their innervation, it is doubtful that these fibres would be colonized by branches from axons of a motor unit that is innervating muscle fibres a long distance away from them. This would have to be the case if the 'linked potentials' were related to sprouting, since the delay is often as long as 50 ms. To account for such delays, the nerve branches would not only have to be very long but also very fine, with extremely slow conduction velocities. The histological examination of muscles from Duchenne patients argues against the presence of such nerve fibres (Woolf, 1969). The presence of 'linked potentials' may be due to synchronous firing of motoneurones, or some other abnormalities of the neuromuscular system.

3.4. Current ideas on the pathogenesis of Duchenne dystrophy

In his editorial review Rowland (1976) discussed three theories of the pathogenesis of muscular dystrophy: (a) vascular, (b) neurogenic, and (c) surface membrane abnormality.

He criticised the vascular hypothesis on the grounds that patients suffering from Duchenne dystrophy have little or no alterations in their muscle blood flow or distribution of capillaries (See Rowland, 1976), and he argued against the neurogenic theory because of the criticisms concerning the findings of McComas and his colleagues (1977) regarding the reduced number of motor units in muscles from patients with Duchenne dystrophy, which was the most compelling argument in favour of the neurogenic hypothesis. This evidence was challenged by several workers, most notably Ballantyne and Hansen (1974), who were unable to confirm the results of McComas and his colleagues, and using a modification of their method, could not find a reduced number of motor units in muscles from Duchenne children. However, the method used by both groups of research workers is flawed, for it is based on the assumption that axons of different motor units have different thresholds to electrical stimulation and that it is therefore possible to discern individual units by the stepwise increments in twitch tension with increasing stimulus strength. It would only take a few axons with similar thresholds to alter the estimate of the number of motor units in a muscle and thus produce an incorrect result.

It was the third hypothesis that was favoured by Rowland (1976), and has since become very popular. The idea that abnormal properties of the surface membrane are the cause of the disease was first based on findings of increased levels of muscle specific enzymes in the plasma of patients suffering from primary myopathies (Heyck et al., 1966). However, such enzymes were also found in other conditions where severe muscle damage took place, such as trauma and cardiac muscle damage (see Rowland, 1976) so that these changes are not unique to Duchenne dystrophy, and could be secondary to the ongoing destruction of muscle tissue. The idea of a 'leaky membrane' is based predominantly on circumstantial evidence, but it recently gained momentum when it was reported that the plasmalemmal membrane is defective in muscle fibres from patients suffering from Duchenne dystrophy (Mokri and Engel, 1975; Schotland et al., 1977) and that free Ca^{2+} can be detected in a proportion of muscle fibres from Duchenne patients (Bodensteiner and Engel, 1978). Since these changes were found in muscle fibres that did not show any other signs of abnormality it was assumed that they are the cause and not the consequence of the disease process. The ultrastructural studies of membranes are interesting, but their significance is difficult to assess, mainly because of the problem of sampling. The presence of free Ca^{2+} on the other hand is now well documented and is indicative of a serious insufficiency of the muscle fibre. This insufficiency however need not be attributed to a change in the surface membrane properties, particularly since there is no evidence of an increase in Ca^{2+} conductance of membranes of muscle fibres or other cells from patients suffering from Duchenne dystrophy. The presence of free Ca^{2+} could be accounted for by the repeatedly confirmed finding that the ability of the sarcoplasmic reticulum to take up Ca^{2+} in muscles from Duchenne patients is very much reduced (Samaha and Gergely, 1969; Takagi et al., 1973; Peter et al., 1974). The 'leaky membrane' hypothesis is difficult to reconcile with the finding that, in muscles from Duchenne children, the fast glycolytic fibres are affected by the disease more than the oxidative ones (Buchthal et al., 1971). Yet these glycolytic fibres normally have a more efficient Ca^{2+} uptake mechanism (Margreth et al., 1974), and if it was to function properly this system could probably take up all the excess Ca^{2+} that entered the cell. The presence of free ionized Ca^{2+} in the muscle fibres of Duchenne patients, together with the preferential involvement of fast, glycolytic muscle fibres during the disease, suggests that it is the sarcoplasmic reticulum, i.e., the internal membrane system rather than the surface membrane that is abnormal in the diseased muscle. Fast glycolytic muscle fibres are normally activated by their motoneurones at high frequencies so that more Ca^{2+} is released within a short interval; an efficient uptake system is therefore needed to prevent the accumulation of Ca^{2+} inside the cell. It is known that in muscles from Duchenne patients, the efficiency of this system is very much reduced (Peter et al., 1974). It could therefore be expected that the fast glycolytic fibres that are more dependent on this system will be most affected by its inefficiency. It would be extremely interesting to identify the type of muscle fibre that contains the free Ca^{2+}.

From the available information it appears that it is indeed the muscle itself that is abnormal as Rowland (1976) suggested, and this abnormality is most damaging to the muscle when it is required to perform certain specific functions. With the maturation of the child's motor performance the motoneurones activate the muscle fibres they supply with different activity patterns than in early infancy, and it is during this period of the child's development that the symptoms of the disease become clearly pronounced.

3.5. Duchenne dystrophy viewed as a disturbance of nerve–muscle interaction

During postnatal development mammalian muscles undergo a great number of changes. These changes seem to develop in response to the increasing functional demands imposed on the muscles by the developing nervous system, in particular the gradual development of more elaborate motor reflexes (Kendrick-Jones and Perry, 1968; Vrbová, 1980).

Skeletal muscles of newborn mammals contract and relax slowly, and develop their maximal tetanic tension at low rates of firing (see Close, 1964). Moreover, at higher rates of firing they are unable to maintain tension for longer than 500 ms

(Handyside et al., 1982). At this stage of development the firing rate of the motoneurones is low (Bursian and Sviderskaya, 1971) and the activity of each motor unit lasts for only a short time, less than 1 sec, so that the mechanical properties of the muscles are well adjusted to the firing pattern of motoneurones (Navarrete and Vrbová, 1980, 1982). With increasing age the mechanical and biochemical properties of skeletal muscle fibres change and prospective fast muscle fibres develop properties that enable them to develop relatively large forces and contract and relax more rapidly (see Fig. 4). Together with these changes, or perhaps preceding them, the pattern of activity of the motor units alters in that the motoneurones to muscle fibres that are destined to become fast start firing transiently at high rates, and motoneurones to muscle fibres destined to become slow will fire at low-rates for long periods of time (Navarrete and Vrbová, 1980, 1983).

There are many indications that some early stages of muscle maturation, i.e., the initial increase of the speeds of contraction and relaxation of both future slow and fast muscles (Buller et al., 1960a,b) are induced by the relatively small amounts of low frequency activity that muscles of newborn animals receive, for when muscles are deprived of this activity by denervation, their speeds of contraction and relaxation fail to increase. However, the change normally seen during development can take place, even in the denervated muscles if they are activated by electrical stimulation. Thus, when denervated neonatal muscles were stimulated at low frequencies, their speeds of contraction and relaxation increased like during normal development (Brown, 1973). From these results it appears that the small amounts of low frequency activity that occur during early development are probably necessary for inducing the maturation of the muscle. An important step in this process of maturation is the development of the internal membrane system. It is known that this system is responsible for the release of Ca^{2+} into the cell and for its rapid removal from it (see Endo, 1977). It was shown that the ability of the sarcoplasmic reticulum (SR) to take up Ca^{2+} is low during the early stages of muscle development and increases with age. This change like that of mechanical properties depends on activity, for it does not take place if activity is prevented by paralysing muscles with curare (Martonosi et al., 1977). Thus, even after the synaptic connections between muscle and nerve are established and properly organized the two systems, i.e., the motoneurone and muscle, continue to interact, but at this later stage of development it is mainly the muscle that depends on its innervation for continued growth and development.

This dependance continues throughout life, for when a mammalian muscle is denervated it undergoes atrophy and, finally, after long periods of time, degenerates (Gutmann, 1962). Nevertheless, if slow or fast muscles of adult animals are temporarily denervated and the nerve fibres find their way back to the muscles they originally supplied, the muscles recover completely and the size of the motor units returns to normal (see Gordon and Stein, 1980). This is not so in muscles of young, developing animals. Denervated muscle fibres of young animals (1 to 2 weeks after birth) survive denervation and do not seem to degenerate more rapidly than denervated muscle fibres of adults (Vrbová, 1952; Zelená, 1962). Nevertheless, when reinnervated by the nerves that originally supplied them, the muscles of young animals are unable to recover their usual strength or histochemical properties. An insight as to why the recovery of these young muscles is so poor is given by the finding that slow muscles, such as soleus, recover, while fast muscles of the hindlimb remain permanently weak (see Fig. 5). Moreover, it

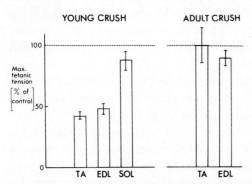

Fig. 5. Decrease of maximal tetanic tension developed by reinnervated muscles 3 to 4 months after nerve crush at 5 to 6 days (Young crush) and at 6 to 8 weeks of age. The dotted line represents 100% (i.e., tension developed by the control muscles) and the change is expressed as % of this value. TA = tibialis anterior; EDL = extensor digitorium longus; SOL = soleus.

was found that their weakness is caused by a selective loss of their large glycolytic fibres, and that the selective destruction of potential fast glycolytic fibres takes place after reinnervation (Lowrie and Vrbová, 1980, 1982).

This finding could be explained by the following suggestion: during normal development the muscle is gradually maturing and acquiring new properties that enable it to respond to the demands imposed by the developing motoneurone. If the muscle is disconnected from its motoneurone during this critical stage of its maturation it will fail to develop these characteristics. In the meantime the nervous system continues its development, and when the motoneurone reinnervates the muscle a few days later, it will impose on the muscle an activity pattern typical of an older animal. The muscles, with their arrested development, may not be capable of this performance, and those muscle fibres which would normally become highly specialized for rapid production of force, i.e., the fast muscle fibres, are not ready for their task. Since these more mature motoneurones no longer activate the muscle fibres with the less mature activity pattern, i.e., low frequencies for short periods of time, some muscle fibres are destroyed, and degenerate.

The similarities between fast muscles that had their nerves crushed during early postnatal life and that of muscles from patients suffering from Duchenne dystrophy suggested the possibility that the cause of muscle wasting in this primary muscle disease may be due to a slower rate of maturation of the diseased muscles due to a genetic defect of the muscle fibres. The altered enzyme profiles of dystrophic muscles also indicates that the maturation of the diseased muscle is abnormal, for, in muscles of Duchenne children, the activities of those enzymes that normally increase with the increased use of the muscles do not do so, and their enzymes retain their immature characteristics. Thus if the rate of maturation could be increased by slow frequency activity, it might be possible to reduce the degree of muscle wasting.

This possibility has been tested in an animal model of a genetically determined muscle dystrophy, the BL57 dy^{2J}/dy^{2J} mouse. The results of this study showed that, when fast muscles of these animals were stimulated for 2 to 4 weeks at low frequencies, their rate of deterioration was reduced, the stimulated muscles developed more tension and contained more muscle fibres than the untreated control muscles (Luthert et al., 1980). Other abnormal features of the diseased mouse muscles were also reversed towards normal by the treatment. Like in muscles from patients suffering from Duchenne dystrophy, in this dystrophic mouse, the activities of glycolytic enzymes characteristic of fast muscles are low and the isoenzymic pattern of lactate dehydrogenase (LDH) is that typical of immature muscles. Low frequency activity reversed these changes and induced an increase of the enzyme activities and the transformation of the isozymic forms of the LDH from an immature type to that of the adult muscle (Reichmann et al., 1981).

Thus, at least in animal models of dystrophy, it appears that the rate of maturation can be enhanced by imposing the appropriate activity pattern. Moreover, the progress of the disease can be halted using this procedure.

In Duchenne dystrophy, muscle fibres display characteristic features of immature muscles at a time when the child's motor activity increases and the development of coordinated movement is more or less completed. It would be during this period of development that the motoneurones to fast muscles can be expected not only to increase, but also change their activity pattern from a low to a higher firing rate. If, in these children, the muscle fibres have not yet developed those characteristics that enable them to respond to high frequency activity, this activity may cause them to degenerate. There are several indications that muscle fibres from Duchenne patients are not well equipped to follow activity at high frequencies; the ability of the SR to take up Ca^{2+} is less than that in muscles from normal children (Peter et al., 1974), and there is evidence that Ca^{2+} accumulates in the muscle fibres and initiates the degeneration process (Bodensteiner and Engel, 1978; Cullen and Fulthorpe, 1975).

Why proximal muscles of limbs are affected earlier than the rest of the musculature is unclear, but there may be a cranio-caudal progression of the maturation of motor reflexes, so that the most proximal muscles would be used a little earlier than the more distal ones, leaving them even less time to complete their maturation, before normal locomotor activity is imposed upon them.

Even if the first symptoms of the disease could be explained by a slower than normal maturation of the muscle fibres, additional factors must contribute to the relentless progress of the disease. There may be a number of these but they could all be a consequence of the initial lesion. EMG recordings have revealed more or less normal frequencies of motoneurone discharge in muscles from dystrophic children. This, and the findings that the time course of contraction is slower than normal and the muscles contain more 'slow' type fibres, indicates a 'mismatching' between the activity pattern and muscle properties.

In addition, the high frequency activity of the fast motoneurones is probably unfavourable not only to the diseased muscle fibres, but also to the regenerating, newly-developed fibres. These too are immature and exposure to this type of activity, could prevent them from surviving. Moreover, the non-uniform involvement of different muscle groups leads to deformities, and these alter the lengths of many muscles. It is known that muscle length is important for growth and development of muscles so that the growth of some muscles could be reduced simply as a result of having an inappropriate length (Goldspink, 1981). However, even fully developed muscles will undergo degeneration if they are activated in a shortened position. This applies particularly to slow muscle fibres; if the soleus muscle is tenotomized its muscle fibres rapidly degenerate, and this degeneration can be reduced by denervation and enhanced by activity (McMinn and Vrbová, 1961, 1967). Thus the increased usage of shortened muscles will aggravate the symptoms of the disease. While the possibility of a slower rate of maturation of muscle fibres in Duchenne dystrophy could be a useful working hypothesis in the study of the pathogenesis of this disease, the genetic defect that might cause this reduced rate of maturation remains a mystery.

Conclusions

In this chapter, two examples of genetically dertermined neuromuscular diseases are considered: spinal muscular atrophy and Duchenne muscular dystrophy. The possibility that some features of these diseases can be explained as disturbances of nerve–muscle interaction during development is discussed.

In the case of spinal muscular atrophy, it is suggested that the normally occurring reduction of connections between muscle and nerve proceeds too rapidly, so that many motoneurones lose contact with the muscle fibres they supply, and die as a result of this loss of contact. Results obtained from experiments show that many motoneurones die during normal development of vertebrates and that these can be made to survive if they are allowed to maintain connections with the muscle. On the other hand, if more connections are lost a greater than normal number of motoneurones die. Mechanisms that determine the maintenance and disruption of nerve muscle contacts are suggested.

It is argued that the initial disturbance in Duchenne dystrophy is due to a slower rate of maturation of skeletal muscle fibres. Normally, the maturation of the motor reflexes is followed by maturation of the muscle fibres. The possibility is considered that, if muscle fibres mature more slowly than normal, they are not able to withstand the type of activity imposed upon them by the mature motoneurone. This applies mainly to the high frequency activity imposed on the fibres of motor units that are destined to become 'fast'. Experimental evidence suggesting such a mechanism is presented.

Acknowledgements

I am grateful to Professor V. Dubowitz, Dr. J.A.R. Lenman and Dr. R. O'Brien for their help in discussing the ideas included in this chapter.

References

Adams, R.D. (1969) in: Disorders of Voluntary Muscle (Walton, J.N., ed.), pp. 143–202, Churchill, London.
Bagust, J., Lewis, D.M. and Westerman, R.A. (1973) J. Physiol. (London) 229, 247–255.
Ballantyne, J. and Hansen, D. (1974) J. Neurol. Neurosurg. Psychiatry, 37, 1195–1201.
Betz, H., Bourgeois, J.P. and Changeux, J.P. (1980) J. Physiol. (London) 302, 197–219.
Bodensteiner, J.B. and Engel, A.G. (1978) Neurology 28, 439–448.
Brown, M.C., Jansen, J.K.S. and Van Essen, D. (1975) Acta Physiol. Scand. 95, 3–4A.
Brown, M.C., Jansen, J.K.S. and Van Essen, D. (1976) J. Physiol. (London) 261, 387–422.
Brown, M.D. (1973) Nature (London) 244, 178–179.
Buchthal, F. and Olsen, P.Z. (1970) Brain 93, 15–30.
Buchthal, F., Schmalbruch, H. and Kamieniecka, Z. (1971a) Neurology 21, 58–67.
Buchthal, F., Schmalbruch, H. and Kamieniecka, Z. (1971b) Neurology 21, 131–139.
Buchthal, F., Kamieniecka, Z. and Schmalbruch, H. (1974) in: Explanatory Concepts in Muscular Dystrophy II (Milhorat, A., ed.) pp. 526–551, Excerpta Medica, Amsterdam.
Buller, A.J., Eccles, J.C. and Eccles, R.M. (1960a) J. Physiol. (London) 150, 399–416.
Buller, A.J., Eccles, J.C. and Eccles, R.M. (1960b) J. Physiol. (London) 150, 417–439.
Bursian, A.V. and Sviderskaya, G.E. (1971) J. Evol. Biol. Fiziol. 7, 309–317.
Byers, R.K. and Banker, B.W. (1961) Arch. Neurol. Chic. 5, 140–164.
Campbell, N.J. and Liversedge, L.A. (1981) in: Disorders of Voluntary Muscle (Walton, J., ed.) pp. 725–751, Churchill and Livingstone, London.
Close, R. (1964) J. Physiol. (London) 173, 74–95.
Chou, S.M. and Fahadej, H.V. (1977) J. Neuropath. Exp. Neurol. 30, 368–378.
Cullen, N.J. and Fulthorpe, J.J. (1975) J. Neurol. Sci. 24, 179.
Desmedt, J.E. (1981) in: Progress in Clinical Neurophysiology (Desmedt, J.E., ed.) Vol. 9, pp. 250–354, Karger, Basel.
Desmedt, J.E. and Borenstein, S. (1976) Arch. Neurol. 33, 642–650.

Desmedt, J.E. and Borenstein, S. (1977) in: Motor Neurone Disease (Rose, ed.) pp. 112–220, Pitman, London.
Diamond, J. and Miledi, R. (1962) J. Physiol. (London) *162*, 393–408.
Dreyfus, J.C., Demos, J., Schapira, F. and Schapira, G. (1962) C.R. Acad. Sci. (Paris) *254*, 4384.
Dubowitz, V. (1969) in: Some Inherited Disorders of Brain and Muscle (SSIEM Symposium No. 5) (Allan, J.D. and Raine, D.N., eds.) pp. 32–46, Livingstone, London.
Dubowitz, V. (1978) Muscle Disorders in Childhood, Saunders, London.
Dubowitz, V. (1981) in: Disorder of Voluntary Muscle (Walton, J., ed.) pp. 261–295, Churchill and Livingstone, London.
Dubowitz, V. and Brooke, M.H. (1973) Muscle Biopsy: A Modern Approach, Saunders, London.
Duchenne, G.B. (1868) Archive Générales de Médicine 6 ser. *11*, pp. 5, 179, 305, 421, 552.
Duxson, M. (1982) J. Neurocytology *11*, 395–408.
Edds, M.V. (1953) Quart. Rev. Biol. *28*, 260–276.
Endo, M. (1977) Physiol. Rev. *57*, 71–109.
Erb, W.H. (1884) Deutsches Archiv. Für Klinische Medizin. *34*, 466–481.
Erb, H.W. (1891) Deutsche Zeitschrift für Nervenheilkunde *1*, pp. 13, 94, 173, 261.
Gardner-Medwin, D. and Walton, J.N. (1969) in: Disorders of Voluntary Muscle (Walton, J.N., ed.) pp. 411–499, Churchill, London.
Goldspink, G. (1980) in: Development and Specialization of Skeletal Muscle (Goldspink, D.F., ed.) pp. 19–37, Cambridge University Press.
Gordon, T., Perry, R., Tuffery, A.R. and Vrbová, G. (1974) Cell Tiss. Res. *155*, 13–25.
Gordon, T. and Stein, R.B. (1980) in: Plasticity of Muscle (Pette, D., ed.) pp. 283–297, de Gruyter, Berlin.
Gutmann, E. (1962) in: The Denervated Muscle (Gutmann, E., ed.) Czech. Acad. Sci. Publishing House, Prague.
Hamburger, V. (1975) J. Comp. Neurol. *160*, 535–546.
Handysides, N.S.P., Navarrete, R. and O'Brien, R.A.D. (1982) J. Physiology (London) (in press).
Heyck, H., Laudahn, G. and Carsten, P.M. (1966) Klin. Wschr. *44*, 695.
Hudgson, P., Pearce, G.W. and Walton, J.N. (1967) Brain, *90*, 565–576.
Huizar, P., Kuno, M., Kudo, N. and Miyata, Y. (1977) J. Physiol. (London) *265*, 175–192.
Jedrzejowska, H., Johnson, A.G. and Woolf, A.L. (1965) Acta. Neuropath. *5*, 225–242.
Jeruzalem, F., Engel, A.G. and Gomez, M.R. (1974b) Brain, *97*, 12–130.
Kendrick-Jones, J. and Perry, S.V. (1967) Biochem. J. *103*, 207–214.
Kugelberg, E. and Welander, M. (1956). Arch. Neurol. Psychiatry *75*, 500–509.
Kuno, M. and Llinas, R. (1970a) J. Physiol. (London) *210*, 807–821.
Kuno, M. and Llinas, R. (1970b) J. Physiol. (London) *210*, 823–838.
Laing, N.G. and Prestige, M.C. (1978) J. Physiol. (London) *282*, 33P.
Lieberman, A.R. (1971) Int. Rev. Neurobiol. *14*, 50–115.
Lieberman, A.R. (1974) in: Essays on the Nervous System. A Festschrift for Professor J.Z. Young (Bellairs, R. and Gray, E.G., eds.) pp. 71–105, Oxford University Press.
Lipton, B.H. (1978) in: Muscle Regeneration (Mauro, A., ed.) pp. 31–41, Raven Press, New York.
Lowrie, M.B. and Vrbová, G. (1981) J. Physiol. (London) *307*, 19P.
Lowrie, M.B., Subramaniam Krishnan and Vrbová (1982) J. Psysiol. (London) *331*, 51–66.
Luthert, P., Vrbová, G. and Ward, K.M. (1980) J. Neurol. Neurosurg. Psychiatry, *43*, 803–809.
Margreth, A., Salviati, G., Dalla Libera, B.R., Biral, D. and Salviati, S. (1980) in: Plasticity of Muscle (Pette, D., ed.) pp. 193–209, de Gruyter, Berlin.
Margreth, A., Salviati, G., Mussini, J. and Carraro, V. (1974) Exploratory Concepts in Muscular Dystrophy II (Milhorat, A.T., ed.) pp. 406–416, Excerpta Medica, Amsterdam.
Martonosi, A., Roufa, D., Boland, R. and Reyes, E. (1977) J. Biol. Chem. *252*, 315–321.
McComas, A.J. (1977) Neuromuscular Function and Disorders, Butterworths, London.

McComas, A.J., Sica, R.E. and Currie, S. (1971) J. Neurol. Neurosurg. Psychiatry 34, 461–468.
McMinn, R.M.H. and Vrbová, G. (1964) Q. J. Exp. Physiol. 49, 424–428.
McMinn, R.M.H. and Vrbová, G. (1967) Q. J. Exp. Physiol. 52, 411–415.
Meryon, E. (1852) Med. Chirurg. Trans. 35, 73.
Milner-Brown, H.S., Stein, R.B. and Lee, R.G. (1974) J. Neurol. Neurosurg. Psychiatry 37, 670–676.
Mokri, B. and Engel, A.G. (1975) Neurology 25, 1111–1120.
Navarrete, R. and Vrbová, G. (1980) J. Physiol. (London) 305, 33–34P.
Navarrete, R. and Vrbová, G. (1983) Deu. Brain Res., 8, 11–20.
Nyström, B. (1968) Acta Neurol. Scand. 44, 405–439.
O'Brien, R.A.D., Östberg, A.J.C. and Vrbová, G. (1978) J. Physiol. (London) 282, 571–582.
O'Brien, R.A.D., Östberg, J.A. and Vrbová, G. (1980) Neuroscience 5, 1367–1379.
O'Brien, R.A.D. and Vrbová, G. (1980) in: Plasticity of Muscle (Pette, D., ed.) pp. 271–238, de Gruyter, Berlin.
O'Brien, R.A.D. and Vrbová, G. (1978) Neuroscience 3, 1227–1230.
Oppenheim, R.W. (1982) in: Studies in developmental neurobiology. Essays in Honour of Victor Hamburger (Cowan, W.M., ed.) Oxford University Press.
Oppenheim, R.W. and Nunez, R. (1982) Nature (London) 295, 57–59.
Perry, S.V. (1971) J. Neurol. Sci. 12, 289–306.
Peter, J.B., Worsfold, M. and Fiehn, W. (1974) in: Exploratory Concepts in Muscular Dystrophy II (Milhora, A.T., ed.) pp. 491–498, Excerpta Medica, Amsterdam.
Pittman, R. and Oppenheim, R.W. (1979) J. Comp. Neurol. 787, 425–446.
Reichmann, H., Pette, D. and Vrbová, G. (1981) FEBS Lett. 128, 55–58.
Richardson, A.T. and Borwick, D.D. (1969) in: Disorders of Voluntary Muscle (Walton, J., ed.) pp. 873–842, Churchill, London.
Romanes, G.J. (1966) J. Anat. 80, 117–131.
Rowland, L.P. (1976) Arch. Neurol. 33, 315–321.
Sacco, G., Buchthal, F. and Rosenfalck, P. (1962) Arch. Neurol. 6, 366–377.
Samaha, F.J. and Gergely, J. (1969) N. Engl. J. Med. 280, 184.
Schotland, D.L., Bonilla, E. and Van Meter, M. (1977) Science 198, 1005–1007.
Srihari, T. and Vrbová, G. (1978) J. Neurocytol. 7, 529–540.
Takagi, A., Schotland, D.I. and Rowland, L.P. (1973) Arch. Neurol. 28, 380–384.
Van Harreveld, A. (1952) J. Comp. Neurol. 97, 385–407.
Vrbová, G. (1952) Physiol. Bohemoslov. 1, 22–26.
Vrbová, G., Gordon, T. and Jones, R. (1978) Nerve–muscle Interaction, Chapman and Hall, London.
Vrbová, G. (1980) in: Development and Specialisation in Skeletal Muscle (Goldspink, D.F., ed.) pp. 37–51, Cambridge University Press.
Werdnig, G. (1891) Archiv. Psychiatr. Nervenkr. 26, 706–744.
Wohlfart, G., Fex, J. and Eliasson, S. (1955) Acta Psychiatr. Neurol. Scand. 30, 395–406.
Woolf, A.L. (1960) Cerebral Palsy Bull. 2, 19.
Woolf, A.L. (1969) in: Disorders of Voluntary Muscle (Walton, J.N., ed.) pp. 203–237, Churchill, London.
Zelena, J. (1962) in: The Denervated Muscle (Gutmann, E., ed.) pp. 103–126, Publishing House Czech. Acad. Sci. Prague.
Zubrzycka-Gaarn, E. and Sarzala, M.G. (1980) in: Plasticity of Muscle (Pette, D., ed.) pp. 209–225, de Gruyter, Berlin.

Index

A23186 262
Acetylcholine (ACh) 26, 42, 157, 186, 367, 368
Acetylcholine receptors (ACh-R) 93, 122, 123, 203, 205, 254, 256, 312, 368, 370
  aggregation 129
  channel open times 131
  clusters 127, 129, 135, 138
  regulation 144
  turnover 131
Acetylcholinesterase (AChE) 122, 157, 202, 206, 345, 368
  localization of 132, 136
  regulation 143
Achalasia 333, 334, 335
ATP 260
Adrenergic
  cross-interactions 258
  fibres 346
  receptors 256, 257, 258, 267
  see also: Neurone
Angiotension II 259
Autonomic nerve 35–53, 327–351
  cholinergic innervation 201
  development 35, 185
  innervation pattern 201
  sympathetic development 186
Axon
  atrophy 304, 305, 306
  branching 71, 92, 94
  conduction velocity 309
  crushed 308
  diameter 303, 306, 309, 316
  growth 122, 300
  impulse traffic 306
  injury 301
  motor 305
  neuroma 310
  outgrown 66, 71
  profiles in nerve terminals 75
  recovery 313
  sensory 305
  size 316, 317, 318
  sprouting 92, 305
  terminals 367
  viable 305
  see also: Reinnervation; Differentiation; Regeneration
Axonal transport 290, 291, 295
  axoplasmic flow 298–301, 305
Axotomy 300, 301–302, 303, 313, 317
  atrophic effects 313
see also: Motoneurones; Neurones

Basal lamina 123, 127, 141, 157
Bowel
  aganglionic 330
  dilated 342
  large 344–345
  small 327, 343
  small bowel dysfunction 342
  wall 328
Bradykinin 259

Calcium (Ca^{2+}) 144, 256, 261, 263, 292
  permeability 257
Cardiac muscle 48–51
Cell death
  column of Terni 97

function 102
histogenetic 61
induction 61
prevention of 79
programmed 60
sympathetic 97
Chagas' disease 330, 335–336, 338
  American trypanosomiasis 348–349
  trypanosomiasis 341
Chloride (Cl$^-$) 263
  permeability 256
Cholecystokinin 340
Choline acetyltransferase (CAT) 26, 187, 189, 202, 203, 205
Cholinergic
  cross-interactions 258
  receptors 267, 367
Cholinesterase 312, 329
Chromatolysis 292, 301
Chronic stimulation 286
Coated vesicles 126
Concanavalin A (con A) 116
Competition hypothesis 79
Crohn's disease 330, 343
Cyclic AMP 144, 230, 261, 263, 264
Cyclic GMP 144, 256, 263–264
Cytoskeleton 299, 300

Dedifferentiation 298
Degeneration
  nerve 306
  process 379
  Schwann cell 310
  sympathic nerve fibre 309
  Wallerian 290–292
Denervation
  atrophy 39, 308
  end-organs 312
  muscle fibres 378
  supersensitivity 257, 266, 267
  sympathetic 40
Dentanyl 262
DNA
  synthesis 111
Desensitization 269-271
Diabetes mellitus 350
Dibenamine 254
Differentiation 12
  adrenergic cell 12
  axonal 89

neural crest cells 23, 28
neuronal 28
neurotransmitter 24, 44, 52
  of muscle 120
  of skeletal muscle 281
  peripheral ganglionic cell 18
  sympathetic neurones 40, 42
Duchenne muscular dystrophy 360, 370–381
  as a disturbance of nerve-muscle interaction 376
  ideas on the pathogenesis 375
Duodenum 328, 341–342
  duodeno-jejunal anastomosis 341

Electrical stimulation of chick embryos 94
EMG 363, 365, 380
  changes 374
  Werdnig–Hoffmann disease 365
Endorphins 260
Enzymes
  glycolytic 361
  oxidative 361, 363, 365
  proteolytic 174, 367, 370
Epidermal growth factor (EGF) 119
Fast muscle 377
  maturation 373
Fibroblast growth factor (FGF) 119
Fibronectin 6, 116
Fusion of muscle cells 111, 116, 117
  effects of viral tranformation on 117–118
  involvement of membrane lipids 114–115
  lectins as mediators 116

Ganglion
  blocking drugs 188
  ciliary 301
  dorsal root (DRG) 215, 216, 217, 218, 219
  intermyenteric 329
  myenteric 340
  sympathetic 191, 218, 259, 260, 301
  sympathetic development 186
  vagal 335
Ganglion precursor cells 2
  automatic 18, 20
  dorsal root 10, 18, 21, 96
  enteric 8
  migration 2
  nodose 22
  Remak 8
  sensory 6

spinal 65
Gap junction 113
Gastric secretion 333
Gastrin 336
  release 335
Gene expression 112
Glial cells 215, 217
Glucocorticoids 45, 119
Glycolytic fibre 378
Grafting 18
  back-transplantation 18, 22
Growth cone 298
Gut 327–351
  anal sphincter 344–345
  colon 328
  gastrointestinal tract 327
  ileum 328
  intestinal obstruction 342
  peristalsis 327, 334, 337
  stomach 337
  see also: Bowel; Oesophagus; Duodenum

Heterotopic grafting 11
Hirschsprung's disease 327, 347
  aganglionosis 346
  congenital aganglionosis 341, 345–347
Histamine 258, 259
Histochemical pattern 363
Hollander tests 334
Horseradish peroxidase (HRP) 66, 130
Hyaluronic acid 6
6-Hydroxydopamine 269
Hyperinnervation 94, 168
Hyperplasia experiments 79
Hypertensive 264
Hypoganglionosis 347, 348

Innervation
  automatic 350
  development 153, 191
  multiple 154, 158, 278
  of nerve–muscle junction 92
  of smooth muscle 253
  pattern 204
  polyneuronal 71, 153
Insulin 119
Insulin hypoglycaemia 333
Irritable colon 349

Lactate dehydrogenase 379

Lateral motor column (LMC) 61, 63, 94
  lumbar 86
  motoneurons 79
Latissimus dorst
  anterior (ALD) 67, 68
  posterior (PLD) 67, 68
Limb-bud removal 63
Lysosomes 294

Mast cell 259
Melanocytes 4, 5
Melanoblasts 2
Membrane
  chorio-allantoic 15
  folding 312
  surface 375
  surface properties 376
Methacholine 256
Microtubules 230, 300
Motoneurone 301, 307, 316, 317, 318, 374, 376
  alpha 313
  chromatolytic 365
  disease 332, 365, 366–370
  electrophysiological changes in axotomized
    motoneurones 306–307
  fast 306, 307, 380
  lower motoneurone disease 368–370
  properties after low and high frequencies of
    stimulation 318
  slow 306, 307
Motoneurone death 57, 69, 81, 360, 366, 367,
    368, 381
  ALD and PLD 68
  error correction hypothesis 104
  genetic factors 60
  neuromuscular blockade 81, 88, 93, 102,
    104, 124, 127, 135, 143, 166
  programmed 60
  redundancy hypothesis 103
  spinal 59, 93
  transient/provisional hypothesis 103
Motor axon 305, 314
Motor fibre 315
Motor nerve 319
  endings 363
  see also: Reinnervation
Motor unit 169, 173, 280, 314, 315, 316, 374,
    375
  activity 319, 377
  fast 359

fast-fatiguable 316
fast-fatigue-resistant 316
potential 363, 364, 365, 367
size 364
slow 316, 359
see also: Reinnervation
mRNA 120
Muscle 312
activity 4, 317
atrophy 315
clusters 77
development 75, 88, 91, 109
dystrophic 125
fast twitch (PLD) 68
hypertrophy 317
maturation 377
sarcoplasmic reticulum 373, 376, 377, 379
satellite cells 77
slow tonic (ALD) 68
see also: Denervation; Differatiation; Fusion of muscle cells; Reinnervation
Muscle fibre
atrophy 312
deformities 380
development 155
properties after reinnervation 315
smooth 327–351
see also: Denervation
Myelin 306
Myelination 71
Myoblasts 110
Myotubes 110

Nerve
activity patterns 284–287
atrophy 315
development of parasympathetic nerves 43
diameter 308
embryonic 310
endings 365, 366
extrinsic 329–330
growth factor 260
hypertrophy 327
inhibitory 332
injury 303, 304, 307, 310
intrinsic 337
parasympathetic 330
peptide-containing 346
postganglionic sympathetic 329
properties after reinnervation 315

recovery after crush injury 314
section 316
serotonergic 347
somatic 309
sprouting 159, 365
stimulation 317
sympathetic 39, 48–51, 329–330, 337, 340
see also: Reinnervation; Degeneration; Regeneration; Plexus
Nerve fibres
crush injury 310
growth 45, 47
hypertrophy 318
mechanical barriers to regenerating fibres 310–312
sprouts 318
see also: Degeneration; Regeneration
Nerve growth factor (NGF) 17, 42, 47, 81, 99, 185, 187, 188, 189, 190, 191, 192, 193, 194, 198, 200, 201, 207, 213, 216, 218
anti- 199
distribution 217
ionic hypothesis for 231
Na^+/K^+ pump 233, 236, 239, 242
radioimmunoassay 195
receptor 225, 236
release 196
RNA protein synthesis 227–228
synthesis 196, 217
transport of 194, 197, 226
Nerve–muscle
connections 368
interaction 366–370, 376
mismatching 30
neuromuscular activity 165
Nerve–muscle junction
formation 48–52
ultrastructure 51
Neural cell adhesion molecule (N-CAM) 6, 124
Neural crest 10
autonomic structures 10
phenotypic expression 12
removal 65
Neural crest cells 2, 8
chromaffin cell 219
migration 2, 8, 185
see also: Differentiation
Neurofibromatosis 348
Neurofilament 300, 316

microfilament 230
protein 299, 300, 305, 317
Neurogenic hypothesis 375
Neurone 42, 43, 305, 328
  adrenergic 329
  axotomized 302, 305, 316
  cell body 317
  development of enteric neurones 43
  disease 343
  dorsomedial (DM) 218
  dysplasias 354
  intrinsic 330, 334
  loss of 334
  migratory 42
  myenteric 336, 338
  myenteric cholinergic 335
  neurite adhesion 205
  neurite growth 222, 241
  neurite orientation 218
  neurite outgrowth 189, 190, 196, 199, 203, 219
  neurite promoting factors 204, 207
  parasympathetic 203
  parasympathetic, development of 202
  polyornithine-binding neurite promoting factors (PNPFs) 223
  survival 206
  sympathetic 188, 189, 192
  ventrolateral (VL) 218
Neurone death 214, 220–221, 238–240
  parasympathetic 203
  sympathetic 188, 190, 192, 197, 199
Neuromuscular
  activity 318
  blockers 295–301
  junction 317
  see also: Motoneurone; Innervation
Neuromuscular diseases 359, 380
  mopathies 359
  myogenic 359
  neurogenic 359
Neuropathy
  autonomic 341, 350
Neurotensin 260
Neurotransmitter 12
  phenotype expression 12
  release 168
  shift 18
Neurotropic 335
Non-adrenergic 259, 344

Non-cholinergic 259, 344
Noradrenaline 185, 191
Notochord 15
Nucleus ambiguus 332

Oesophagus 327, 331–336
  oesophageal pathology 333
  oesophageal sphincter 331
Oestrogen 261
Opiates 260, 270
Ornithine decarboxylase (ODC) 228–229

Parasympathetic fibres 331
Pentagastrin 332, 340
Peptic ulceration 337
Peptides 259
  neurotransmitters 259
  see also: Vasoactive intestinal polypeptide; Nerve
Phosphatidylinositol (PI) 231, 256, 258
  'PI effect' 231
Pigment cells – see Melanocytes
Plexus
  myentric (Auerbach's) 331, 332, 334, 337, 341, 348, 350–351
  nervous 327, 328, 335
Poliomyelitis 332
Polyneuronal innervation 71, 153
  elimination of 160, 171
  techniques for estimation of 176
Posterior nerotic zone (PNZ) 60
Potassium (K^+) 262, 263
  permeability 256, 257
Preganglionic cholinergic fibres 332
Presynaptic cholinergic autoreceptors 255
Prolactin 260
Propylbenzilcholine mustard 254
Prosencephalic crest cells 12
Prostaglandins 261
Protein
  fibrillar 299, 308
  see also: Neurofilament
Putrescine 297
Pyloric
  hypertrophic 339
  innervation 338
  motility 340
  pathology 338–340
  stenosis 338, 340
Pylorus 339

Quantal mitosis 111
Quinuclidinyl benzilate 254, 268

Reflex
  monosynaptic 302
  polysynaptic 302
Regeneration 307–319
  of axon 190, 298, 299, 305, 306, 308, 309, 310, 311, 317
  of nerve 167, 292, 294, 296, 297, 299, 306
  of nerve fibres 311
  trophic role 310
Reinnervation 303, 312, 317
  axon 316
  cross- 278–281, 282, 283, 284
  motor nerve 314
  motor unit 316
  muscle 315
  nerve 319
  self- 280
Reserpine 269
Retrophin 185

Schwann
  cell 310
  Sheath 311
  tube 308, 310, 311, 312
  see also: Degeneration
Sciatin 141
Scleroderma 336, 341
Secretin 340
Sensory fibre 314
Skeletal muscle
  contractile properties 277, 280, 282, 283
  electrical activity of 285
  see also: Differentiation
Slow muscle 377
Small intestine 340
  jejuno-ileum 342–344
Smooth muscle 35, 48–51
  development 35–53
  phenotypic modulation 36, 39
  properties 253–275
  receptors 254
  regulation of 265
  see also: Innervation
Sodium (Na^+)
  permeability 256
Sodium-potassium ATPase 263
Sodium-potassium pump 269
Somatomedins 119
Spasm
  diffuse 336
Spikes
  dendritic 302, 307
Spinal cord
  transection 318
Spinal musular atrophies 361, 380
  severe 360
  Werdnig–Hoffman type 360–370
Sprouts 308, 311, 316
  see also: Nerve; Axon; Nerve fibre
Substance P 259, 260
Sympathetic chains 5
Sympathetic cholinergic neurones 17
Sympathoblasts 17
Synapse
  ectopic 312
  formation 121, 125, 186
  monosynaptic 302
  onset of 125
  postjunctional thickening 301
  postsynaptic chemosensitivity 301, 302
  postsynaptic specialisation 200, 312
  regulation 135
  regulation of formation 199
  specificity 123
  transmission and 301–302
  ultrastructure 126

Tetrodotoxin (TTX) 135
Thymidine ^{3}H 37
  incorporation of 37
Thyrotrophic-releasing hormone 259–260
Trophic 39, 45, 65
  agent 173
  ciliary Neurotrophic Factors (CNTFs) 214, 222
  effect 264
  factors 200
  neurotrophic factors 214–216
Trypanosoma cruzi 335
Tubulin 299, 300
Twitch/tetanus ratio 282
Tyrosine hydroxylase 186, 187, 191, 193, 194, 198, 201

Ulcerative colitis 349

Vanadium 263
Vasoactive intestinal polypeptide (VIP) 260, 330, 343

Wheat germ agglutinin (WGA) 116